Edexcel GCSE

Statistics

Written by

Gill Dyer, Jane Dyer, Keith Pledger, David Kent, Brian Roadnight and Gordon Skipworth

edexcel
advancing learning, changing lives

D1493597

Series editor: Keith Pledger

A PEARSON COMPANY

Published by Pearson Education Limited, a company incorporated in England and Wales, having its registered office at Edinburgh Gate, Harlow, Essex, CM20 2JE. Registered company number: 872828

Edexcel is a registered trademark of Edexcel Limited.

Text © Gill Dyer, Jane Dyer, Keith Pledger, David Kent, Brian Roadnight, Gordon Skipworth, 2009

© First published 2009

13 12 11 10
10 9 8 7 6 5 4

British Library Cataloguing in Publication Data is available from the British Library on request.

ISBN 978 1 846904 547

Edited by Nicola Morgan and Alex Sharpe
Typeset by Tech-set Ltd
Original illustrations © Pearson Education Ltd
Illustrated by Tech-set Ltd
Cover design by Siu Hang Wong
Picture research by Chrissie Martin
Cover photo/illustration © Masterfile / Dreams Stock
Printed in Malaysia (CTP-VP)

Acknowledgements

Additional material provided by Jim Newall

The author and publisher would like to thank the following individuals and organisations for permission to reproduce photographs:

Alamy / Jim Wileman p1; Shutterstock / Susan McKenzie p29; Shutterstock / Stephen Coburn p78; Pearson Education Ltd / Clark Wiseman, Studio 8 p116; Image Source p169; Getty Images / Taxi p208; Getty Images / Photographers Choice p241; Alamy / Robert Ashton / Massive Pixels 293; Shutterstock / Barbro Bergfeldt p6; Rex Features / Micha Theiner p13; Shutterstock / Tomasz Trojanowski p19; Shutterstock / Sandro Donda p33; Shutterstock / EcoPrint p38; Shutterstock / Geko p43; Shutterstock / Danilo p54; Shutterstock / Jan van den Hoeven p63; Image Source Ltd p68; Getty Images / Photographers Choice p81; Shutterstock / Morgan Lane Photography p88; Shutterstock / Rachael Grazias p92; Getty Images / Stone p99; Shutterstock / P Kruger p112; Alamy / John Cooper p122; Shutterstock / Silver-John p128; Shutterstock / C. p131; Getty Images / Oli Scarff p140; Ardea / Robert T. Smith p144; Shutterstock / Photography Perspectives - Jeff Smith p153; Alamy / Paul Baldesare p158; Alamy / Christiane Nehring p173; Alamy / Alex Segre p177; Shutterstock / Schmid Christophe p185; Getty Images / PhotoDisc p198; Alamy / Image Source Pink p210; Shutterstock / STERS p214; Shutterstock / Hannamariah p217; Shutterstock / Malibu Books p225; Getty Images / PhotoDisc p243; Shutterstock / 8781118005 p250; Shutterstock / Matsonashvilli Mikhail p256; Shutterstock / Galina Borskaya p259; Alamy / Rob Watts p270; Alamy / Hugh Threlfall p277; Getty Images / Stone p297; Alamy / Phil Wills p301; Shutterstock / Cathy Keifer p307; iStockPhoto.com / Efendi Kocakafa p322; iStockPhoto.com / Alex Slobodkin p322; iStockPhoto.com / ZoneCreative p322; iStockPhoto.com / Stockphoto4u p328; Getty Images / PhotoDisc p331; Getty Images / PhotoDisc p332; Pearson Education Ltd / Rob Judges p333, Alexander Caminada p335

Every effort has been made to contact copyright holders of material reproduced in this book. Any omissions will be rectified in subsequent printings if notice is given to the publishers.

Disclaimer

This Edexcel publication offers high-quality support for the delivery of Edexcel qualifications.

Edexcel endorsement does not mean that this material is essential to achieve any Edexcel qualification, nor does it mean that this is the only suitable material available to support any Edexcel qualification. No endorsed material will be used verbatim in setting any Edexcel examination/assessment and any resource lists produced by Edexcel shall include this and other appropriate texts.

Copies of official specifications for all Edexcel qualifications may be found on the Edexcel website – www.edexcel.com

43231

How to use this book

This book is designed to give you the best preparation possible for your GCSE Statistics examination:

- This is Edexcel's own course for the GCSE Statistics specification.
- It covers the whole specification for the Foundation and Higher tiers, and clearly shows which is which.
- Written by senior examiners.

Finding your way around the book

Detailed contents list, clearly showing which topics are also covered in GCSE Mathematics.

Brief chapter overview, plus an example of a real-life application of the statistics in the chapter.

Every chapter ends with a review exercise. Star ratings show the level of challenge for each question.

Questions marked edexcel past paper question are from past exam papers

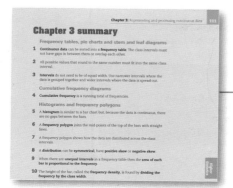

Every chapter ends with a summary of key points and a 'test yourself' quiz.

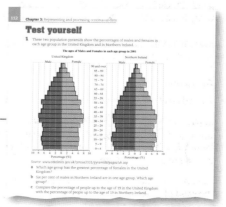

How to use this book

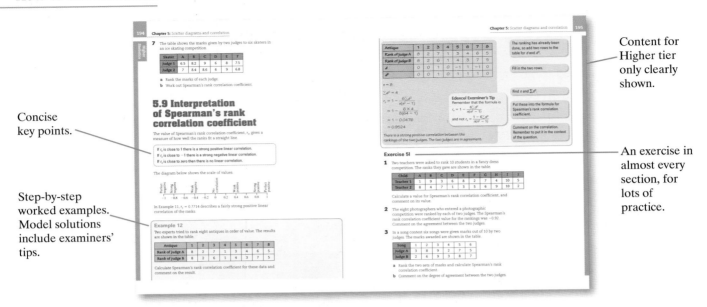

Concise key points.

Step-by-step worked examples. Model solutions include examiners' tips.

Content for Higher tier only clearly shown.

An exercise in almost every section, for lots of practice.

We've broken down the six stages of revision to ensure you are prepared every step of the way.

Zone In: How to get into the perfect 'zone' for revision.

Planning Zone: Tips and advice on how to effectively plan revision.

Know Zone: The facts you need to know and memory tips.

Don't Panic Zone: Last-minute revision tips.

Exam Zone: What to expect on the exam paper and in the controlled assessment and exam-style practice papers.

Zone Out: What happens after the exam.

ResultsPlus

These features are based on how students have performed in past exams. They draw on expert advice and guidance from examiners to show you how to achieve better results.

There are three different types of ResultsPlus features throughout this book.

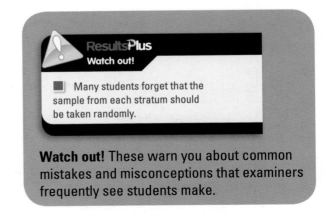

Watch out! These warn you about common mistakes and misconceptions that examiners frequently see students make.

ResultsPlus
Build Better Answers

Question: A town council plans to build a swimming pool. It is going to carry out a survey to find out what people think of the plan.
(ii) What type of **statistical** diagram could the council use to show the results of the survey?
Give a reason for your answer

■ **Zero marks**
A common incorrect answer was 'tally chart'. This is not a diagram, it is a data collection sheet.

● **Good 1 mark answer**
Choose a diagram that shows the different groups of responses clearly, for example a bar chart or pie chart.

▲ **Excellent 2 mark answer**
Give a reason for your choice, explaining why the diagram you suggest will present the information so that it is easy to read. An example of an excellent answer was: 'Bar chart, because it shows the results clearly'.

Build better answers gives an opportunity to answer exam-style questions. They include tips for what a basic or incorrect ■, good ● and excellent △ answer will contain.

ResultsPlus
Exam Question Report

Tyson took a Statistics test and a Maths test.
Both tests were marked out of 100.
The table gives information about Tyson's marks.
It also shows the mean mark and standard deviation for the group that took the test.

Test	Tyson's mark	Group mean mark	Group standard deviation
Statistics	55	52	15
Maths	48	45	12

(a) i)   Work out Tyson's standardised score for Statistics.
    ii) Work out Tyson's standardised score for Maths.
                                                     (3 marks)
(b) Write down the subject in which Tyson did better.
     Give a reason for your answer.   (2 marks)
(c) Comment on the group's performance in the two tests.
                                                     (2 marks)
                                              (Total 7 marks)

How students answered

Poor   37%

Around one-third of students answered this question poorly. Some candidates wrote down the calculation for (a) i) as $55 - \frac{52}{15} (= 51.5)$, instead of $\frac{55 - 52}{15} (= 0.2)$.

Good   19%

Some students calculated the standardised scores correctly in part a), but then in part b) stated that Tyson did better in statistics because he got a higher mark, without using the standardised scores that compared Tyson's results with the group.

Excellent   44%

These students answered parts a) and b) correctly. For part c), some commented that statistics was an easier test, but few mentioned the spread or variability of the marks.

Exam question reports show an exam question and how well students answered it, with examiners's comments, so you can see common errors and what an excellent answer should include.

Contents

- Topics that are fully covered in Higher GCSE maths (but not in Foundation GCSE maths).
- Topics that are not fully covered in Foundation or Higher GCSE maths.
- ▨ Topics that are in Higher GCSE Statistics only.

Chapter 1:
Collecting data

After completing this chapter you should be able to

- understand what is meant by statistics
- specify a line of enquiry and form a suitable hypothesis
- recognise the difference between quantitative and qualitative variables
- know the difference between discrete and continuous quantitative data and that the measurement of continuous data is subject to some error
- understand the meaning of bivariate data
- know the difference between primary and secondary data

- understand the meaning of the term population
- know what is meant by a census
- understand the reason for sampling and that sampling is used to estimate values in the population
- understand the terms 'random', 'randomness' and 'random sample'
- understand and use a variety of sampling methods
- know how to collect data using questionnaires, interviews and observation.

Who is better at spelling, boys or girls?

To find out, first suggest a hypothesis such as: 'Girls are better at spelling than boys'.

Then collect information, known as data, to test the hypothesis.

Think about:
- what sorts of data to collect
- where to collect them
- how to collect them
- how accurate the data are.

After reading this chapter you should be able to form a suitable hypothesis and collect data to prove or disprove the hypothesis.

1.1 The meaning of statistics

> **Statistics** are a way to answer questions using information. The information has to be observed or collected, ordered, represented and then analysed.

The information collected is called raw data. The data collected will depend upon the question you are trying to answer. This question is often called a **hypothesis**.

> A **hypothesis** is an assumption made as a starting point for an investigation. It may or may not be true.

'How does the price of a second-hand car change as the car gets older?' The hypothesis to help answer this question could be: 'The price of a second-hand car goes down as it gets older'.

The data needed to answer this question would be the age and price of second-hand cars.

In this chapter you will learn about the different types of data, how to choose samples and how to collect data.

1.2 Types of data

Raw data are obtained by collecting information such as the number of brothers or sisters a person has, or by measuring things such as people's heights.

The things being observed are known as variables because they vary from observation to observation, for example one person's eyes may be blue and the next person's may be brown.

Shoe size, height and eye colour are all variables.

Variable	Observation or measurement
Shoe size	5, 6, 7, 8, 9
Height	169 cm, 178 cm, 183 cm, 185 cm, 179 cm
Eye colour	blue, brown, green, brown, blue

Shoe size and height are numerical measurements (they are written as numbers). They are examples of **quantitative variables**.

Eye colour cannot be written as numbers. It is a quality that people possess. Eye colour is an example of a **qualitative variable**.

> **Quantitative variables** are numerical observations or measurements.
>
> **Qualitative variables** are non-numerical observations.

Quantitative variables (numerical data) can be **continuous** or **discrete**.

The length of a piece of string could take any value on this scale. It is continuous data.

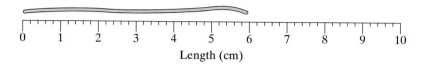

The number of eggs laid by a chicken can only take particular values. So it is discrete data

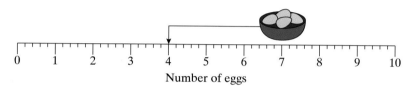

> **Continuous data** can take any value on a continuous numerical scale.
>
> **Discrete data** can only take particular values on a continuous numerical scale.

Weight, length, time, temperature and speed are all examples of continuous data.

Number of sisters, number of doors and shoe size are all examples of discrete data.

Example 1

Mr Jones is going to buy a new car.

He will consider these variables before deciding which car to buy:

miles per gallon	colour	number of passenger seats	engine size
number of doors	length	type of car	

Write the variables in three lists: qualitative variables, discrete quantitative variables and continuous quantitative variables.

Ask yourself:

Qualitative: colour, type of car.

Which cannot be described using numbers?

Discrete: number of doors, number of passenger seats.

Which can only take certain numerical values?

Continuous: miles per gallon, engine size, length.

Which can take any numerical values?

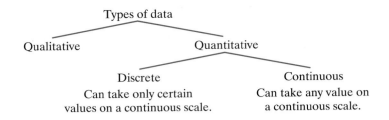

1.3 Categorical and ranked data

To make raw data easier to handle and easier to display they may be gathered or ordered in a particular way.

Categorical data and **ranked data** are examples of ordered data.

> A set of data is **categorical** if values or observations belonging to it can be sorted into different categories.

Each piece of categorical data is put into one of a set of non-overlapping categories.

Single-coloured shoes in a cupboard can be sorted according to colour. The characteristic 'colour' can have non-overlapping categories: 'black', 'brown', 'red' and 'other'.

Numerical data can be put into size categories. The characteristic 'engine sizes of cars' could have non-overlapping categories:

$$e \leqslant 1000 \text{ cc}, \quad 1000 \text{ cc} < e \leqslant 2000 \text{ cc}, \quad e > 2000 \text{ cc}.$$

> **Ranked data** have values/observations that can be ranked (put in order) or have a rating scale attached. Ranked data can be counted and ordered, but not measured.

The categories for a ranked set of data have a natural order. If judges rank ten dogs on a scale of 1 to 10, 1 would represent the best example of the breed and 10 the worst.

1.4 Bivariate data

In many statistical investigations pairs of variables are examined to find out how they are related or how changes in one variable affects the other variable.

Here are some examples:
- The price and age of second-hand cars.
- The distance and times taken for train journeys.

This is known as **bivariate data**.

> **Bivariate data** are pairs of related variables.

Example 2

Are each of the following categorical data, ranked data or bivariate data?

  a Students' year groups.

  b The ages and heights of students.

  c The league positions of football teams.

a categorical	The students are grouped by age, so the category is age.
b bivariate	Two sets of data are observed: age and height.
c ranked	The teams are ordered according to the number of points they scored.

1.5 Accuracy of continuous data

Continuous data are usually measured and rounded to the nearest sensible unit.

Numbers are usually rounded to the nearest 10, 100, or 1000 depending what is being measured.

The question is whether to round up or down? This is the easy bit!

If the last number is 5 or above, round up.

If the last number is less than 5, round down.

For example, the length of a field would probably be measured to the nearest metre. So, if the exact length is 235.3 m, it would be acceptable to say it is 235 m long.

> A measurement given correct to the nearest whole unit can be inaccurate by up to $\pm\frac{1}{2}$ unit.

Age is treated in a slightly different way. A person's age is usually given as their age at their last birthday. So, a person who is 16 years and 11 months old would be called 16 years old. The actual age of a person could be nearly 1 year more than their age at their last birthday, but could not be less.

Example 3

John cycles $4\frac{1}{2}$ miles to the nearest half mile.

What are the longest and shortest distances (upper and lower limits) that John's journey could actually be?

> The maximum error will be $\pm\frac{1}{4}$ mile.
>
> The longest distance $= 4\frac{1}{2} + \frac{1}{4} = 4\frac{3}{4}$ miles.
>
> The shortest distance $= 4\frac{1}{2} - \frac{1}{4} = 4\frac{1}{4}$ miles.

> John measures to the nearest half mile, so the maximum error will be half of this: $\pm\frac{1}{4}$ mile.
>
> He actually rides any distance between $4\frac{1}{2} \pm \frac{1}{4}$ miles, i.e. $4\frac{1}{4}$ up to $4\frac{3}{4}$ miles.

Exercise 1A

1 A food processing company thinks that males are the main buyers of their products.

It decides to investigate this.

Write a hypothesis the company could use.

2 A researcher wants to investigate whether Drug A has a better cure rate than Drug B.

Write a hypothesis he could use.

3 Are these discrete data or continuous data?

  A   The weight of a dog.

  B   The number of flowers in a bouquet.

  C   The time it takes to bake a cake.

4 Are these quantitative data or qualitative data?

  A   The makes of cars.

  B   The number of acorns on an oak tree.

  C   The weight of acorns on an oak tree.

5 Julita sold raffle tickets at a village fair.

The tickets were red, green, blue and yellow.

discrete   continuous   qualitative   quantitative

Which of the above can be used to describe:

  a   the number of tickets sold

  b   the colour of tickets sold?

6 Which of these could be ranked?

  A   The marks gained in a test by a group of students.

  B   The position of dogs in a dog show.

  C   The colours of sweets.

7 Write down **two** ways in which cars could be categorised.

8 Bivariate data are data that are related in some way.

Choose a word that could be used to make each pair bivariate data.

a Height and _____ of people.

b Hours of work and _____.

c Car age and _____.

9 Here are the ages, in years, of some people in a village.

32	40	17	34	58	60	15	14	22	29
44	18	26	31	36	42	18	23	25	38
31	33	28	47	65	72	19	77	30	34
37	34	58	56	60	63	42	15	82	17
40	61	33	42	21	31	42	72	16	22

Group these ages according to the categories: 0 to <10, 10 to <20, 20 to <30, 30 to <40, 40 to <50, 50 to <60, 60 to <70, 70 to <80, 80 to <90.

10 Karen says that her age is 16 years.

Explain why this response is actually *continuous* data but may appear to be *discrete*.

11 a A journey is 54.2 km. What is this to the nearest kilometre?

  b The weight of tomatoes picked from one plant is 4.7 kg. What is this to the nearest kilogram?

  c A water butt holds 60.5 litres. What is this to the nearest litre?

12 Find the upper and lower limits for:

  a a TV that costs £200, correct to the nearest £5

  b a car that costs £10 000, correct to the nearest £500

  c the height of a wall labelled as 4 m, correct to the nearest metre

  d a tree whose height is given as 5 m, correct to the nearest metre

  e a railway journey of 125 miles, correct to the nearest mile

  f a boy's height of 175 cm, correct to the nearest centimetre

  g the age of a man of 63 years

  h the distance from London to Manchester, which is given as 200 miles, correct to the nearest 5 miles.

1.6 Populations and sampling

The first thing to do in an investigation is to identify the **population**.

> A **population** is everything or everybody that could possibly be involved in an investigation.

Populations can vary in character as these two examples show.

The owner of Highfield Motor Company wants information about the number of miles travelled by the cars in the company garage. The population is all the cars in the garage.

Some people who are waiting in a queue for a shop to open are to be asked questions about their spending habits. The population is all people in the queue.

Census data

Sometimes, the entire population will be sufficiently small to include the entire population in a study. This type of research is called a **census** study because data are gathered on every member of the population.

> **Census** data contain information about every member of the population.

Workers for central and local government, health authorities, police and such organisations often use census data to help allocate resources or plan services for people.

The best known census is the Government's National Census. This takes place every ten years. All people who own or rent a house are identified and asked certain questions, including how many people live in their house.

Sample data

Usually, the population is too large to survey all of its members. A small, but carefully chosen, sample can be used to represent the population. The **sample** must reflect the characteristics of the population from which it is drawn.

> A **sample** contains information about part of the population.

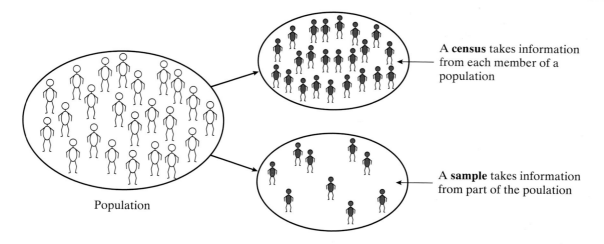

A **census** takes information from each member of a population

A **sample** takes information from part of the poulation

Population

	Advantages	Disadvantages
Census	Unbiased Accurate Takes the whole population into account	Time-consuming Expensive Difficult to ensure the whole population is used Lots of data to handle
Sample	Cheaper Less time consuming Less data to be considered	Not completely representative May be biased

The people or items in the population are called **sampling units**.

To take a sample, the population is formed into a list called a **sampling frame**.

> The **sampling units** are the people or items that are to be sampled.
> The **sampling frame** is a list of the people or items that are to be sampled.

There are two questions to be asked about a sample:

- How big does the sample need to be?
- How is a sample taken so that it represents the population accurately?

Sample size can vary but usually the larger the sample, the more reliable the results. It is important to remember what the sample is being used for and to balance the size of the sample with the accuracy needed. In opinion polls, the views of the whole nation are often found by using a sample of only 1500 people.

1.7 Random sampling

To represent the population accurately, the sample should be taken so that it is free from bias. This simply means that the results are not distorted in any way.

Bias can occur in many ways. Examples are:

- using an unrepresentative sample
- poor or misleading questions
- external factors affecting the data collection
- not correctly identifying the whole population.

> A **random sample** is chosen without a conscious decision about which items from the population are selected.

Random sampling methods include simple random sampling and stratified sampling.

Simple random sampling

Simple random sampling is the purest form of sampling. Each sample of size n has an equal chance of being selected.

There are many ways of taking a simple random sample. Each item of the sampling frame could be given a number and then the items to be included could be selected by:

- using a random number table
- using a random number generator on a calculator
- using a computer to choose numbers
- putting the numbers in a hat and then selecting however many you need for your sample.

Example 4

This is an extract from a random number table.

33 52 21 17 04 51 78 62 73 41

53 27 15 82 38 59 48 20 82 34

Starting at 33, and working across the rows, use the table to give eight numbers between 1 and 50.

> Start with 33 and take pairs of digits.
>
> Ignore any number larger than 50.
>
> 04 counts as the number 4.

33   21   17   04   41   27   15   38

Stratified sampling

There may often be factors that divide the population into sub-populations (known as 'strata') such as gender (male/female), age, earnings, etc.

This has to be considered when selecting a sample from the population to ensure it is representative of the population.

> **Examiner's Tip**
>
> 'Stratum' means just one sub-population. 'Strata' is the plural word, it means more than one sub-population.

To take a stratified sample from each stratum, or sub-population, a simple random sample is taken.

The size of each sample must be in proportion to the relative size of the stratum from which it is taken.

Example 5

The headteacher of a school of 1000 students wants to take a sample of 60 students. Here are the numbers of students in each year.

	Year 7	Year 8	Year 9	Year 10	Year 11
Students	250	250	200	150	150

How many students should be included from each year?

The sample for Years 7 and 8 will be $\dfrac{250}{1000} \times 60 = 15$.

> In Year 7 there are 250 out of the total 1000 students, so take $\frac{250}{1000}$ of 60.

The sample for Year 9 will be $\dfrac{200}{1000} \times 60 = 12$.

The sample for Years 10 and 11 will be $\dfrac{150}{1000} \times 60 = 9$.

> Calculate other numbers (you do not need to calculate Year 8 because it is the same as Year 7, or Year 11 because it is the same as Year 10).

Check: $15 + 15 + 12 + 9 + 9 = 60$

> Check the answer by totalling the answers from all year groups.

1.8 Non-random sampling

These are sampling methods in which the selection of the items is not random.

Methods include cluster sampling, quota sampling and systematic sampling.

Cluster sampling

Cluster sampling is used when the population being sampled splits naturally into groups or clusters.

A sample of the groups is randomly selected and the required information is collected from the sampling units within each selected group.

One version of cluster sampling is area sampling. Clusters consist of geographical areas such as towns.

Quota sampling

Quota sampling is a method of sampling widely used in opinion polling and market research. Each member of the population will have certain characteristics such as age, gender, etc. Instructions are given about the quota (or number) to be sampled from each section of the population who have a particular combination of these characteristics.

An interviewer might be told to interview 20 men over twenty years of age, 20 women over twenty years of age, 15 teenage girls and 15 teenage boys.

Systematic sampling

To take a systematic sample, choose a starting point from the sampling frame at random, and then choose items at regular intervals.

This example shows how to sample 20 students from a year group consisting of 100 students.

$\frac{100}{20} = 5$, so every fifth student is chosen after a random starting point between 1 and 5 has been chosen. If the random starting point is 4, then the students selected are numbers 4, 9, 14, 19, ..., 94 and 99.

Systematic sampling is used when the population is very large. It is simple to use, but is not always truly representative.

Sampling summary

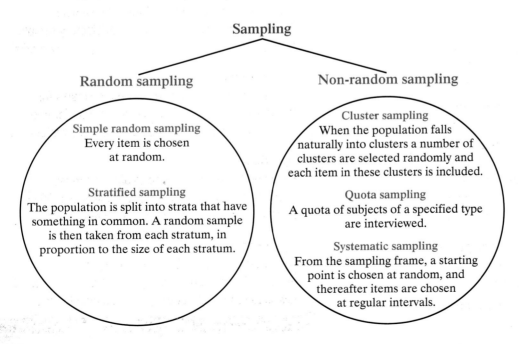

Exercise 1B

1 A new canteen is going to open at Edewell College.
The canteen manager wants to find out what students would like
on the menu. He decides to ask the students.

  a Write down the population he should use.

  b Describe a sampling unit.

  c Give **one** advantage and **one** disadvantage of him using a census.

2 An estate agent wants to get information about house prices
in the city where she works.

  a What is the population she will use?

  b Why would she not use a census of the house prices?

  She decides to use a sample. She also decides to use the prices
of all houses on her list of houses for sale.

  c Give a reason why this might be a poor sample.

3 A college decides to investigate the number of
hours that students study per week.

  a Describe the sampling frame the college will
use.

  b Describe a sampling unit.

4 **a** Give **two** advantages of using a sample rather
than a census.

  b Give **two** advantages of using a census rather
than a sample.

5 Jack wishes to find out how much people in Britain
are prepared to spend on a weekend break. He
asked people in his village.

  a Identify Jack's population.

  b Explain why Jack's sample is likely to be biased.

6 Write the name of the sampling method that is
being used in each of these cases.

  A Yves needs a sample of 20 people from a
numbered list of 100. He generates 20 random
numbers and uses those numbered people.

  B A factory manager requires a sample of 20
from his work force of 60 men and 40 women.
He randomly selects 12 men and eight women.

7 A nursery school has three age groups. The first age
group has 60 children, the second has 40 children,
and the third has 20 children.

Describe how you would get a sample of 30
children, stratified by age.

ResultsPlus
Build Better Answers

Question: A market research company is going to do
a national opinion poll.
They want to find out what people think about the
European Union.
The company is going to do a telephone poll.
First they will pick 10 towns at random.
Then they will pick 10 telephone numbers from the
telephone book for each town.
They will ring these 100 telephone numbers.
The people who answer will form the sample.
Discuss whether this will form a satisfactory sample
for the poll.   (2 marks)

■ Zero marks
Over a quarter of candidates scored no marks for
this question. The examiners were looking for two
reasons why the sample would not be satisfactory.

● Good
For a good answer you need to give one reason, for
example point out that the number involved (100 or
less) is too small to represent the whole country.

▲ Excellent
For an excellent answer you need to give two
reasons. A second reason might be that using only 10
towns to represent the country could lead to bias.

8 A university decides to investigate the use of the common room facilities.

It wants to ask a sample of 50 students in total from three year groups.

Year group 1 has 540 males and 420 females.

Year group 2 has 600 males and 660 females.

Year group 3 has 360 males and 420 females.

It decides to use a stratified sample.

ResultsPlus

Watch out!

■ Many students forget that the sample from each stratum should be taken randomly.

  a Describe the strata it will use.

  b Work out the number of males and females in each stratum that will be used.

  c Describe how it will choose the individual members of the strata.

9 Here is an extract from a table of random numbers.

335217	045178	627341	532715	823859	482082	342173
451739	936415	526338	127642	137284	463919	394821
264519	143857	012653	628491	558317	316832	229103

  a Select 10 random numbers each less than 50. Start at the top left-hand corner and work across from left to right.

  b Select 10 random numbers each less than 50. Start at the top left-hand corner and work down in pairs, and from left to right.

10 Tim is asked to take a random sample of 25 students from the registration roll at his school.

He attempts to do so by:

  - listing all their names in order
  - rolling a dice
  - selecting the student shown by the number on the dice (i.e. if the dice shows a 4, he selects the student numbered four in the list)
  - rolling the dice again. If it shows a 3, he selects the 4th + 3rd = 7th name on the list, and so on.

Give **two** reasons why Tim's method will not give him a random sample.

11 Write the name of the sampling method that is being used in each of these cases.

  A A health centre is interested in which of their facilities are most appreciated by patients. They send a questionnaire to every 20th person on their patient list starting at a random number between 1 and 20.

  B A market research company wants some information about the use of parking bays in a supermarket car park. They question 20 people in total from four different age groups of the population.

  C A company director wants to know what his workers think about the company pension plan. There are 20 departments in the factory. He asks people in eight of the departments.

1.9 Collecting data

Primary and secondary data

> **Primary data** are collected by, or for, the person who is going to use them.
> **Secondary data** have been collected by someone else.

Examples of **primary data** include:

- measuring hand spans of students in your class
- observing and tallying the colours of all the cars passing your house on a certain morning.

Secondary data may be collected from

- websites
- magazines, newspapers
- databases
- research articles.

Both forms of data collection have advantages and disadvantages.

Data	Advantages	Disadvantages
Primary	You know how the data were obtained. Accuracy is known.	Time-consuming. Expensive.
Secondary	Easy to obtain. Cheap to obtain.	Method of collection unknown. The data might be out of date. May contain mistakes.

Surveys

Primary data may be collected using a **survey**.

> A **survey** is the collection of data from a given population. The data are used to analyse a particular issue.

There are a number of methods of collecting primary data in a survey. The main methods are

- questionnaires
- interviews
- observations

Other methods include

- experiments
- data logging

Pilot surveys

> A **pilot survey** is conducted on a small sample to test the design and methods of that survey.

A **pilot survey** should identify any problems with questions, such as wording, likely responses, etc.

Questionnaires

Questionnaires are a popular means of collecting data, but are difficult to design. Many rewrites are often needed to produce an acceptable questionnaire.

Sometimes the questions can be factual, such as 'How old are you?', or they might ask for an opinion, such as 'What is your favourite colour?'.

> A **questionnaire** is a set of questions designed to obtain data.

Anyone answering a questionnaire is called a respondent.

It is important to follow these rules for writing a questionnaire.

- Keep the questions short, simple and to the point.
- Use words and phrases that are easily understood.
- Avoid leading questions that suggest a certain answer. For example, 'Don't you agree that Sudso is the best washing powder?' is a leading question that is asking for the answer 'Yes'.
- Write questions that only address a single issue. For example, questions such as 'Does your car run on diesel fuel?' should be broken down into two stages. Firstly find out if the respondent has a car and then secondly, if they own a car, find out if it is a diesel car.
- State units required but do not aim for too high a degree of accuracy. Use an interval, such as £10 to £15, 3–5 or to the nearest metre, rather than an exact figure.
- Avoid embarrassing words.

There are two types of question to use in a questionnaire: **open questions** or **closed questions**.

> An **open question** is one that has no suggested answers.

An open question could reveal an answer or response not considered by the person asking the question.

The main problem with open questions is that many different answers have to be summarised to enable them to be analysed.

ResultsPlus
Exam Question Report

A town council plans to build a swimming pool. It is going to carry out a survey to find out what people think of the plan.
The council should carry out a pilot study (pre-test).
(e) Give **two** reasons why.

How students answered

Poor   51%

Most students answered this question poorly. Common incorrect answers were: 'to check the results from the main survey'; 'to see if it is worth doing the actual survey'.

Good   31%

Giving one good reason gained one mark. For example: 'to see if there are any problems with the question wording or response boxes'.

Excellent   18%

Not many students gave excellent answers. Giving two good reasons gained the full 2 marks. A second good reason is: 'Makes sure the survey is designed and planned to collect the information needed'.

A **closed question** has a set of answers for the respondent to choose from.

A closed question has the advantage that it is easier to summarise the data.

Closed questions will often use an **opinion scale**.

For example:

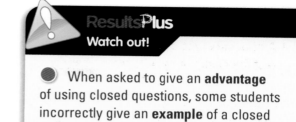

Read the following statement and then tick to show whether you strongly agree, agree, disagree or strongly disagree with the statement.

	Strongly agree	Agree	Disagree	Strongly disagree
Statistics is an easy subject				

One problem with opinion scales is that most people will answer somewhere near the middle. They are unlikely to indicate a strong opinion either way as they do not wish to seem extreme.

Sometimes the respondent may be asked to tick a box.

For example:

Tick one box to indicate your age group.

☐   ☐   ☐   ☐

  Under 21   21 to 40   41 to 60   over 60

Make sure that the possible answers are clear, do not overlap and cover all possibilities.

Example 6

State what is wrong with each of the following questions.

a   'How many brothers do you have?'

☐   ☐   ☐   ☐

  1   2 to 3   3 to 4   5

b   'You do support the idea of a school uniform, don't you?'

c   'Do you take drugs?' Tick the correct box.

☐   ☐   ☐

  Sometimes   Usually   A lot

d In what age range are you? Tick the correct box.

☐ ☐ ☐ ☐

under 10   10 to 20   20 to 30   40 to 60

a There isn't a box for someone who has no brothers or more than five brothers. The option for three brothers appears twice.

b This is forcing the respondent to answer 'Yes'.
There isn't a space for an answer.

c This is a sensitive question and people are unlikely to give a correct answer.
The terms are vague.
There isn't a box for 'Never'.

d 20 can be ticked in two boxes.
There isn't a box for 31 to 39.
There isn't a box for over 60s.

Look for:

- boxes that do not cover all possibilities
- boxes that cover one option more than once
- biased questions that try to persuade you to 'agree'
- questions that people are unlikely to answer honestly
- open questions that allow for personal opinions and do not have tick boxes

Interviews

Interviews may be conducted in different ways, for example:

- Sending questions to people by post or email.
- Calling people on the telephone.
- Face-to-face personal questioning.

Each of these has advantages and disadvantages.

- Postal surveys are cheap but the least likely to provide a response.
- Emails are cheap but they may often be ignored if people think they are spam messages.
- Telephone surveys cost more but the response rate is better than for postal surveys. There is a small possibility that the interviewer may cause bias.
- Personal interviews are the most costly and there is a greater chance of the interviewer causing bias. However, it is more suitable for complex questions where the interviewer can explain the question to the respondent.

Investigations and experiments

Data can be collected from experiments or by direct observation.

For example, flipping a coin many times and seeing how many times it came down heads would identify if the coin is biased.

Here are five possible ways of completing an experiment as part of a statistical investigation.

Before and after experiments

For example, looking at the number of accidents on a stretch of road before and after a 30 miles per hour speed restriction is enforced.

Control groups

A control group is often used to test things such as the effectiveness of drugs. People often feel better because they are receiving attention rather than because of the effect of the drug they are given. The control group and the group to be tested are randomly selected. The control group is then given an inactive substance and the other group is given the actual drug. The effect of the drug can then be assessed by comparing the two groups.

Matched pairs

Two groups of people are used to test the effects of a particular factor. Each individual in one group is paired with an individual in the second group who has everything in common with him/her except the factor being studied. Identical twins can be very important in this type of experiment.

Data logging

This is a mechanical or electronic method of automatically collecting primary data. The instrument is programmed to take readings at set intervals.

Capture–recapture method

This is a method for estimating the size of a population.

Suppose the population is of size N, so that N is the number to be estimated.

The first step is to capture M members of the population, mark (or tag) them and release them back into the population.

After waiting some time so that captured members have had time to mix, n members of the population are then captured, and the number, m, of these that are marked is recorded.

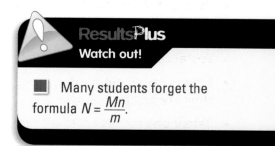

ResultsPlus
Watch out!

■ Many students forget the formula $N = \dfrac{Mn}{m}$.

The ratio of the members originally captured and marked, M, to the whole population, N, is assumed to be the same as the ratio of marked members who were recaptured, m, to the total number of members who were recaptured, n.

So $\dfrac{M}{N} = \dfrac{m}{n}$.

The population size can be found using the formula $N = \dfrac{Mn}{m}$.

The following assumptions are made in the capture-recapture method:

i   The population has not changed, that is there have been no members entering the population or leaving the population and no births or deaths between the release and recapture times.

ii   The probability of being caught is equal for all individuals.

iii Marks (or tags) are not lost and are always recognisable.

Example 7

Twenty birds from a bird colony were captured, ringed and released. Subsequently 100 birds were caught, of which 10 were already ringed. Estimate the number of birds in the colony.

$$\frac{20}{N} = \frac{10}{100}$$

$$N = \frac{20 \times 100}{10} = 200 \text{ birds}$$

Examiner's Tip
Use whichever formula you can remember. They both give the same answer.

Replication

Experimental replication means repeating experiments. If similar results are obtained each time, it is reasonable to rely on them more.

Direct observation

Investigations can also be done by direct observation. This involves recording the behaviour patterns of people, objects and events in a systematic manner. For example, recording the number of cars passing in every 10-minute interval.

Example 8

James wishes to see which class in Years 10 and 11 is better at arriving on time in the mornings.

How should he collect suitable data?

He could use class registers to record how many students were late.

Example 9

A creative design company is given a contract to design and market a lifestyle magazine for younger women.

a Explain how they could use both primary and secondary data.

b Explain a possible method of collecting primary data.

a Primary – survey, in the form of a questionnaire, for younger women to find out features of a magazine that they find most and least attractive.

Secondary – refer to other publications to see what they contain.

b By post, email, personal interview or telephone.

Exercise 1C

1 Are the data collected in these examples primary data or secondary data?

A Banji decides to investigate the amount of rainfall his garden gets in one month. He uses a measuring cylinder to collect the rainfall each day.

B A research student decides to investigate the sales of books. He collects data from several websites.

C A student decides to do a project on the milk yield of dairy cows. He gets his data from a local farm's records.

D A council decides to investigate the use of a waste disposal site. A council member goes to the site and collects data by questioning the people using the site.

2 James and Colin wish to predict the winners of the next Football World Cup.

James looks at the World Cup results from 2006, when Italy were the winners.

Colin looks at the table of all World Cup results starting from when the competition began in 1930 to the present day.

a What types of data are they considering?

b Whose opinion would you trust the most and why?

3 Kerry is writing a questionnaire about people's ages.

In it she asks the question 'How old are you?'

Young ☐   Middle-aged ☐   Old ☐

a What is wrong with the question and answers?

b Rewrite the question and answers to improve them.

4   Leslie is carrying out a survey about cricket teams.

He uses this question 'How often do you watch a cricket match?'

Never ☐   Once a week ☐   Whenever I can ☐

   a   This is not a good question. Explain why.

   b   Rewrite the question in a better form.

5   Which of these questions are open and which are closed?

   A   Do you like the food?   Yes/No

   B   What do you think about the proposed new town hall?

   C   How old are you?   Under 30 ☐   30 to 60 ☐   Over 60 ☐

6   A hotel leaves a questionnaire in the hotel rooms for guests to complete.

One of the questions on the questionnaire is 'Do you agree that this hotel has an excellent dinner menu?'

What is wrong with this question?

7   Write **three** things you need to think about when writing a question for a questionnaire.

8   Write **two** advantages and **two** disadvantages of using an interview to get information.

9   A large chemist company with 170 shops wants to obtain information about sales.

They decide to send out a questionnaire to all shops, but first they carry out a pilot survey.

What are the advantages of doing a pilot survey?

10   You are carrying out a survey to see how much money people will spend buying a car. Give **one** reason why you would choose to conduct a personal interview rather than a postal survey. Also explain why you might not choose to conduct personal interviews.

11   Twenty birds in a large aviary are caught and tagged. They are then returned to the aviary. Later, 40 birds are caught and two are found to have tags. Estimate the number of birds in the aviary.

12   Forty fish in a lake are caught, marked and returned to the lake. A second sample of 100 fish is later caught. Of these 100 fish, 10 are marked.

   a   Estimate the number of fish in the lake.

   b   Give **two** assumptions you made before estimating the number of fish in the lake.

13   Describe what is meant by a control group.

ResultsPlus
Watch out!

● When asked to comment on a question for a questionnaire, some students don't state clearly whether the question is unsuitable or not, and lose marks.

Chapter 1 review

1   There are three horses in a field.

Use **one** of these words to complete each sentence.

discrete   quantitative   qualitative   continuous   cumulative

   a   The colour of the horses is _____ data.

   b   The number of horses is _____ data.

★☆★★★
challenge

2   Which of these are continuous data and which are discrete data?

   A   Time

   B   Number of dogs

   C   Quantity of milk

★☆★★★
challenge

3   Which of these are qualitative data and which are quantitative data?

   A   Number of pets

   B   Height

   C   Make of car

★☆★★★
challenge

4   Which of these are primary data and which are secondary data?

   A   Data collected from a car magazine.

   B   Data from the BBC website.

   C   Data collected by asking questions of people at a supermarket.

★☆★★★
challenge

5   A council included this question in a questionnaire:

'Do you agree that the new roundabout is an improvement?'

Describe what is wrong with the question.

★☆★★★
challenge

6   Give **one** disadvantage of each of these:

   A   using a census

   B   using a sample.

★☆★★★
challenge

7   The headteacher in a primary school took a random sample of 10 boys and 12 girls from all the children in the school.

   a   What is the sampling frame used by the headteacher?

The headteacher asked each of these 22 children this question as part of a questionnaire:

'You go to bed before 9 pm, don't you?'

   b   Give **one** reason why the headteacher should not have asked the question in this way.

★★★★★
challenge

8   A hospital decides to investigate whether they have more women than men visiting their Accident and Emergency Department.

Write a suitable hypothesis they could use for the investigation.

★★★★★
challenge

9 Here are some examples of different types of data.

A The number of people in a café.

B The time it takes to go to work.

C The colour of a dress.

a Which **one** of these is continuous data?

b Which **one** of these is qualitative data?
In a survey it is decided to use secondary data.

c Write **one** advantage and **one** disadvantage of using secondary data.

edexcel ⠿ *past paper question*

10 A hotel owner wants to give his guests information about the number of hours of sunshine they can expect in June.

a Write **one** way that he can collect the information if he wants to use primary data.

b Write **one** way that he can collect the information if he wants to use secondary data.

c Are the data he collects qualitative or quantitative?

edexcel ⠿ *past paper question*

11 In a study about smoking, a doctor selected a sample of adult patients from those registered with him. He looked at their records to see to what degree they claimed to smoke.

a What is the population being studied?

b The doctor wishes to get the number of males and the number of females in proportion to the numbers on his register. What sampling method should he use?

> **Examiner's Tip**
> Don't forget to say how you pick within each stratum.

c The doctor's information on smoking was obtained by asking the patients at the time of setting up a database. Give **two** reasons why the data obtained may be unreliable.

edexcel ⠿ *past paper question*

12 A manufacturer makes two types of ropes – Twineasy and Plasuper.

The manager of the company thinks that Twineasy is the stronger rope. He decides to investigate this.

a Write a suitable hypothesis he could use.

b What would form his population?

c Why would he use a sample to test the hypothesis?

d Describe a sampling unit he will use.

13 The table shows the number of students in each of the four Year 11 maths classes in a school.

challenge

Maths class	Number of pupils
Class 1	35
Class 2	25
Class 3	20
Class 4	10

A sample of size 30 is to be taken from Year 11. Omar suggests that three of the classes are chosen at random and 10 students selected at random from each class.

a Would this method give a random sample? Explain your answer.

Nesta suggests a stratified sample of size 36 from the whole of Year 11 using classes as the strata.

b How many students from Class 1 should be in the sample?

edexcel ⠿ *past paper question*

Chapter 1 summary

Data

1 **Statistics** are a way to answer questions using information. The information has to be observed or collected, ordered, represented and then analysed.

2 A **hypothesis** is an assumption made as a starting point for an investigation. It may or may not be true.

3 **Quantitative variables** are numerical observations or measurements.
Qualitative variables are non-numerical observations.

4 **Continuous data** can take any value on a continuous numerical scale.
Discrete data can only take particular values on a continuous numerical scale.

5 A set of data is **categorical** if values or observations belonging to it can be sorted into different categories.

6 **Ranked data** have values/observations that can be ranked (put in order) or have a rating scale attached. Ranked data can be counted and ordered, but not measured.

7 **Bivariate data** are pairs of related variables.

8 A measurement given correct to the nearest whole unit can be inaccurate by up to $\pm \frac{1}{2}$ unit.

Sampling

9 A **population** is everything or everybody that could possibly be involved in an investigation.

10 **Census** data contain information about every member of the population.

11 A **sample** contains information about part of the population.

12 The **sampling units** are the people or items that are to be sampled.
The **sampling frame** is a list of the people or items that are to be sampled.

13 A **random sample** is chosen without a conscious decision about which items from the population are selected.

14 **Primary data** are collected by, or for, the person who is going to use them.
Secondary data have been collected by someone else.

Surveys

15 A **survey** is the collection of data from a given population. The data are used to analyse a particular issue.

16 A **pilot survey** is conducted on a small sample to test the design and methods of that survey.

17 A **questionnaire** is a set of questions designed to obtain data.

18 An **open question** is one that has no suggested answers.

19 A **closed question** has a set of answers for the respondent to choose from.

Test yourself

1 *continuous   discrete   quantitative   qualitative   primary   secondary*
Which of the above words can be used to describe the following data?

   a Height

   b Colour

   c Number of aunts

   d Time

   e Census information on a website

   f A tally you make of car types.

2 Give **two** advantages and **two** disadvantages of using:

   a primary data

   b secondary data.

3 Describe briefly the meaning of:

   a population

   b census.

4 Write **two** advantages of using a sample rather than a census.

5 What is the name given to a sample that allows everyone or everything to have an equal chance of selection?

6 Is this a closed or an open question?

   'What do you think about the new hall?'

   Give a reason for your answer.

7 Explain **two** ways of selecting a set of random numbers.

8 Write **two** advantages of using a pilot survey.

9 Explain what is meant by a 'control group'.

10 Write the name of the sampling method for each of these.

   a Maeve forms her sample by picking boys and girls from her class in proportion to the numbers of boys and girls.

   b Jack has a list of 50 students. He uses every 5th student to form a sample.

Chapter 2:
Representing and processing qualitative and discrete data

After completing this chapter you should be able to

- construct tally charts, frequency tables and cumulative frequency tables
- group data into classes including classes of varying width and open ended classes
- construct two-way tables

- read and interpret data presented in tabular or graphic form
- construct, draw, use and understand pictograms, bar charts including multiple and composite bar charts, vertical line graphs, stem and leaf diagrams, pie charts and comparative pie charts.

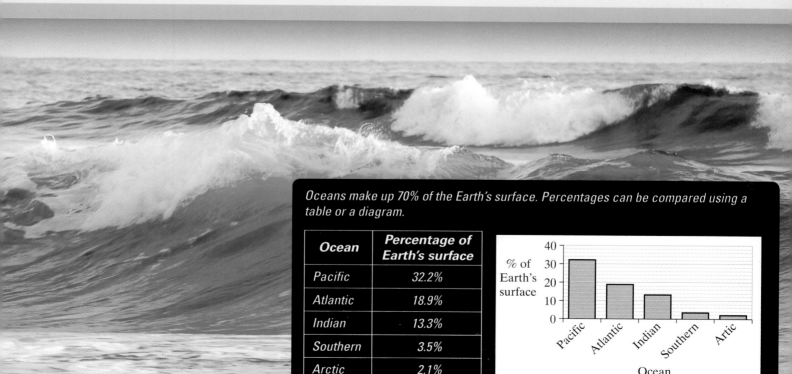

Oceans make up 70% of the Earth's surface. Percentages can be compared using a table or a diagram.

Ocean	Percentage of Earth's surface
Pacific	32.2%
Atlantic	18.9%
Indian	13.3%
Southern	3.5%
Arctic	2.1%

When you have completed this chapter you will be able to represent and process qualitative data and discrete quantitative data, like the ocean data. You will know how to order this sort of data, draw diagrams like the one shown and make comparisons.

2.1 Tally charts and frequency tables

Before data are processed they are known as raw data.

Here are some raw data: the colours of the front doors of 30 houses chosen at random.

They are qualitative data.

blue	brown	red	yellow	red	red
brown	red	green	green	yellow	red
green	blue	red	green	blue	brown
yellow	blue	blue	brown	blue	green
blue	red	brown	blue	blue	red

It is not easy to see the pattern or trend of these raw data.

To make these data easier to interpret they are sorted into a **frequency table**.

Colour	Tally	Frequency
blue	ЖЖ IIII	9
brown	ЖЖ	5
green	ЖЖ	5
red	ЖЖ III	8
yellow	III	3

> Go through the raw data and each time you see 'blue' put a tally mark in the table next to blue. Make groups of five tally marks with the fifth tally mark drawn through the other four ЖЖ . This makes them easier to count.
> Repeat for the other colours.
> Add up the tally marks for each colour to find the frequencies.

- The first column shows the colours of the front doors.
- Each door is represented by one tally mark in the second column.
- The frequency column shows the number of times each colour occurs.

The most common colour is blue, with a frequency of 9. The least common colour is yellow, with a frequency of 3.

The same idea may be used for discrete data, as Example 1 shows.

> A **tally chart**, or **frequency table**, can be used to process raw data. They make it easier to spot patterns.

Example 1

These are the numbers of bedrooms in each of 40 houses chosen at random.

5 6 3 4 5 2 6 4 5 7 3 8 4 4 3 5 4 2 7 8
6 3 4 5 3 3 5 5 3 4 6 2 4 5 3 6 4 4 3 4

Sort the data into a frequency table and comment on it.

Number of bedrooms	Tally	Frequency
2	|||	3
3	ⵜHT ||||	9
4	ⵜHT ⵜHT |	11
5	ⵜHT |||	8
6	ⵜHT	5
7	||	2
8	||	2

Start with the top row of data and work from left to right. Make a tally mark for each house next to the appropriate number of bedrooms.

Count up the tallies for each number of bedrooms. Write these in the frequency column.

Four-bedroom houses are the most frequent. Houses with 7 and 8 bedrooms are the least frequent.

Look to see which had the highest frequency and which had the lowest frequency.

2.2 Cumulative frequency tables

Sometimes quantitative data are presented in a **cumulative frequency** table.

The **cumulative frequency** of the value of a variable is the total number of observations that are less than or equal to that value.

Example 2

This frequency table gives information about the number of matches in each of 50 boxes.

Number of matches	Frequency	Cumulative frequency
47	3	
48	10	
49	18	
50	12	
51	7	

Examiner's Tip
The final cumulative frequency must be the same as the total number of observations.

Complete the table by filling in the cumulative frequencies.

Number of matches	Frequency	Cumulative frequency
47	3	3
48	10	13
49	18	31
50	12	43
51	7	50

Add the frequency of an item to the cumulative frequency for the previous item to get its cumulative frequency.
- There are 3 boxes with 47 or fewer.
- There are 3 + 10 = 13 boxes with 48 or fewer.
- There are 13 + 18 = 31 boxes with 49 or fewer.
- There are 31 + 12 = 43 boxes with 50 or fewer.
- There are 43 + 7 = 50 boxes with 51 or fewer.

2.3 Grouping discrete data

Sometimes data can have a wide range of values, with few values that are the same. Such data may be sorted into groups or **classes**.

> If data are widely spread they should be grouped into **classes**.

Example 3

Sixty computer game scores were collected.

0	7	34	40	52	53	52	24	48	32
56	3	2	35	42	55	56	14	23	34
51	54	6	6	49	48	55	57	12	29
62	63	73	4	22	45	44	45	47	5
71	79	70	77	3	28	44	47	49	2
72	78	71	72	75	9	27	43	46	9

Sort the 60 pieces of data into classes.

Score	Frequency
0–9	12
10–19	2
20–29	6
30–39	4
40–49	14
50–59	10
60–69	2
70–79	10
Total	60

Decide on the classes. Each class in this table covers ten different integers. The intervals 0–9, 10–19, etc. are called **class intervals**.

Make sure your class intervals don't overlap. For example, if you had 0–10 and 10–20, which class interval would 10 be in?

Using class intervals makes it easier to see the distribution of the data and to spot any patterns.

Tally marks do not need to be included in every frequency table.

When grouping data, think about the number of class intervals and the width of these intervals:

- If there are not enough classes, important detail may be lost.
- If too many classes are used, the classes will be very small which could hide any pattern.

Exercise 2A

1 Here are the colours of 50 cars in a car park.

red	white	white	blue	white	yellow	white
blue	white	red	blue	red	yellow	white
blue	blue	red	black	white	white	black
white	white	blue	red	white	white	black
yellow	blue	white	red	white	red	red
red	red	black	red	blue	black	blue
blue	red	black	white	white	black	white
white						

Copy and complete the frequency table to show the colours of the cars.

Colour	Tally	Frequency
black		
blue		
red		
white		
yellow		
Total		

2 Last season, Woodbank football team played 30 matches.

This is a list of the number of goals scored in each match.

2	3	0	1	2	4	3	0	2	1
2	2	1	3	0	2	4	0	2	1
0	5	3	2	1	4	1	2	0	1

Design and complete a frequency table to show the distribution of the number of goals scored.

3 A company sells drawing pins in boxes. They claim that there are 50 drawing pins in each box.

Des checks the contents of 20 boxes. The number of drawing pins in each box is shown below.

48	52	51	50	50	50	49	50	50	51
53	50	51	52	49	50	50	52	49	50

a Design and complete a frequency table to show the number of drawing pins in each box.

b Do you think that the company's claim that 'there are 50 drawing pins in each box' is reasonable?

4 The length of a word can be measured by counting the number of letters it contains. Count the length of each of the 34 words in this question. Record your answers in a frequency table.

5   A mathematics test is marked out of 100. Here are the marks for 60 students.

71	62	40	72	59	63	43	81	44	23
55	52	55	58	66	31	45	54	57	59
63	61	54	42	35	47	33	62	41	73
57	82	26	71	52	48	38	65	52	56
68	36	49	63	57	53	77	65	27	88
41	62	35	47	63	39	62	43	46	51

a   Copy and complete the frequency table to show the students' marks.

Mark	Tally	Frequency
20–29		
30–39		
40–49		
50–59		
60–69		
70–79		
80–89		
Total		

b   The pass mark for this test was 40 out of 100.

How many students passed the test?

6   A newsagent recorded the number of newspapers sold on each day in January:

40	62	67	40	49	52	57	42
46	44	48	55	53	51	56	58
58	59	60	44	52	63	48	49
42	53	57	56	53	61	51	

a   Draw and complete a frequency table, using class intervals 40–44, 45–49, and so on.

b   In order to cut costs, the newsagent decides that he will stock only 60 newspapers each day. In January, on how many days would he have sold out of newspapers?

7   Forty people took part in a competition flying model aeroplanes.

The competitors included experts and beginners.

Each competitor was given a score out of 120.

Here are the scores awarded:

111	97	36	41	115	15	112	99	56	105
73	71	47	33	46	105	109	22	56	52
109	43	36	95	17	48	85	107	42	35
56	28	103	59	57	116	38	29	53	61

  a Draw a frequency table, using class intervals of 1–10, 11–20,
     and so on.

  b Using the same data, draw a new frequency table, using class
     intervals of 1–20, 21–40, and so on.

  c Using the same data, draw a third frequency table, using class
     intervals of 1–40, 41–80 and 81–120.

  d Which of the three frequency tables gives the best information
     about the distribution of scores? Explain your answer.

  e Explain the limitations of the other two frequency tables.

8 **Activity:** you will need three coins.

   Flip all three coins at the same time, and count how many show
   'heads'.

   Conduct this experiment 25 times.

   Record your answers in a frequency table.

2.4 Classes of varying width and open-ended classes

Data do not always have to be grouped into class intervals of equal
width. Many of the grouped frequency distributions appearing in
government publications have unequal class widths.

> When there is an uneven spread of data across the range, class intervals
> may vary in width.

A grouped frequency distribution can include one or two class intervals
that are open-ended. For an interval 0–9 you could use less than 9 (<9)
and the last interval could, for example, be over 70 (>70).

> When the extreme values of data are not known the first and/or last
> intervals may be left open.

Example 4

In a competition, students were loaded with exercise books until they dropped one.

The raw data below show the number of exercise books balanced by the first 50 students.

7	37	41	33	44	31	49	15	32	40
69	45	42	40	38	52	27	42	31	37
36	31	39	53	46	82	41	47	43	42
38	39	40	43	41	37	32	51	11	33
32	66	45	42	39	35	47	58	3	44

Sort these data into groups.

Number of exercise books held	Frequency
0–29	5
30–34	8
35–39	10
40–44	14
45–49	6
50–59	4
60–	3
Total	50

First decide on the classes. Although there is a large range, there are few data below 30 or over 50 so use wider class intervals at the extremes.

Most of the data are between 30 and 49 so these class intervals need to be smaller.

The maximum number of books a person can hold without dropping one is unknown, so leave this end class open.

2.5 Two-way tables

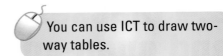
You can use ICT to draw two-way tables.

A **two-way table** allows you to show information about two related categorical variables, such as a student's gender and the class they are in.

These are called bivariate data.

A **two-way table** displays frequencies for two categories of data.

Example 5

The two-way table shows information about the gender and tutor group of students in Year 11 at Hitchen High School. Complete the table.

	11A	11B	11C	11D	Total
Boys	18	16	13	14	
Girls	12	17	14	19	
Total					

The table shows the frequencies of boys and girls in each tutor group.

There are 14 boys and 19 girls in tutor group 11D.

	11A	11B	11C	11D	Total
Boys	18	16	13	14	61
Girls	12	17	14	19	62
Total	30	33	27	33	123

Find row and column totals by simple addition.

$18 + 16 + 13 + 14 = 61$

$14 + 19 = 33$

Sometimes you may be asked to fill in missing information in a two-way table.

Example 6

This table shows the results of a survey of houses on an estate. Complete the table.

| | Type of house | | | |
Accommodation	Detached	Semi-detached	Terraced	Total
2 bedrooms	1	3		10
3 bedrooms	4		12	24
4 bedrooms		4	2	
Total	11		20	

Examiner's Tip
Look for rows or columns with only one figure missing.

| | Type of house | | | |
Accommodation	Detached	Semi-detached	Terraced	Total
2 bedrooms	1	3	6	10
3 bedrooms	4	8	12	24
4 bedrooms	6	4	2	12
Total	11	15	20	46

1 Row 10 − (3 + 1) = **6**

2 Row 24 − (12 + 4) = **8**

3 Column 11 − (4 + 1) = **6**

4 Row 6 + 4 + 2 = **12**

5 Column 3 + 8 + 4 = **15**

6 Row or column
10 + 24 + 12 = **46**
11 + 15 + 20 = **46**
Check totals are the same.

Exercise 2B

1 Henry asked 30 students in his class how much money they had in their pockets.

These are the amounts.

£1.20	£2.43	76p	0p	£1.63	£1.42
£2.09	£1.80	£1.36	£10.50	37p	£1.28
£2.61	£1.60	£1.50	£2.00	£1.22	£1.55
£3.50	£2.32	£1.40	£1.50	£2.00	£1.87
£1.75	£2.50	£1.35	£1.40	£1.59	£2.05

Copy and complete the frequency table.

Amount	Tally	Frequency
£0–£1.00		
£1.01–£1.50		
£1.51–£2.00		
£2.01–£3.00		
£3.01–		
Total		

2 Here are the batting scores for 50 cricket players.

30	24	15	31	23	28	32	29	33	48
31	37	42	18	20	34	40	25	36	31
29	32	26	33	25	27	32	22	29	28
21	35	34	29	30	34	26	32	22	31
29	35	19	28	24	33	27	50	32	27

a Write down the lowest score.

b Write down the the highest score.

c Design and complete a frequency table for these data.
Use classes of varying width.

3 Eighty people were asked how many television programmes they
had watched in one week.

The results of the survey are shown in two different frequency
tables.

Table 1 Equal class intervals

Number of programmes	Frequency
0–9	4
10–19	54
20–29	16
30–39	4
40–49	1
50–59	0
60–69	1
Total	**80**

Table 2 Varied class intervals

Number of programmes	Frequency
0–8	3
9–12	21
13–16	8
17–20	32
21–30	11
31–40	3
41–	2
Total	**80**

a Give two limitations of Table 1.

b In Table 2, why has the last class been left open?

c Only one person watched over 60 programmes in one week.
What reason could there be for this not happening more
often?

d What extra information can you read from Table 2 that was
hidden in Table 1?

4 Sue and John conducted a survey into the ages of 125 people at a
classical music concert. They used the same data but drew
different frequency tables.

These are their frequency tables.

Sue's table

Age	Frequency
0–9	1
10–19	0
20–29	2
30–39	55
40–49	56
50–59	8
60–69	2
70–79	0
80–89	1
Total	125

John's table

Age	Frequency
0–29	3
30–34	18
35–39	37
40–44	43
45–49	13
50–59	8
60–	3
Total	125

a   What is the main difference between the two frequency tables?

b   Explain why John has made such a wide class interval for people below the age of 30.

c   Which frequency table shows more detail about the most common age ranges? Explain your answer.

d   Why did John leave the last interval open?

e   How could Sue have improved her frequency table? Give **two** ways.

5   Charles asked 87 adults and children whether they were right-handed or left-handed.

	Adults	Children	Total
Right-handed	32		
Left-handed		22	
Total	47		87

a   Copy and complete the two-way table.

b   How many right-handed children were in the sample?

Charles thinks that a child is more likely to be left-handed, than an adult.

c   Do these data support his view?

6   Copy and complete the two-way table to show drinks chosen by 28 children at a birthday party.

	Lemonade	Orange juice	Total
Girls		9	
Boys	10	6	
Total			

Hint: Fill this one first!

7 Victoria conducted an experiment to see whether or not a piece of buttered toast was more likely to land 'butter-side' down.

Each time, she either dropped the toast or threw it.
- She conducted the experiment 82 times in total.
- She dropped the toast 37 times in total.
- When the toast was thrown, it landed 'butter-side down' on 21 occasions.
- When the toast was dropped, it landed 'butter side-up' on 11 occasions.

a Copy and complete the two-way table.

	'Butter-side down'	'Butter-side up'	Total
Dropped			
Thrown			
Total			

b In the experiment, which way up did the toast land most frequently?

8 In a survey, teachers of different subjects were asked how they preferred to travel to work. Some of the results are shown in this table.

	Car	Bus	Cycle	Walk	Other	Total
English		4	0	1	0	
Games	3	1	18		3	32
Geography	8		1	18		32
Maths	28	3	1		1	
Science	16	5	7	6		
Total		17		33	9	156

a Copy and complete the two-way table.

b How many science teachers were involved in the survey?

c In total, how many teachers travelled by car?

d How many maths teachers preferred to walk to school?

2.6 Other tables and databases

A **database** is an organised collection of information.

Computers can store huge **databases** that can be easily selected, sorted and ordered at the touch of a button.

A **summary table** shows data that have been sorted and summarised. It is easier to interpret than the original data.

Example 7

The table gives information about world production of nuclear electricity in 1995.

Producer	Percentage of world total	Installed capacity (Gigawatts)	Percentage of total domestic electricity
Canada	4.2	15	17.7
France	16.2	60	77.1
Germany	6.6	22	28.9
Japan	12.5	44	29.7
Korea	2.9	9	36.3
Russia	4.3	20	11.6
Sweden	3.0	10	47.6
Ukraine	3.0	12	36.3
UK	3.8	14	26.7
USA	30.6	102	20.1
Rest of world	12.9	46	5.6
World total	100	354	17.7

Source: International Energy Agency

a What percentage of the world total was produced by the UK?

b Which countries had the capacity to produce more than 50 Gigawatts of nuclear electricity?

c Which country produced less than 4% of the world total of nuclear electricity but more than 40% of its total domestic electricity was nuclear?

Examiner's Tip
There could be several answers to part **b** so check through the whole table.

a 3.8%

Look down the 'Percentage of world total' column to where it crosses the UK row and read off the percentage.

b France and the USA

Look down the 'Installed capacity' column for values greater than 50 and read off the names of countries in those rows.

c Sweden

Look down the 'Percentage of world total' column and note those that are less than 4%. Look across to the 'Percentage of total domestic electricity' to see if any generated more than 40% of their total by nuclear electricity.

Exercise 2C

This exercise shows a variety of tables that you may come across in everyday life.

1

In the table, all distances are given in kilometres.

a What is the distance between
   i) Liverpool and Dover
   ii) Newcastle and Edinburgh
   iii) London and Cardiff?

b Mrs McQueen travels from Dover to London, and then on to Newcastle. How far does she travel altogether?

c Mr King is a travelling salesman. In one week he travelled from London to Cardiff, to Liverpool, to Newcastle and finally returned to London. How far did he travel in that week?

2 Here is part of a train timetable.

Train timetable – Andwich to Elchester									
	X			**X**	**S**				**S**
Andwich	07:30	08:45	10:40	12:55	13:15	15:30	17:25	19:50	23:30
Balstone	07:42	08:57	10:52	13:07	13:27	15:42	17:37	20:02	23:42
Ciffingham	07:58	09:13	11:08	13:23	13:43	15:58	17:53	20:18	23:58
Dilsbury	08:09	09:24	11:19	13:34	13:54	16:09	18:04	20:29	00:09
Elchester	08:20	09:35	11:30	13:45	14:05	16:20	18:15	20:40	00:20
	X service not available on Saturday								
	S Saturday only								

a What time is the earliest train from Andwich to Elchester on a Monday?

b What time is the earliest train from Andwich to Elchester on a Saturday?

c Winston catches the train from Ciffingham at 11:08.
   What time does he arrive at Dilsbury?

d Cathy catches the train from Balstone at 20:02. What time does she arrive at Elchester?

   e How long does the train take to travel between
      i) Andwich and Balstone
     ii) Balstone and Ciffingham
    iii) Ciffingham and Dilsbury
    iv) Andwich and Elchester?

   f Daniel arrives at Dilsbury at 13:30 on a Saturday.
   What time is the next train to Elchester?

3 This table gives information about the number of egg-laying hens
   and pullets in England over three years.

Type	Number (in thousands)		
	June 2005	June 2006	June 2007
Caged hens	14 712	12 935	12 091
Barn hens	1278	1172	992
Free-range hens	7294	8457	7982
Pullets	8823	7628	6354

Source: Defra Survey of Agriculture, June 2007

   a How many barn hens were there in June 2006?
   b Describe what has happened to the number of caged hens over
   the three years.
   c Work out how many more pullets there were in June 2005 than
   June 2007.
   d Work out the percentage change in the number of free-range
   hens between 2006 and 2007.

4 The price of car insurance is dependent on many factors. These
   include the gender of the driver, and the area in which the driver lives.
   The table shows the price of insurance for a particular make and
   size of car.

Age	Gender	Area				
		A	B	C	D	E
17–25	M	£484	£366	£633	£500	£558
	F	£387	£293	£506	£400	£446
26–35	M	£397	£300	£519	£410	£458
	F	£315	£238	£411	£325	£363
36–50	M	£242	£183	£317	£250	£279
	F	£194	£146	£253	£200	£223
51–	M	£266	£201	£348	£275	£307
	F	£266	£201	£348	£275	£307

   a How much would the insurance cost for
      i) Arthur, a 42-year-old male who lives in area C
     ii) Amrita, a 20-year-old female who lives in area E
    iii) Michael, a 70-year-old male from area A
   b In which age group do males and females pay the same for
   their insurance?

    c How much more would insurance cost a 33-year-old man than a woman of the same age, both living in area B?

    d Describe a person who pays the most for insurance according to this table.

    e Describe a person who pays the least for insurance according to this table.

5 This table gives information about ten countries.

Country	Capital city (official)	Population density (People per km²)	Life expectancy (years)	Birth rate (births/1000)	Death rate (deaths/1000)
Denmark	Copenhagen	126	77.9	11.13	10.43
Italy	Rome	193	79.81	8.72	10.30
Belgium	Brussels	341	78.77	10.38	10.22
Greece	Athens	84	79.24	9.68	10.15
Ireland	Dublin	59	77.56	14.45	7.85
Austria	Vienna	98	79.07	8.74	9.70
Spain	Madrid	85	79.65	10.06	9.63
Norway	Oslo	12	79.54	11.46	9.45
France	Paris	110	79.73	11.99	9.08
Switzerland	Bern	176	80.51	9.71	8.48

Source:www.doheth.co.uk

    a Write down the name of the capital of Austria.

    b Write down the country that has the highest life expectancy.

    c Write down the names of countries that have a higher death rate than birth rate.

    d Write down the name of the country that is least populated.

    e Work out which country has the greatest difference between its birth rate and death rate.

6 The summary table shows eight hotels and the facilities they offer.

    a Which two hotels have a fitness suite?

    b Which hotel has both a kids' play area and bicycle hire?

    c Which hotels could you go to if you wanted both a wake-up call and room service?

    d Raj enjoys swimming and cycling and likes to have a wake-up call in the morning. Which hotels would you recommend?

    e Which hotel does not have a bar?

	Evening meal	Swimming pool	Bicycle hire	Fitness suite	Bar	Laundry	Room service	Wake-up call	Satellite TV	Kids' play area
The HILLSTONE	✓	✓	✓		✓	✓		✓		✓
FIVEWAYS	✓		✓		✓	✓		✓	✓	
ROLLING HILLS	✓		✓		✓				✓	
PORTENDALES	✓	✓		✓		✓	✓	✓		
The MARION	✓	✓	✓		✓		✓	✓		
The TOWN	✓	✓		✓	✓	✓	✓			✓
WALKERSTONES	✓	✓	✓		✓		✓			
The RED TIGER	✓				✓			✓	✓	✓

7 This table shows the numbers of road accident casualties in thousands. The accidents were in Northern Ireland in the years 1986 to 2000 and all involved illegal alcohol levels.

Year	Fatal injuries	Serious injuries	Slight injuries	Total casualties
1986	1.03	6.57	19.60	27.20
1987	0.93	6.01	17.99	24.93
1988	0.81	5.18	17.25	23.24
1989	0.84	4.92	17.05	22.81
1990	0.80	4.23	16.01	21.04
1991	0.69	3.72	14.00	18.41
1992	0.69	3.40	13.28	17.37
1993	0.57	2.82	12.25	15.63
1994	0.54	2.95	12.26	15.75
1995	0.56	3.10	12.89	16.56
1996	0.60	3.13	13.93	17.67
1997	0.57	3.07	13.90	17.55
1998	0.49	2.68	13.25	16.42
1999	0.48	2.60	14.64	17.72
2000	0.56	2.71	15.75	19.02

Source: Department for Transport, Royal Ulster Constabulary

ResultsPlus
Watch out!

■ In this type of question, very few students mentioned that the numbers given are 'thousands' so they have been rounded – which can lead to inaccuracies.

a Write down the number of total casualties in 1989.

b The total number of casualties in 1996, found by adding together the Fatal, Serious and Slight injuries columns comes to 17.66 thousand.

The number of casualties in the total casualties column is 17.67 thousand.

Give a reason for this difference.

c Describe the trend in the total numbers of fatal injuries in the years

　i) 1986 to 1991

　ii) 1993 to 2000.　　　　　edexcel ▦ *past paper question*

8 **Activity:** ask 10 students in your class for the following data.
- the month of their birthday
- their height in centimetres
- their gender
- their shoe size.

Design a summary table to show these data.

2.7 Pictorial representation

Data that are presented in the form of tables may not be as clear or easy to understand as the same data presented as pictures or diagrams.

Pictorial representation lets you:

- show clearly any relations or patterns in the data
- show how big or small the differences are between the data items
- quickly and accurately show the important information.

Frequencies can be represented in two ways.

- By the length of a line or bar.
- By an area.

There are many ways of representing data pictorially. You need to be able draw and interpret pictograms, bar charts, line graphs and pie charts.

2.8 Pictograms

A **pictogram** is used to show the frequency of qualitative data. Pictograms should have a **key**.

> A **pictogram** uses symbols or pictures to represent a certain number of items.
>
> It must have a **key** to tell you the number of items represented by a single symbol or picture.

Example 8

This pictogram shows the number of computers in the art department in each of four schools.

Woodridge High

Ecliffe Secondary

Hursley Comprehensive

Caslehey High

> In pictograms, each picture is related to what it represents. In this case it is a computer picture.

Key: 　represents 4 computers

a　Which art department had the most computers?

b　How many computers are there in the art department at Woodridge High School?

c　How many computers are there in the art department at Hursley Comprehensive?

a   Caslehey High has most computers.

Look to see which school has the greatest number of symbols.

Each picture represents 4 computers so it is easy to see that one quarter of a picture represents 1 computer.

b   16

Look at Woodridge High, it has 4 complete symbols, each representing 4 computers.

$4 \times 4 = 16$ computers

c   10

Hursley Comprehensive has $2\frac{1}{2}$ complete symbols.

$2\frac{1}{2} \times 4 = 10$ computers

When drawing a pictogram, make sure that:
- each picture is the same size
- the picture can be divided easily to show different frequencies
- the spacing between the pictures is the same in each row
- you give a key.

Exercise 2D

1   Woodridge High School organised an activity day. The pictogram shows how the students decided to spend their day.

    a   Which activity was the most popular?

    b   Write down how many students went bowling?

    c   Write down how many students went to the theme park.

    d   Write down how many students took part in the least popular activity.

    e   How many students took advantage of the activities offered?

Theme Park	☺ ☺ ☺ ☺ ☺ ☺ ☺ ☺ ☺
Bowling	☺ ☺ ☺ ☺ ☺
Cinema	☺ ☺ ☺ ☺
Sport	☺ ☺ ☺ ☺ ☺ ☺ ☺ ☺

Key: ☺ represents 48 students

2   This pictogram shows the number of days in January with more than one hour of sunshine in three cities.

The information for Cardiff is not shown on the pictogram.

Cardiff had eight days with more than one hour of sunshine in January.

    a   Copy and complete the pictogram.

    b   Write down the city that had the most days with more than one hour of sunshine?

    c   Write down the number of days with more than one hour of sunshine in Edinburgh.

London	◯ ◯ ◯ ◖
Edinburgh	◯ ◔
Belfast	◯ ◯
Cardiff	

Key: ◯ represents 4 days

edexcel ⠿ past paper question

3   Five primary schools were asked to estimate the number of
library books they kept in each classroom. Their responses are
shown in the frequency table.
Draw a pictogram to display this information.

School	Frequency
Parkfield	50
Easedale	30
Whinton	35
Graymans	45
Harris	40

Examiner's Tip
Use a rectangle to represent five books.

4   Forty customers at a supermarket were asked which of the local
towns they came from. Their responses are shown in the
frequency table.
Draw a pictogram to display this information.

Town	Frequency
Upshaw	5
Bunton	10
Chuckleswade	18
Newtown	5
Shenford	2

Examiner's Tip
Use a cylinder shape to represent
5 customers

5   The table below shows the amount of money that is spent on
Education, Health, Transport and Emergency services in the city
of Suncastle.
Draw a pictogram to display this information.

Area of spending	Amount of money
Health	£57 000 000
Education	£62 000 000
Transport	£15 000 000
Emergency services	£34 000 000

Examiner's Tip
Use a rectangle to represent £5 000 000.

2.9 Bar charts

Bar charts are used to show trends for qualitative or discrete data.
The bars may be horizontal or vertical.

In a **bar chart**, bars of equal width are drawn for each category or quantity.
The length of a bar is equal to the frequency of the category it represents.
The gaps between bars must be of equal width.

Example 9

This bar chart shows the results of a survey of 75 people.
Each person was asked which newspaper they read.

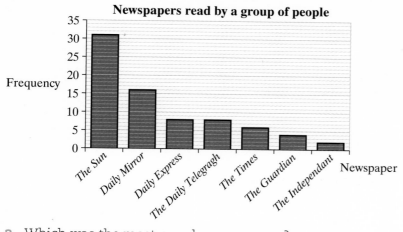

Examiner's Tip

With qualitative variables it is usual to put the highest frequency on the left and the lowest on the right.

a Which was the most popular newspaper?

b How many people read *The Times*?

c Which **two** papers had the same number of readers?

a *The Sun*	Look to see which is the highest bar.
b 6	Find *The Times* on the horizontal axis. Go to the top of its bar, follow the top across and read off the vertical scale.
c *The Daily Express* and *The Daily Telegraph*.	Look for the two bars that are the same height and read off the papers from the horizontal scale.

Example 10

This bar chart shows students' marks out of 10 scored in a class spelling test.
Students were given one mark for each correct spelling.

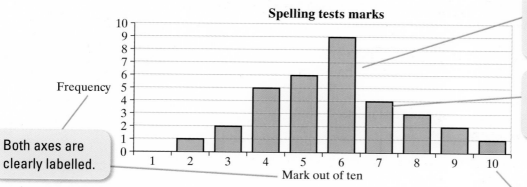

Bars are of equal width and separated by equal gaps.

Four people scored a mark of seven out of ten.

Both axes are clearly labelled.

Although the marks are discrete, they are ordered.

a What is the range of marks in this test?

b How many students scored full marks?

c Which is the most common mark?

d How many students are in the class?

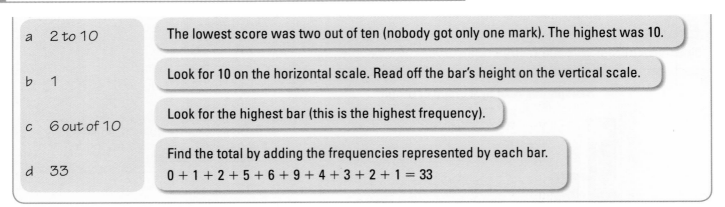

a   2 to 10   The lowest score was two out of ten (nobody got only one mark). The highest was 10.

b   1   Look for 10 on the horizontal scale. Read off the bar's height on the vertical scale.

c   6 out of 10   Look for the highest bar (this is the highest frequency).

d   33   Find the total by adding the frequencies represented by each bar.

0 + 1 + 2 + 5 + 6 + 9 + 4 + 3 + 2 + 1 = 33

This horizontal bar chart shows the number of copies of *Daily Stats* sold in the UK in one week.

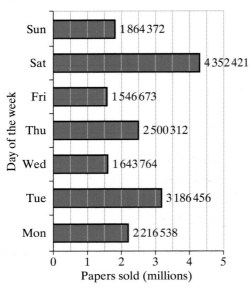

Sun   1 864 372
Sat   4 352 421
Fri   1 546 673
Thu   2 500 312
Wed   1 643 764
Tue   3 186 456
Mon   2 216 538

Day of the week

0   1   2   3   4   5
Papers sold (millions)

On this chart the bars are horizontal.

It is not possible to read off how many newspapers were sold on each day with any accuracy so the actual numbers are shown next to each bar.

2.10 Vertical line graphs

A vertical line graph is similar to a bar chart except that the bar width is reduced to a pencil line thickness.

This graph shows the data from Example 10 on a vertical line graph. The heights of the lines are exactly the same as the heights of the bars in Example 10.

 Bar charts and line graphs can be drawn using ICT.

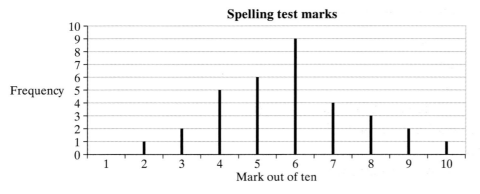

Spelling test marks

Frequency

Mark out of ten

Example 11

This vertical line graph shows the number of people in each car that passed the gates of a school between 9 am and 10 am.

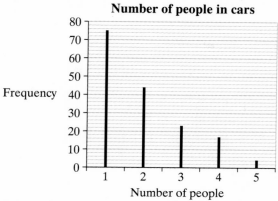

Number of people in cars

a What was the most frequently occurring number of people?

b Estimate the number of cars that contained two people.

c Why do you think there were no cars with more than five people in them?

a 1

> Look for the highest line.

b 44

> Find the line for two people and read off its height from the vertical scale.

c There are not many cars that can hold more than five people.

> This answer has to relate to the data given.

Exercise 2E

1 The vertical line graph shows the shoe sizes of students in class 11G.

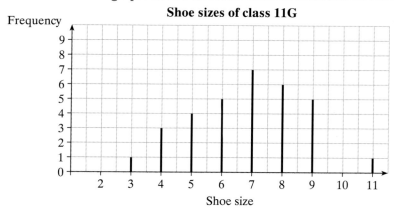

Shoe sizes of class 11G

a How many students wear size 5 shoes?

b Which is the most common shoe size?

c Which size of shoe does the student with the smallest feet wear?

d Which two shoe sizes have identical frequencies?

e How many students were in 11G?

f Construct a pictogram to show the same information.

2 The bar chart shows the number of pets owned by members of class 11H.

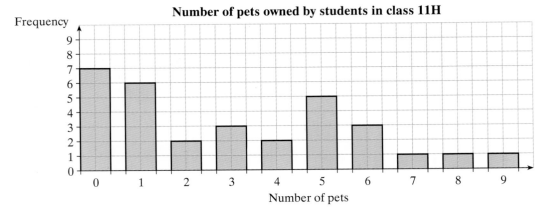

a How many students did not have any pets?

b What is the largest number of pets owned by a member of this class?

c How many students are in 11H altogether?

d How many pets is a student chosen at random most likely to have?

e Is this a good way to show these data? Explain your answer.

3 This frequency table shows the number of children in 30 families.

Number of children	0	1	2	3	4	5
Frequency	4	7	10	5	3	1

Draw a bar chart to display this information.

4 A book was opened at random and the lengths of 1000 words were counted. The results were as follows.

Number of letters	1	2	3	4	5	6	7	8	9	10
Frequency	35	132	306	183	123	96	62	41	14	8

a Draw a pictogram to display this information.

b Is the pictogram a good way to show these data? Explain your answer.

c Draw either a bar chart or vertical line graph for these data. Explain your choice.

5 **Activity:** conduct a survey to find out the number of mobile phones owned by the families of your classmates. Choose a frequency diagram to show this information. Explain your choice.

2.11 Stem and leaf diagrams

A **stem and leaf diagram** shows data distribution in the same way as a bar chart but it retains the details of the data.

A **key** shows how the stem and the leaves are combined to form a number.

Examiner's Tip
Do not forget the key. Students often do and lose a mark.

Example 12

Here is information about the number of motorists who bought diesel fuel on each of 25 randomly selected days.

19	34	41	26	18
14	8	36	33	25
18	30	37	19	40
25	31	43	21	35
22	33	13	10	23

Draw an ordered stem and leaf diagram to show these data.

Unordered

Stem	Leaves
0	8
1	9 8 4 8 9 3 0
2	6 5 5 1 2 3
3	4 6 3 0 7 1 5 3
4	1 0 3

First, draw a vertical line.

To the left of this line, write the first figures of the observations in increasing order. This is called the stem (like the stem of a plant).

To the right of the stem, write down the remaining figures in each observed value. These are the leaves.

Ordered   Key 2 | 1 means 21

Stem	Leaves
0	8
1	0 3 4 8 8 9 9
2	1 2 3 5 5 6
3	0 1 3 3 4 5 6 7
4	0 1 3

Re-draw the diagram with the numbers in each row in order.

Add a key.

Exercise 2F

1 Thirty members of a fitness club were asked how many sit-ups they could do in a minute. These are the results.

12	15	16	23	26	26	27	28	29	29
32	33	33	33	35	37	38	39	40	40
41	42	45	48	53	59	68	72	75	239

a Explain why the club manager decided to ignore the final result of 239.

b Draw a stem and leaf diagram to show these data. Use steps of 10. Leave out the final result of 239.

c Which number of sit-ups was the most common?

2   A shop manager records details of the customers during the first half hour that his shop is open.

The stem and leaf diagram shows the ages of the customers.

```
0 | 1   2   4
0 | 5   6   6   8   9
1 | 1   1   2   2   2   3   4   4
1 | 5   5   6   7   8   8
2 | 2   3   3   4
2 | 5   5   8
3 | 2
```

Key
1	8 = 18 years old

a   How many customers visited the shop in this time?

b   What was the age of the oldest customer?

c   What was the most common age of customer?

d   How many customers were 6 years old?

e   Draw a frequency table from these data.

f   Redraw the stem and leaf diagram with steps of 10 between the stems.

3   Every day a dairy farm records the milk yield, to the nearest litre, for each of its herd of 50 cows. Here are the data for one particular day.

```
26   35   23   15   35   32   13    9   42   36
40   34   39   34   25   17   19   21   31   16
41   23   32   28   26   25   19   25   22   24
18   24   28   27   26   18    9   14   24   25
34    5   25   39   23    7   29   34   26   25
```

Draw a stem and leaf diagram to show this information.

4   This stem and leaf diagram shows the weekly number of complaints received by a media company.

```
0 | 5   6
1 | 1   1   3   4   7
2 | 2   2   3   3   3   6   6   9   9
3 | 1   2   2   4   8
4 | 0   5   5   6
5 | 2
```

Key
2	6 means 26

a   Work out the number of weeks that are represented on the diagram.

b   Write down the greatest number of complaints.

c   Write down the most common number of complaints.

5 Carol asked 60 of her friends to name a whole number between 100 and 200.

These are their answers.

100	107	134	140	152	153	152	124	148	132
162	163	173	104	122	145	144	145	147	105
156	103	102	135	142	155	156	114	123	134
172	178	171	172	175	109	127	143	146	109
151	154	106	106	149	148	155	157	112	129
171	179	170	177	103	128	144	147	149	102

Draw a stem and leaf diagram to show this information.

2.12 Pie charts

> A **pie chart** is a way of displaying data when you want to show how something is shared or divided.
>
> A pie chart uses area to represent frequency.
>
> The **angles** at the centre of a pie chart add up to 360°.

This **pie chart** shows the types of housing in Showtown.

Types of housing in Showtown

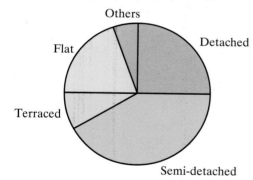

Each sector of this pie chart represents a type of housing. The proportion of each type of housing is clear.

The most common type is semi-detached houses.

About one quarter are detached houses.

Example 13

This frequency table shows what 24 people in a hotel had for breakfast.

Choice of breakfast	Frequency
Cereal	6
Full English	11
Continental	5
Fruit	2
Total	24

Draw a pie chart to show this information.

Method 1

Cereal: $\frac{6}{24} \times 360° = 90°$

Full English: $\frac{11}{24} \times 360° = 165°$

Continental: $\frac{5}{24} \times 360° = 75°$

Fruit: $\frac{2}{24} \times 360° = 30°$

Or

Method 2

24 guests are represented by 360°.

One guest is represented by $\frac{360°}{24} = 15°$.

Cereal: $6 \times 15° = 90°$

Full English: $11 \times 15° = 165°$

Continental: $5 \times 15° = 75°$

Fruit: $2 \times 15° = 30°$

$90° + 165° + 75° + 30° = 360°$

First calculate the angles for each sector.

6 out of 24 of the guests having cereal is $\frac{6}{24}$ths of total. So, it needs to be $\frac{6}{24}$ths of 360°.

Calculate the other angles in the same way.

Examiner's Tip
Try both methods and then use the one you prefer.

Add the angles to check that they total 360°.

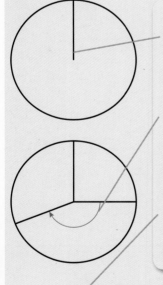

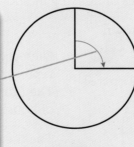

Draw a circle. Mark the centre.

Draw a radius from the centre of the circle to the circumference.

For the first sector, measure and draw an angle of 90°.

For the second sector, measure and draw an angle of 165°.

For the third sector, measure and draw an angle of 75°.

Add labels for each sector and shade them to make the proportions clearer.

Add the title.

Examiner's Tip
Measure the final sector to check that you have measured the other sectors correctly.

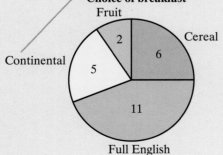

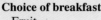

Choice of breakfast

Examiner's Tip
It is a good idea to include the actual figures on pie charts.

 You can use ICT to draw pie charts.

Exercise 2G

1 A research company recorded the country of manufacture of 120 cars. It will use a pie chart to show the data. This table shows the information.

Country	Frequency	Angle
UK	20	60°
France	15	
Germany	48	
Italy	5	
Japan	32	96°

  a Show how the angle for the UK is calculated.

  b Copy and complete the table.

  c Draw and label the pie chart.

> **Examiner's Tip**
> Remember that the sector angles in a pie chart must add up to 360°

2 This table shows the amount of money, in millions of pounds, to be spent on Education, Health, Transport and Emergency services in Tadcastle. A pie chart is needed to show the information.

Area of spending	Amount (£million)	Angle
Health	59	
Education	67	
Transport	17	
Emergency services	37	

  a Copy and complete the table by filling in the angles.

  b Draw and label the pie chart.

3 A dentist recorded the numbers of each type of treatment she carried out in a single week. She wants to draw a pie chart to show the information in this table.

Treatment	Frequency
Check up	24
Filling	22
Clean/Scale	10
Cap	4

  a The angle for a Cap is 24°. Show how the angle was calculated.

  b Copy and complete the following table.

Treatment	Angle in pie chart
Check up	
Filling	
Clean/Scale	
Cap	24°

  c Draw and label the pie chart.

4 The pie chart shows the 240 g of ingredients used to make a cake.

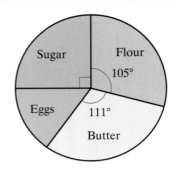

Ingredients	Weight
Flour	
Butter	
Eggs	
Sugar	

Use the pie chart to complete the table.

5 The pie chart shows the way Edward spent the last 24 hours.

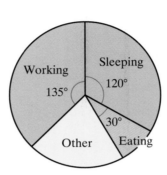

a How many degrees represents each hour in the day?

b What angle represents 'Other activities' in the pie chart?

c How many hours did Edward sleep yesterday?

d How many hours did Edward spend working yesterday?

6 Some research was done into the popularity of different carpet colours in offices. A number of offices were asked to fill in a questionnaire. A pie chart is to be drawn to show the information collected. The carpet colour results are shown in the table.

Colour	Frequency	Angle
Grey	22	
Green	16	
Patterned	10	50°
Red		30°
Other		90°

a Copy and complete the table.

b Work out how many offices filled in the questionnaire.

c Write down the most popular carpet colour.

d Draw a pie chart to show these data.

7 **Class activity**

a Draw a frequency table for the number of children in each student's family.

b Draw a pie chart for the information collected.

2.13 Advantages and disadvantages of each type of diagram

Pictograms

- Although similar to a bar chart, they are not very accurate.
- Suitable for people who require a simple, but not accurate, representation.

Bar charts/line charts

- Easy to draw.
- Fairly accurate.
- Values can, in many cases, be read from a scale.
- Information more easily seen than on pie charts.

Pie charts

- Not as accurate as bar charts.
- Not easy to draw.
- Values cannot be read off, but must be given.
- Show the relation of parts to the whole better than bar charts.

Stem and leaf diagrams

- Give visual comparison like bar charts.
- Store a large amount of data in a smaller space.
- Retain details of the data.
- Not all sets of data can be displayed as stem and leaf diagrams.

2.14 Cumulative frequency step polygons

Another way of displaying frequencies is to use a cumulative frequency step polygon.

Example 14

The cumulative frequency table gives information about the number of matches in each of 50 boxes.

Number of matches	Frequency	Cumulative frequency
47	2	2
48	10	12
49	18	30
50	10	40
51	10	50

Draw a cumulative frequency step polygon for these data.

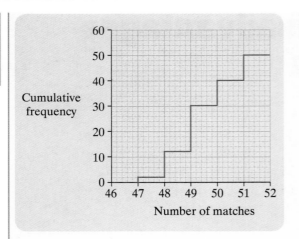

The cumulative frequency is zero until 47 is reached, it then jumps up to 2. Draw a vertical line from 0 to 2 to represent this jump.

The value of the cumulative frequency from 47 up to, but not including 48, remains constant at 2. It then jumps up to 12.

Draw a horizontal line at 2 from 47 to 48, then a vertical line from 2 to 12 to represent this jump.

Then draw a horizontal line at 12 between 48 and 49, then a vertical line up to 30.

Continue in this way.

2.15 Using bar charts to make comparisons

Multiple bar charts and **composite bar charts** can be used to make comparisons.

Multiple bar charts

Multiple bar charts have more than one bar for each class.

Example 15

This multiple bar chart gives information about the shoe sizes of 40 boys and 40 girls.

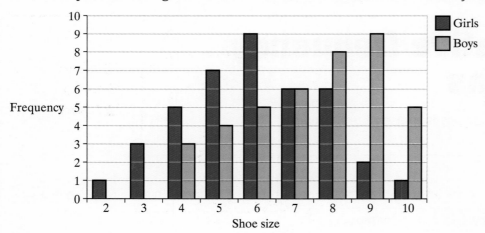

a   Which is the most common shoe size for girls?

b   Did more boys or girls wear size 8 shoes?

c   Which shoe size was worn by an equal number of boys and girls?

d   Only one child had size 2 shoes. Was this a boy or a girl?

e   Janet thinks boys have larger feet than girls. Do you think she is correct? Give a reason for your answer.

a   Size 6

> The green bars represent the girls' shoe sizes. Look for the highest green bar.

b   More boys

> Compare the boys' orange bar for size 8 with the girls' bar for size 8. Which is the highest?

c   Size 7

> Look for a shoe size that has green and orange bars the same height.

d   A girl

> Look at the colour of the bar for size 2.

e   Janet's theory seems to be correct, but she would have to assume that the boys' and girls' ages are similar.

> Compare the range of sizes and the most common sizes.

The bar chart shows the girls' shoes range in size from size 2 to size 10. The most common size is 5 or 6.

The chart shows the boys' shoes range from size 4 to size 10. The most common size is 8 or 9.

Composite bar charts

> Each bar in a **composite bar chart** shows how the total frequency for each category is made up from the separate component groups.
>
> The total frequencies and the frequencies of each of the component groups can be compared.

Example 16

This composite bar chart gives information about the numbers of computers and printers sold by a shop over a three-year period.

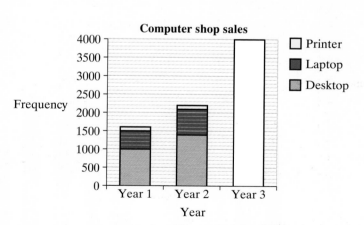

In the third year the shop sold 1500 desktop computers, 2000 laptop computers and 200 printers.

a   Copy and complete the chart by filling in the bar for Year 3.

b   In which year was the total number of sales highest?

c   How many laptops were sold in Year 1?

d   Describe how the sale of laptops changed over the three years.

a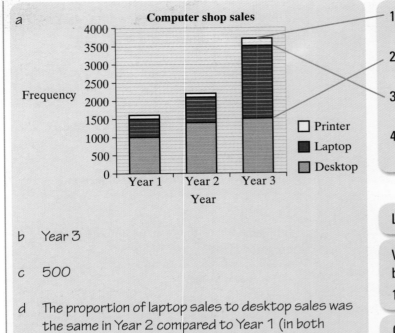

1. The total sales are 1500 + 2000 + 200 = 3700. Draw a rectangle 3700 high.

2. 1500 desktop computers were sold so draw a line across at 1500.

3. 1500 + 2000 = 3500 laptops and desktop computers were sold so draw a line at 3500.

4. Shade the sections of the Year 3 bar in the correct colours.

b   Year 3

Look to see which bar is highest.

c   500

Work out the height of the laptop section of the bar for Year 1.

1500 − 1000 = 500

d   The proportion of laptop sales to desktop sales was the same in Year 2 compared to Year 1 (in both years laptops accounted for approximately one third of the total computer sales), but increased in Year 3 (when laptops accounted for over half the total computer sales).

Compare the laptop sections of the bar to see how total laptop sales have varied for the different years. Use the desktop sections and the laptop sections of the bars to make comparisons about the proportion of sales.

 You can use ICT to draw both multiple and composite bar charts.

Exercise 2H

1   Paul sells scarves in three colours. Each scarf is black or red or blue. The multiple bar chart below shows his sales over one week.

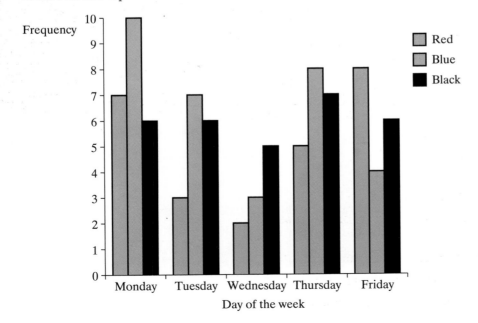

a What colour scarf was most popular on Monday?

b Which colour had the most consistent sales over the whole week?

c How many scarves were sold on Wednesday?

d How many red scarves were sold over the whole week?

e On which day did Paul sell the fewest scarves?

2 Joel has collected some data on the types of animals taken to a veterinary surgery. He wants to show his data on a diagram but cannot decide whether to use a bar chart or a pie chart.

Give the advantages and disadvantages of using each type of diagram.

3 A nursery collects some data on the number of tomatoes produced by each of their plants. They decide to draw a stem and leaf diagram to represent these data.

Write down the advantages of using a stem and leaf diagram.

4 Roland owns a market stall, from which he sells fruit and vegetables. One day he compared the types of fruit that he sold in the morning and afternoon.

His results are shown in the composite bar chart.

ResultsPlus
Build Better Answers

Question: A town council plans to build a swimming pool.
It is going to carry out a survey to find out what people think of the plan.
(ii) What type of **statistical** diagram could the council use to show the results of the survey?
Give a reason for your answer

■ **Zero marks**
A common incorrect answer was 'tally chart'. This is not a diagram, it is a data collection sheet.

● **Good 1 mark answer**
Choose a diagram that shows the different groups of responses clearly, for example a bar chart or pie chart.

▲ **Excellent 2 mark answer**
Give a reason for your choice, explaining why the diagram you suggest will present the information so that it is easy to read. An example of an excellent answer was: 'Bar chart, because it shows the results clearly'.

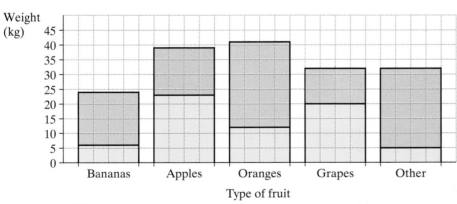

a Which fruit sold the most in weight overall?

b Which fruits sold more by weight in the morning than the afternoon?

c Estimate the weight of apples sold in the morning.

d Estimate the weight of grapes sold in the afternoon.

e Estimate the weight of bananas sold in the whole day.

5   This table is part of a larger table that shows the percentages of male and female students achieving GCSEs in the UK in 2001–2.

UK student GCSE achievements in their last year of compulsory education 2001–2

	Percentage of students				Total number of students (=100%) (thousands)
	5 or more grades A*–C	1–4 grades A*–C	Grades D–G only	No graded GCSEs	
Male	45.7	24.6	23.1	6.5	372.1
Female	56.5	23.6	15.5	4.4	357.6

Source: Department of Education and Skills

a   The percentages for male students do not add up to 100%. Explain why.

b   Copy the following grid. Draw a composite bar chart to show the data given in the table.

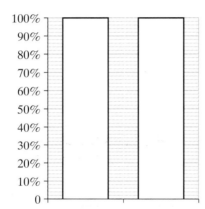

Key	
5 or more A*–C grades	
1–4 A*–C grades	
Grades D–G only	
No graded GCSEs	

ResultsPlus

Watch out!

⬤ When asked to draw a composite bar representing three data values, some students added them up incorrectly and so drew the bar the wrong height. Others read the scale wrongly, so their bars were also the wrong height.

6   Use the data in Question 1 to construct a composite bar chart.

7   Use the data in Question 4 to construct a multiple bar chart.

2.16 Using pie charts to make comparisons

It is possible to use pie charts to compare two sets of similar data. Often the comparison is made between the data for one year and the same set of data for a different year.

Pie charts show proportions. The area of any sector is proportional to the frequency of the category it represents. The whole pie chart area is therefore proportional to the total frequency.

Often two sets of data have different total frequencies. Drawing two pie charts the same size would be misleading.

This problem can be avoided by drawing scaled pie charts. This means the areas of the pie charts are in the same ratio as the two frequencies.

> **Comparative pie charts** can be used to compare two sets of data.
>
> The areas of the two circles should be in the same ratio as the two total frequencies.
>
> To compare the total frequencies, compare the areas.
>
> To compare proportions, compare the individual angles.

Calculating the radii of scaled pie charts

The area of a circle is calculated using the formula: area $= \pi r^2$, where r is the radius of the circle.

Call the radius of the first circle r_1 and the radius of the second circle r_2.

The ratio of the areas of the two circles is: $\pi r_1^2 : \pi r_2^2$

which is $r_1^2 : r_2^2$ (dividing both sides by π).

If you call the first frequency F_1 and the second frequency F_2,

then $r_1^2 : r_2^2 = F_1 : F_2$ or $r_1 : r_2 = \sqrt{F_1} : \sqrt{F_2}$

We can write this as $\dfrac{r_2}{r_1} = \dfrac{\sqrt{F_2}}{\sqrt{F_1}}$.

Since the first radius, r_1, can be any size, the formula can be rearranged as

$$r_2 = r_1 \frac{\sqrt{F_2}}{\sqrt{F_1}}$$

> **Examiner's Tip**
> You must remember this formula.

Example 17

These tables give information about the number of television sets in households in Appleville and Orangeford.

Orangeford

Number of TVs	Frequency
0	157
1	848
2	415
3	90
more than 3	90
Total	1600

Appleville

Number of TVs	Frequency
0	90
1	420
2	360
3	15
more than 3	15
Total	900

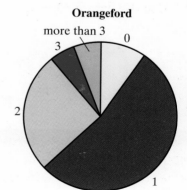

Orangeford

Comparative pie charts are to be drawn to show this information.

The pie chart for Orangeford has already been drawn, and has a radius of 2 cm.

a Work out the radius of the pie chart for Appleville.

b Draw the pie chart for Appleville.

c Compare the number of TVs per household in Orangeford and Appleville.

a $2 \times \dfrac{\sqrt{900}}{\sqrt{1600}} = 2 \times \dfrac{30}{40} = 1.5 \text{ cm}$

Use the formula $r_2 = r_1 \dfrac{\sqrt{F_2}}{\sqrt{F_1}}$

b

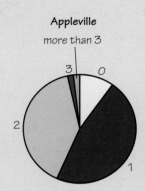

Appleville

Orangeford

Draw a circle with a radius of 1.5 cm, then complete the pie chart in the usual way.

c Orangeford has more TVs. This could be because there are more households in Orangeford. There is a greater proportion of households with 2 TVs in Appleville. The proportion with no TVs is about the same. The proportions of households with 1, 3 or more than 3 TVs in Orangeford are greater than in Appleville.

You only need to compare the angles to compare the proportions.

ResultsPlus
Build Better Answers

Question: Five shops in Whitehaven sell both videos and DVDs. The comparative pie charts show the percentage of each that were sold by the five shops in one week.

(a) Comment on the sales of videos compared to DVDs in Whitehaven. Give reasons for your opinion. (2 marks)

(b) Discuss any relationship between the percentage of videos sold and the percentages of DVDs sold in each of the two shops. (1 mark)

■ **Basic 1 mark answers**
Do not realise that the pie charts are drawn to scale: the larger pie chart represents a larger sample. Answer to (b) talks about numbers rather than percentages of videos and DVDs.

● **Good 2 mark answer**
Compared the percentages of DVDs and videos sold in part (b).

▲ **Excellent 3 mark answer**
Measure the radii of the pie charts to work out that the total number of DVDs is 1.78 times the total number of videos, and use this fact in answer to part (a).

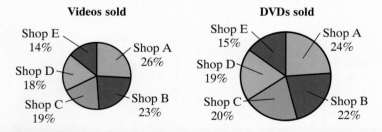

Exercise 2I

1 At Springbank High School, students can choose to study either French or Spanish. The tables give information about their GCSE results.

a Which was the most popular language?

b What grades was a student most likely to achieve in a language GCSE?

c Draw two scaled pie charts to show this information.

French

GCSE result	Frequency
A or A*	12
B or C	36
D or E	24
F or G	9
Total	81

Spanish

GCSE result	Frequency
A or A*	8
B or C	29
D or E	9
F or G	3
Total	49

2 A survey was conducted to find the uses of land in two counties, Southshire and Northshire. The results are shown in the tables.

a What is the most common use of land in Southshire?

b What is the most common use of land in Northshire?

Southshire

Use of land	Land area (acres)
Agriculture	89
Urban	420
Woodland	365
Water	72
Total	946

Northshire

Use of land	Land area (acres)
Agriculture	157
Urban	845
Woodland	403
Water	88
Total	1493

c Draw two scaled pie charts to show this information.

3

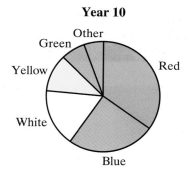

Year 10

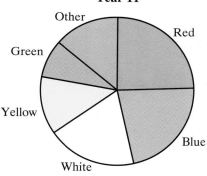

Year 11

The pie charts show the favourite colours of students in two year groups in a school. There are 180 students in Year 10.

a The radii of the two circles are $1\frac{1}{2}$ cm and 2 cm. How many students are there in Year 11?

b How many students liked the most favourite colour in Year 10?

c How many students liked green in Year 11?

d Complete the table below for Year 10.

Colour	Red	Blue	White	Yellow	Green	Other	Total
Frequency							180

e How many students in Years 10 and 11 liked the colour blue?

f Use this information to draw a multiple bar chart to compare the favourite colours of the two year groups.

4 A survey was conducted to find out Europe's most popular classical composer.

The table shows the results of the survey in France and Germany.

Germany

Composer	Frequency
Beethoven	82
Mozart	486
Handel	136
Saint-Saëns	0
Wagner	48
Other	32

France

Composer	Frequency
Beethoven	255
Mozart	321
Handel	189
Saint-Saëns	287
Wagner	36
Other	33

Draw two scaled pie charts to show this information.

5 James and Alex asked the children in their respective classes how they had travelled to school that morning.

The results are shown in these two tables.

James' class

Mode of travel	Frequency
Bus	3
Train	2
Car	5
Bicycle	2
Walk	13
Total	**25**

Alex's class

Mode of travel	Frequency
Bus	3
Train	6
Car	11
Bicycle	1
Walk	15
Total	**36**

They draw pie charts to show their data.

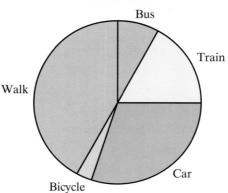

James' class   **Alex's class**

a These pie charts are misleading. It looks as if there were more children in James' class who walked to school. Explain why this is not the case.

b Construct **two** comparative pie charts to show the data in the frequency tables more effectively.

Chapter 2 review

1 This pictogram shows the results of a survey of some people who were asked on which day they preferred to do their shopping.

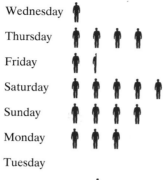

Wednesday

Thursday

Friday

Saturday

Sunday

Monday

Tuesday

Key ♟ = 2 people

Four people stated that they preferred to shop on a Tuesday.

a Copy and complete the pictogram by showing the information for Tuesday.

☆★★★★ *challenge*

b Which day was the most popular day to go shopping?

☆★★★★ *challenge*

c How many people preferred to shop on a Thursday?

☆★★★★ *challenge*

edexcel ⠿ *past paper question*

2 This table is part of a larger table that shows the examination results of students in 2001–2.

UK student GCSE achievements in their last year of compulsory education 2001–2

	Percentage of students				Total number of students (=100%) (thousands)
	5 or more grades A*–C	1–4 grades A*–C	Grades D–G only	No graded GCSEs	
United Kingdom	51.0	24.1	19.4	5.5	729.7
North East	43.9	24.8	24.8	6.5	33.3
North West	48.0	24.5	21.7	5.7	89.6
Yorkshire and the Humber	44.4	24.3	25.2	6.2	63.0
East Midlands	49.1	23.1	22.1	5.7	51.6
West Midlands	47.4	25.0	21.9	5.7	67.9
East	54.0	23.3	18.1	4.6	65.4
London	48.6	26.4	19.5	5.5	77.7
South East	55.5	22.3	17.2	5.0	95.9
South West	54.6	22.8	17.8	4.8	58.9

Source: Department of Education and Skills

a What percentage of students in the North West got five or more grades A*–C?

☆★★★★ *challenge*

b Write down the total number of students in the last year of compulsory education in the United Kingdom in 2001–2.

c Which region shown in the table had the lowest percentage of students with no graded GCSEs?

edexcel ::: *past paper question*

3 This composite bar chart shows information about the metals used to make brass, bronze and pewter.

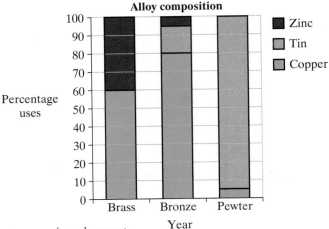

Source: www.gizmology.net

Three of the following statements are true.
Write down the **true** statements.

A   Bronze is made from three different metals.

B   There is a large proportion of tin in pewter.

C   Fifty per cent of brass is zinc.

D   There is copper in brass, bronze and pewter.

E   There is more tin than copper in bronze.

edexcel ::: *past paper question*

4 This multiple bar chart shows information about the average price, to the nearest £1000, of three different types of housing in Newcastle and Bristol in August 2005.

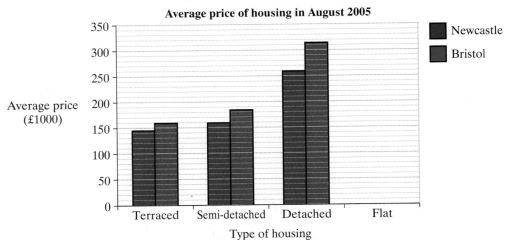

The average price of a flat in Newcastle in August 2005 was £130 000.

The average price of a flat in Bristol in August 2005 was £160 000.

a Copy and complete the multiple bar chart to show the information for flats.

☆☆★★★ *challenge*

b What does the multiple bar chart show you about the average price of detached houses?

☆☆☆★★ *challenge*

c If you had £150 000 to spend on housing in Newcastle in August 2005, which type of housing were you most likely to be able to buy?

 *past paper question*

☆☆☆★★ *challenge*

5 This table gives some information about the number of male and female car drivers who were killed or injured in the UK in 1994.

It also shows the percentage of those killed or injured in each of three age groups.

Males and females drivers killed or injured in the UK in 1994			
Age of driver (years)			Number killed or injured
17–21	22–39	40 and over	
Males	18%	48%	34%	70 100
Females	16%	52%	32%	54 700

Source: Social trends 1996

a Use the information from the table to complete this composite bar chart.

☆☆☆★★ *challenge*

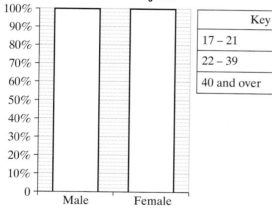

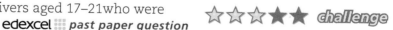

b Calculate the number of male drivers aged 17–21 who were killed or injured.  *past paper question*

☆☆☆★★ *challenge*

6 This pie chart shows information about the area of land used for different crops at High Meadows Farm.

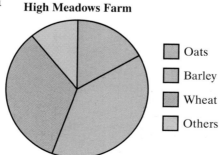

High Meadows Farm

Oats
Barley
Wheat
Others

a Measure and write down the size of the angle for wheat.

☆☆★★★ *challenge*

180 acres of land are used for crops at High Meadows Farm.

b Work out the number of acres used for oats.

The table shows the number of acres of land used for crops at Springfield Farm.

Crops at Springfield Farm			
Oats	Barley	Wheat	Others
40 acres	60 acres	50 acres	30 acres

c Using a cricle radius 13 cm, draw a pie chart to show the information in the table.

Show all your working.

edexcel ::: *past paper question*

7 An estate agent surveyed the number of bedrooms in each house on three streets of Frimmerton. The results of the survey are shown in the charts below.

Squib Street

Number of bedrooms	Frequency
1	3
2	17
3	22
4	11
5	7
Total	60

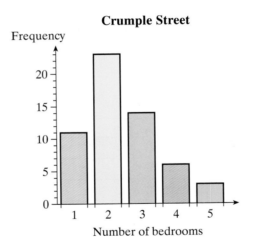

Round Street

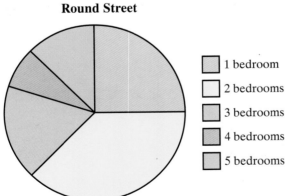

a What is the most common number of bedrooms in Round street?

b How many two-bedroom houses are there in Crumple street?

c How many houses are on Squib Street?

d How many houses are there in total on Crumple Street?

e There are 48 houses on Round Street. How many of them have

   i) 1 bedroom

   ii) 2 bedrooms?

f Which street has the most four-bedroom houses?

g The estate agent said that 'In every street, there are more four-bedroom houses than there are five-bedroom houses.' Is he correct? Explain your answer.

h A house is chosen at random. In which street is it most likely to have exactly two bedrooms?

i Which type of chart do you think shows these data most clearly?

   Explain your answer.

j Show the data for Squib Street using

   i) a bar chart

   ii) a pie chart

   iii) a pictogram.

8 One Saturday, Adrian recorded the ages of the first 40 customers at a supermarket. These are the ages.

25	8	36	29	12	17	33	28	22	36
55	21	27	33	37	48	42	3	35	44
16	22	29	31	36	56	41	24	28	33
46	56	38	25	41	38	11	7	17	26

a Draw a tally chart. Use the class intervals 0–9, 10–19, 20–29, ..., etc.

b Why can you not use the intervals 0–10, 10–20, 20–30 ... for this tally chart?

c Construct a stem and leaf diagram to show these data.

d What information was lost when constructing the tally chart, but could still be seen on the stem and leaf diagram?

9 The number of boys and girls in each year at Finbow High School is shown in this two-way table.

	Year 7	Year 8	Year 9	Year 10	Year 11	Total
Boys	72		71	66		320
Girls	63	75		55		286
Total		122	101			

a Copy and complete the two-way table.

b Draw a multiple bar chart to show the number of boys and girls in each year of Finbow High School.

Chapter 2 summary

Tally charts and frequency tables

1 A **tally chart**, or **frequency table**, can be used to process raw data. They make it easier to spot patterns.

Cumulative frequency tables

2 The **cumulative frequency** of the value of a variable is the total number of observations that are less than or equal to that value.

Grouping discrete data

3 If data are widely spread they should be grouped into **classes**.

Classes of varying width and open-ended classes

4 When there is an uneven spread of data across the range, class intervals may vary in width.

5 When the extreme values of data are not known the first and/or last intervals may be left open.

Two-way tables

6 A **two-way table** displays frequencies for two categories of data.

Other tables and databases

7 A **database** is an organised collection of information.

8 A **summary table** shows data that have been sorted and summarised. It is easier to interpret than the original data.

Pictograms

9 A **pictogram** uses symbols or pictures to represent a certain number of items.

10 It must have a **key** to tell you the number of items represented by a single symbol or picture.

11 When drawing a pictogram, make sure that
- each picture is the same size
- the picture can be divided easily to show different frequencies
- the spacing between the pictures is the same in each row
- you give a key.

Bar charts

12 In a **bar chart**, bars of equal width are drawn for each category or quantity.

13 The length of a bar is equal to the frequency of the category it represents.

14 The gaps between bars must be of equal width.

Vertical line graphs

15 A **vertical line graph** is similar to a bar chart except that the bar width is reduced to a pencil line thickness.

Stem and leaf diagrams

16 A **stem and leaf diagram** shows the data distribution in the same way as a bar chart but it retains the details of the data.

17 A **key** shows how the stem and the leaves are combined to form a number.

Pie charts

18 A **pie chart** is a way of displaying data when you want to show how something is shared or divided.

19 A pie chart uses area to represent frequency. The **angles** at the centre of a pie chart add up to 360°.

Using bar charts to make comparisons

20 **Multiple bar charts** have more than one bar for each class.

21 Each bar in a **composite bar chart** shows how the total frequency for each category is made up from the separate component groups.

22 The total frequencies and the frequencies of each of the component groups can be compared.

Using pie charts to make comparisons

23 **Comparative pie charts** can be used to compare two sets of data.

24 The areas of the two circles should be in the same ratio as the two total frequencies.

25 $r_2 = r_1 \dfrac{\sqrt{F_2}}{\sqrt{F_1}}$ where r_1 = radius of first circle
r_2 = radius of second circle
F_1 = total frequency of first data set
F_2 = total frequency of second data set

26 To compare the total frequencies, compare the areas.

27 To compare proportions, compare the individual angles.

Test yourself

1 A council kept a 30-day record of the number of absentees among its workers. The data are as follows:

5	12	17	27	4	13	32	54	6	13
14	23	24	3	9	5	15	21	7	2
6	8	9	14	14	19	17	18	22	24

Sort these data into groups and draw a grouped frequency table.

2 This two-way table shows ice cream preferences of a group of males and females. Copy and complete the table.

	Vanilla	Chocolate	Strawberry	Total
Male	6	3		20
Female		8	5	18
Total				

3 Write down the difference between a bar chart and a vertical line graph.

4 For 20 days, a garage counts the number of people who buy newspapers when they buy petrol. The data are

26	19	28	17	22	28	32	31	29	41
19	27	9	16	27	27	23	25	30	29

a Draw an ordered stem and leaf diagram to show these data.

b Write down the most common number of people.

c Write down the advantages of a stem and leaf diagram.

5 Clark Garage records the colours of cars visiting the garage. It wants to use the information to help decide the colours of the new cars it orders. A pie chart is to be drawn of these data.

Copy and complete the table for the numbers of cars and for the angles of the pie chart.

Car colour	Frequency	Angle for pie chart
Silver		112°
Red	10	
Blue	6	
White	12	
Black	3	24°

6   Describe the difference between a multiple bar chart and a
composite bar chart.

7   Pie chart 1 represents 80 members of a gym in 2008 and is drawn
with a radius r_1 of 6 cm.

Work out the radius that should be used for pie chart 2, which
represents 160 members in 2009.

Chapter 3:
Representing and processing continuous data

After completing this chapter you should be able to

- select suitable class intervals
- draw and interpret pie charts, frequency diagrams, stem and leaf diagrams, population pyramids, choropleth maps and histograms with equal and unequal class intervals

- understand the terms 'symmetrical', 'positive skew' and 'negative skew', and be able to identify such distributions
- recognise features that make graphs misleading.

Times such as the length of a telephone call are measured on a continuous scale.

After reading this chapter you will be able to organise continuous data, such as lengths of telephone calls, into a grouped frequency table and then represent the data using a histogram or other type of diagram.

3.1 Frequency tables

Continuous data can take any value on a continuous scale and can be sorted into classes. It is important to make sure that the **class intervals** do not overlap. It must also be clear to which classes the class limits belong.

> The class limit 2 is in this class. The class limit 5 is in the next class.

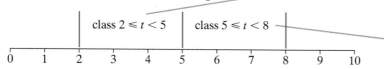

> $5 \leqslant t$ means t is greater than or equal to 5.
> $t < 8$ means t is less than 8.

> **Continuous data** can be sorted into a **frequency table**. The **class intervals** must not have gaps in between them or overlap each other.

Example 1

Twenty students take part in a 400 m race. These are the times taken (in seconds) for each person to complete the race.

| 54.0 | 58.0 | 69.3 | 82.2 | 70.4 | 63.2 | 69.0 | 78.0 | 54.4 | 66.2 |
| 53.0 | 56.2 | 71.4 | 76.3 | 80.0 | 84.0 | 72.2 | 68.4 | 56.4 | 62.3 |

Shaun, Rita and Serguei each tried to sort the data into a grouped frequency table. They each chose different class intervals.

Shaun	Rita	Serguei
Time (s)	**Time (s)**	**Time, t (s)**
50 to 59	50–60	$50 < t \leqslant 60$
60 to 69	60–70	$60 < t \leqslant 70$
70 to 79	70–80	$70 < t \leqslant 80$
80 to 89	80–90	$80 < t \leqslant 90$

a Why are Shaun and Rita's class intervals unsuitable?

b Comment on the suitability of Serguei's class intervals.

Another person runs 400 m in 105.8 seconds.

c How could Serguei change his final class interval to allow for times longer than 90 seconds?

a Shaun's table is unsuitable because there are gaps between his class intervals.

   Rita's table is unsuitable because her class intervals overlap.

> There is nowhere for Shaun to record the number 69.3. It does not fit in the intervals 60 to 69 or 70 to 79.

b Serguei's class intervals do not have gaps and they do not overlap, so they are suitable.

> Rita could record the number 80.0 in the interval 70–80 or in the interval 80–90. It is not clear which class interval it should go into.

c He could use an open interval, for example $t > 80$.

In Example 1, Serguei's class intervals satisfy the criteria of no gaps and no overlapping. However, there is one other issue to consider when designing a table for figures that have been rounded.

> All possible values that round to the same number must fit into the same class interval.

Example 2

a Explain why Serguei's class intervals (in Example 1) are not suitable if the times are rounded to the nearest second.

b Design a frequency table with suitable intervals.

a A time shown as 70 seconds could have been anything in the range $69.5 \leqslant t < 70.5$. It may belong in the class interval $60 < t \leqslant 70$ or $70 < t \leqslant 80$.

Serguei
Time, t (s)
$50 < t \leqslant 60$
$60 < t \leqslant 70$
$70 < t \leqslant 80$
$80 < t \leqslant 90$

The times 69.5 to 70 fit into this group. Not all values that round to 70 fit into this class interval (e.g. 70.2).

b

Time, t (s)
$49.5 \leqslant t < 60.5$
$60.5 \leqslant t < 70.5$
$70.5 \leqslant t < 80.5$
$80.5 \leqslant t < 90.5$

All figures that round to 70 fit into this group.

The data in Example 2 are fairly evenly spread so the class **intervals** can be the same width. If data are unevenly spread class intervals of different widths should be used.

> **Intervals** do not need to be of equal width. Use narrower intervals where the data are grouped together and wider intervals where the data are spread out.

Exercise 3A

1 Gareth records the amount of rainfall each day in Runsby.

Here are the raw data for January (in centimetres).

5.6	4.3	2.1	0	0.8	5.2	3.3	2.8	2.2	1.6	0.4
1.9	3.2	4.2	1.0	3.0	3.6	2.4	1.8	0.4	0	0
3.2	3.5	2.7	1.2	2.1	1.1	5.7	5.2	3.1		

He rounds each measurement down to the nearest centimetre and sorts his results using this frequency table.

Rainfall, r (cm)	Tally	Frequency
$0 \leqslant r < 1$		
$1 \leqslant r < 2$		
$2 \leqslant r < 3$		
$3 \leqslant r < 4$		
$4 \leqslant r < 5$		
$5 \leqslant r < 6$		

a In which class will Gareth put a rainfall of 1 cm?

b Copy and complete the frequency table.

2 Frank has a collection of 34 garden gnomes.

These are the weights in kilograms of each garden gnome to two decimal places.

2.44	1.57	2.35	1.13	2.52	1.59	2.53
0.65	2.56	1.60	2.67	1.22	2.89	1.72
2.99	0.27	3.00	1.77	3.13	1.34	3.22
1.81	0.74	1.88	1.37	1.91	0.48	2.11
1.48	2.36	0.85	2.22	1.53	2.29	

a What is the weight of the heaviest gnome?

b Frank begins to draw a frequency table.

Weight, w (kg)	Tally	Frequency
$0 \leqslant w < 0.5$		
$0.5 \leqslant w < 1$		

Copy and complete Frank's frequency table. Use classes of equal width.

3 Helen and Nigel decide to measure the time (in seconds) that it takes different students to type and send the same text message.

They use these class intervals.

Helen

Time (s)
0 to 0.4
0.5 to 0.9
1.0 to 1.4
1.5 to 1.9
2.0 to 2.9

Nigel

Time (s)
0–0.5
0.5–1.0
1.0–1.5
1.5–2.0
2.0–2.5

a One student took 1.46 seconds. Explain why Helen would have trouble recording this using her class intervals.

b One of the students took exactly 1.5 seconds. Explain why Nigel would have trouble recording this using his class intervals.

c Write the class intervals differently so that these problems would not occur.

4 Here are the weights of 30 boys, rounded to the nearest kilogram.

60	62	51	53	42	52	50	53	48	55
58	59	63	49	52	54	35	53	44	54
46	57	46	67	58	56	48	48	37	41

a What is the range of values that could be represented by the weight 48 kg?

b Kathleen wanted to use class intervals of $35 \leqslant w < 40$, $40 \leqslant w < 45$, $45 \leqslant w < 50$, etc. Explain why this is wrong.

c Choose class intervals of width 5 kg that would suit these rounded data.

d Use your class intervals from part c to create and complete a frequency table for these data.

5 A rugby team held a competition to test the strength of their members. The first event involved holding a heavy object for as long as possible.

The times were measured to the nearest tenth of a second.

Joff, Lee and Ben decided to keep a frequency table to record the results, using these class intervals.

Joff

Time (s)
0 to 9
10 to 19
20 to 29
30 to 39
40 to 49
50 to 59

Lee

Time, t (s)
$0 < t \leqslant 10$
$10 < t \leqslant 20$
$20 < t \leqslant 30$
$30 < t \leqslant 40$
$40 < t \leqslant 50$
$50 < t \leqslant 60$

Ben

Time, t (s)
$0.05 \leqslant t < 10.05$
$10.05 \leqslant t < 20.05$
$20.05 \leqslant t < 30.05$
$30.05 \leqslant t < 40.05$
$40.05 \leqslant t < 50.05$
$50.05 \leqslant t < 60.05$

a The first contestant held the object for 9.7 seconds. Explain why Joff would have trouble recording this in his table.

b The second contestant had his time rounded to 20.0 seconds. Lee did not know which class to record this in. Explain why.

c Ben had more sensible class intervals for rounded data. Before the competition, it was not clear what the longest time would be. How could Ben change his last interval to allow for this?

3.2 Pie charts and stem and leaf diagrams

Pie charts can be used to show continuous data. The area of each sector represents the frequency for that class interval. Other than this the chart is drawn in the same way as for discrete data.

Example 3

This frequency chart shows the reaction times t, in seconds, of 20 students after a gun was fired to start a race.

Draw a pie chart to show this information.

Time, t (s)	Frequency
$0.15 < t \leqslant 0.25$	3
$0.25 < t \leqslant 0.35$	6
$0.35 < t \leqslant 0.45$	4
$0.45 < t \leqslant 0.55$	6
$0.55 < t \leqslant 0.65$	1
Total	20

Method 1

$0.15 < t \leqslant 0.25$: $\frac{3}{20} \times 360° = 54°$

$0.25 < t \leqslant 0.35$: $\frac{6}{20} \times 360° = 108°$

$0.35 < t \leqslant 0.45$: $\frac{4}{20} \times 360° = 72°$

$0.45 < t \leqslant 0.55$: $\frac{6}{20} \times 360° = 108°$

$0.55 < t \leqslant 0.65$: $\frac{1}{20} \times 360° = 18°$

First calculate the angle for each sector.

There are 3 out of 20 students in the interval $0.15 < t \leqslant 0.25$ so this is $\frac{3}{20}$ths of the total, which is $\frac{3}{20}$ths of 360°.

Calculate the other angles in a similar way.

Or

Method 2

20 students are represented by 360°, so one student is represented by $\frac{360°}{20} = 18°$.

$0.15 < t \leqslant 0.25$: $3 \times 18° = 54°$

$0.25 < t \leqslant 0.35$: $6 \times 18° = 108°$

$0.35 < t \leqslant 0.45$: $4 \times 18° = 72°$

$0.45 < t \leqslant 0.55$: $6 \times 18° = 108°$

$0.55 < t \leqslant 0.65$: $1 \times 18° = 18°$

Check: $54° + 108° + 72° + 108° + 18° = 360°$

Calculate how many degrees will represent one student.

There are three students in the interval $0.15 < t \leqslant 0.25$ so multiply 3 by 18°.

Calculate the other angles in the same way.

Time taken

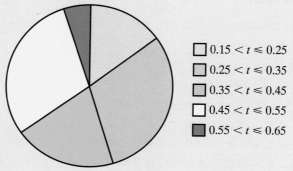

- $0.15 < t \leqslant 0.25$
- $0.25 < t \leqslant 0.35$
- $0.35 < t \leqslant 0.45$
- $0.45 < t \leqslant 0.55$
- $0.55 < t \leqslant 0.65$

Examiner's Tip

Add up the angles for all intervals to make sure they total 360°.

> If there are not too many observations they can be treated as individual numbers and shown as a stem and leaf diagram. The stem shows the first digit(s) of the numbers and the leaves show the last digit.

Example 4

This information represents the amount of rainfall, in centimetres, in Beesie for the 31 days in December.

2.3	1.6	0.7	1.2	3.1	4.5	1.0	3.2
3.5	2.3	1.7	0.4	0.0	0.0	3.0	0.0
2.7	1.2	2.1	1.1	2.8	5.2	1.1	5.7
2.1	0.0	0.8	5.6	4.3	5.2	3.3	

a Draw an ordered stem and leaf diagram.

b Work out the range of the data.

Unordered

a
Stem	Leaves						
0	7	4	0	0	0	0	8
1	6	2	0	7	2	1	1
2	3	3	7	1	8	1	
3	1	2	5	0	3		
4	5	3					
5	2	7	6	2			

Draw a vertical line.

To the left of this line write the first figures of the observations. This is the **stem** (like the stem of a plant).

Write the remaining figure in each observed value to the right of the stem. These are the **leaves**.

Ordered

0	0	0	0	0	4	7	8
1	0	1	1	2	2	6	7
2	1	1	3	3	7	8	
3	0	1	2	3	5		
4	3	5					
5	2	2	6	7			

Key

2 | 1

means 2.1 cm of rainfall.

Order the numbers in each row to the right of the stem.

Add a key.

b Range = 5.7 − 0.0 = 5.7 cm

Range = largest value − smallest value

Exercise 3B

1 Catherine measures the trunk diameters of small trees in a nursery.

When the tree trunks have a diameter of over 6 cm, they are planted in the forest.

Here are the diameters of 50 trees in the nursery (in centimetres).

4.5	2.1	2.5	3.4	5.6	5.7	4.9	2.2	4.7	2.3
5.9	6.0	6.4	5.0	2.6	2.3	2.5	2.7	5.8	5.0
4.2	2.8	3.0	3.4	3.3	3.7	4.6	3.3	3.1	3.5
4.3	3.7	3.7	3.9	3.9	4.3	4.6	4.8	4.9	5.0
5.3	5.3	5.6	3.7	3.3	0.2	5.8	5.9	6.2	6.3

   a Draw an unordered stem and leaf diagram for these data.

   b Use part a to help you draw an ordered stem and leaf diagram
      for these data.

   c How many trees are ready to be planted?

2 This stem and leaf diagram shows the length of time it took people
to answer a general knowledge question.

2	0	1	3	6	8	
3	1	2	2	2	6	9
4	0	1	3	7	7	
5	1	2	3	4		
6	3					

Key

2 | 3 = 2.3 seconds to answer the question

   a Write down the most common amount of time taken to answer
      the question (to the nearest tenth of a second).

   b In total, how many people answered the question?

   c Work out the range of times taken to answer the question.

   d What advantage does this stem and leaf diagram have over a
      frequency table for the same data?

3 As part of a survey, the heights of men visiting a clothes shop
over a period of one hour were recorded. The stem and leaf
diagram shows the heights of the men (in metres).

1.5	1	6	8					
1.6	3	4	6	8	9			
1.7	1	2	2	4	6	7	7	8	9
1.8	2	3	3	4	5	5	7	
1.9	1							

Key

1.5 | 6 = 1.56 metres

   a Write down how many customers visited the shop in this time.

   b Write down the height of the tallest customer.

   c How many customers were at least 170 cm tall?

   d Draw a frequency table for these data.

   e Construct a pie chart to display these data.

4 The pie chart gives information about the frequencies of the
weights, to the nearest gram, of eggs from a flock of chickens.

   a Write down the class in which most eggs belong.

   b Write down the exact limits for each class interval.

   c Which is the smallest class width?

   d Describe how the egg sizes are distributed.

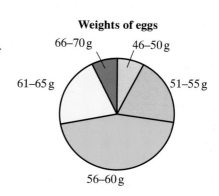

Weights of eggs

5   A delivery of iron ore arrived at a factory in a large lorry. Forty
random samples were taken from the ore and the percentage of
iron in each sample was measured.

This table gives information about these samples.

Iron, p (%)	$29 < p \leqslant 31$	$31 < p \leqslant 33$	$33 < p \leqslant 35$	$35 < p \leqslant 37$
Frequency	8	13	12	7

A pie chart is to be drawn to show these data.

a   Copy and complete this table.

b   Draw a pie chart to show these data.

Frequency	Angle
8	$\frac{8}{40} \times 360° = 72°$
13	
12	
7	

3.3 Cumulative frequency diagrams

A **cumulative frequency** is the total frequency of all values up to and
including the upper value of the class interval being considered. Each
upper class boundary has its own cumulative frequency.

> **Cumulative frequency** is a running total of frequencies.

A cumulative frequency diagram is drawn by plotting the cumulative
frequencies against their corresponding upper class boundaries.

> **Cumulative frequency diagrams** can be used to estimate or predict
> other values.

Example 5

A group of students were asked to complete a
50-piece jigsaw puzzle. The time taken for each
student to complete the jigsaw was recorded in
minutes.

a   Complete the table by calculating the
cumulative frequencies.

b   Draw a cumulative frequency curve for the data.

c   Estimate the number of students who took less
than 12 minutes.

d   Estimate how many minutes it took for 50% of
the students to complete the jigsaw.

e   Estimate how many students took between
8 and 18 minutes.

Time, x (min)	Frequency	Cumulative frequency
$0 < x \leqslant 5$	3	
$5 < x \leqslant 10$	18	
$10 < x \leqslant 15$	32	
$15 < x \leqslant 20$	26	
$20 < x \leqslant 25$	11	
$25 < x \leqslant 30$	4	

a

Time, x (min)	Frequency	Cumulative frequency
$0 < x \leqslant 5$	3	3
$5 < x \leqslant 10$	18	21
$10 < x \leqslant 15$	32	53
$15 < x \leqslant 20$	26	79
$20 < x \leqslant 25$	11	90
$25 < x \leqslant 30$	4	94

Each cumulative frequency = previous cumulative frequency + frequency of the next class interval:

$3 + 18 = 21$

$21 + 32 = 53$

b

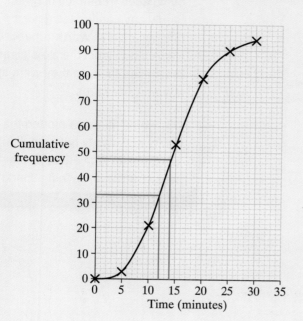

Draw the axes with time on the horizontal axis and cumulative frequency on the vertical axis. Add scales.

Plot the point (0, lowest boundary of first interval), that is (0, 0).

Plot cumulative frequency against the upper boundary of each group, for example (5, 3), (10, 21), etc.

Join the points up with a smooth curve.

Examiner's Tip

Although the question asks for a curve, in an exam the points can be joined by either a curve or straight lines.

c   33

Find 12 minutes on the time axis. Draw a line up to the cumulative frequency curve and then across to the vertical axis. Read the value on the cumulative frequency (vertical) axis.

d   50% of 94 = 47

Therefore 14 minutes.

Find 50% of the total frequency.

Find 47 on the cumulative frequency axis.

Draw a line across to the cumulative frequency curve and then down.

Read off the value on the time axis.

e   There are 70 students ≤ 18 minutes.

There are 11 students ≤ 8 minutes.

So, 70 − 11 = 59 between 8 and 18 minutes.

Do not worry about the difference between ≤ and < for estimates. In this example it is not known if one or more values are exactly 18 minutes or exactly 8 minutes.

Exercise 3C

1 Josh recorded the heights of all the boys in his year at school. The table shows the information he collected.

Height, h (cm)	Frequency
$110 < h \leqslant 120$	5
$120 < h \leqslant 130$	12
$130 < h \leqslant 140$	35
$140 < h \leqslant 150$	40
$150 < h \leqslant 160$	38
$160 < h \leqslant 170$	20

ResultsPlus

Watch out!

■ When asked to draw a cumulative frequency diagram, hardly any candidates gave it a title, which was worth 1 mark. Many didn't label the axes clearly, or labelled them so they went up in unequal size steps.

● Some students plotted the points accurately but then didn't join them up with either straight lines or a curve, so lost a mark.

a Copy and complete the cumulative frequency table.

Height, h (cm)	Cumulative frequency
$110 < h \leqslant 120$	
$120 < h \leqslant 130$	
$130 < h \leqslant 140$	
$140 < h \leqslant 150$	
$150 < h \leqslant 160$	
$160 < h \leqslant 170$	

b Use the information in your table to draw a cumulative frequency diagram.

c Estimate the number of boys whose height is between 148 cm and 152 cm.

2 The table gives information about the time taken, in seconds, to run 100 m by a random sample of 60 members of an athletics club.

Time, t (seconds)	Frequency
$10 < t \leqslant 11$	2
$11 < t \leqslant 12$	15
$12 < t \leqslant 13$	18
$13 < t \leqslant 14$	12
$14 < t \leqslant 15$	13

a Draw a cumulative frequency diagram to represent these data.

b Estimate the number of athletes whose time was greater than 11.4 seconds.

c Estimate the number of athletes who took a time that was between 11.2 seconds and 14.2 seconds.

3   Cigdem caught a bus to school every day. For 100 days she recorded the number of minutes that the bus was late. The table gives information about the lateness of the bus.

Time late, t (min)	Frequency	Cumulative frequency
$0 < t \leqslant 3$	10	
$3 < t \leqslant 5$	31	
$5 < t \leqslant 7$	35	
$7 < t \leqslant 10$	21	
$10 < t \leqslant 15$	3	

   a   Copy and complete the cumulative frequency table.

   b   Use the information in your table to draw a cumulative frequency diagram.

   c   Use your diagram to estimate how many times the bus was more than 8 minutes late.

   d   Use your diagram to estimate the percentage of times the bus was between 4.5 and 6.5 minutes late.

4   The table gives information about the heights, in centimetres, of a group of 100 randomly selected 16-year-olds.

Height, h (cm)	Frequency
$140 < h \leqslant 150$	2
$150 < h \leqslant 155$	15
$155 < h \leqslant 160$	18
$160 < h \leqslant 165$	42
$165 < h \leqslant 170$	15
$170 < h \leqslant 180$	8

   a   Draw a cumulative frequency diagram to represent these data.

   b   Estimate the number of 16-year-olds whose height is less than 162 cm.

3.4 Frequency polygons and histograms with equal class intervals

For qualitative variables the height of a bar in a bar chart represents the frequency. A similar chart is used for continuous data but with two important differences:

- There will be no gaps between the bars. Each bar represents one class interval.

- The area of the bar, not the height, represents the frequency. Class intervals can therefore be different sizes.

A **histogram** is similar to a bar chart but, because the data are continuous, there are no gaps between the bars.

A **frequency polygon** joins the mid-points of the top of the bars with straight lines.

Example 6

A group of students were asked to say the 12-times tables as fast as possible. The time taken by each student was recorded.

The results are shown in this frequency table.

a Draw a histogram for these data.

b Draw a frequency polygon for these data.

Time, t (s)	Frequency
$0 < t \leqslant 10$	1
$10 < t \leqslant 20$	2
$20 < t \leqslant 30$	8
$30 < t \leqslant 40$	12
$40 < t \leqslant 50$	6
$50 < t \leqslant 60$	3

a

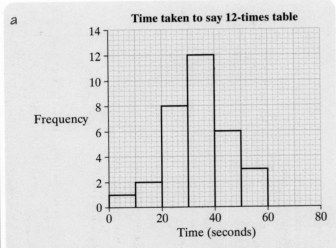

Draw the axes and add scales.

Label the vertical axis 'Frequency' and the horizontal axis 'Time (seconds)'.

Now draw the bars for each interval (e.g. The first bar goes from 0 to 10 and has a height of 1).

Add a title to the histogram.

b

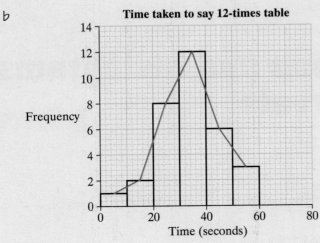

Draw the frequency polygon by joining the mid-points of the top of the bars with straight lines.

3.5 The shape of a distribution

The shape formed by the bars in a **histogram** is known as the shape of the **distribution**.

> A **histogram** shows how the data are distributed across the class intervals.
>
> A **distribution** can be **symmetrical**, or have **positive skew** or **negative skew**.

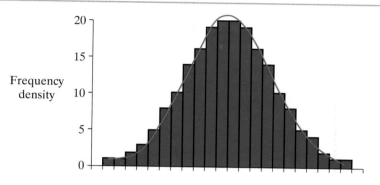

> The shape of a frequency polygon shows the shape of the distribution.

This distribution is **symmetrical**. It has no skew. Example: The lengths of leaves on a tree.

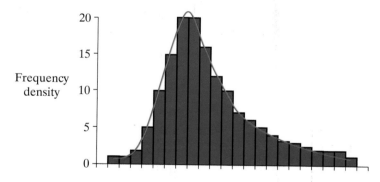

This distribution has **positive skew**. Most of the data values are at the lower end. Example: The age at which a person learns to write.

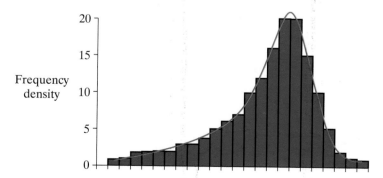

This distribution has **negative skew**. Most of the data values are at the upper end. Example: The age at which a person dies.

Exercise 3D

1 Last season, Gander United played 30 hockey matches away from home. The table gives information about the distances travelled to away matches.

Distance, d (miles)	Frequency
$0 < d \leqslant 20$	6
$20 < d \leqslant 40$	12
$40 < d \leqslant 60$	7
$60 < d \leqslant 80$	4
$80 < d \leqslant 100$	1

Draw a frequency polygon to display this information.

2 Amy decided to measure the height of each of 50 plants of the same variety after they had been growing for a month. The table gives information about the heights of her plants.

Height (cm)	Frequency
up to, but not including 10	3
10 up to, but not including 20	12
20 up to, but not including 30	19
30 up to, but not including 40	10
40 up to, but not including 50	6

Draw a frequency polygon to display this information.

3

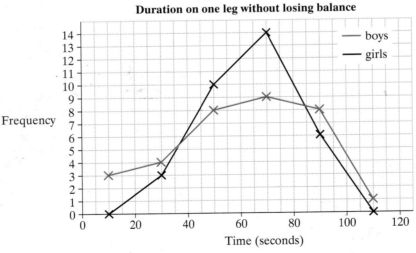

Students were asked to stand on one leg for as long as possible. The frequency polygon gives information about the performance of boys and girls in this task.

a How many girls were able to stand on one leg for between 20 and 40 seconds without losing balance?

b How many boys lost their balance in less than or equal to 40 seconds?

c One student balanced for longer than the others. Was this a boy or a girl?

d Copy and complete this frequency table.

Time, t (s)	Girls' frequency	Boys' frequency
$0 < t \leqslant 20$		
$20 < t \leqslant 40$		
$40 < t \leqslant 60$		
$60 < t \leqslant 80$		
$80 < t \leqslant 100$		
$100 < t \leqslant 120$		

e Were there more boys or girls in the experiment?

f Were boys or girls better at balancing? Explain your answer.

g Draw a histogram to represent the boys' data.

h Comment on the skew of the distribution.

4 This table shows some information about the heights of some Year 7 and 8 students.

Height, h (cm)	Year 7 frequency	Year 8 frequency
$120 < h \leqslant 130$	1	0
$130 < h \leqslant 140$	5	3
$140 < h \leqslant 150$	18	12
$150 < h \leqslant 160$	20	22
$160 < h \leqslant 170$	8	19
$170 < h \leqslant 180$	2	6

a Draw a histogram to show Year 7 heights.

b Draw a histogram to show Year 8 heights.

c On your histograms draw frequency polygons for these data.

5 Thirty students were asked to time their journey to school to the nearest minute. These are the results.

6	18	29	55	7	34	28	56	33	4
2	41	33	23	7	43	26	53	44	41
32	46	16	17	3	26	17	47	22	17

a Design and complete a frequency table to sort these data. Use class intervals of equal width. (Remember: the data are rounded.)

b Draw a histogram to display your results.

3.6 Histograms with unequal class intervals

This section explains why it is necessary for the area of each bar in a histogram, rather than the height, to represent the frequency.

Look carefully at this frequency table.

These data could be represented in a histogram.

Time, t (s)	Frequency
$10 < t \leqslant 20$	8
$20 < t \leqslant 40$	8

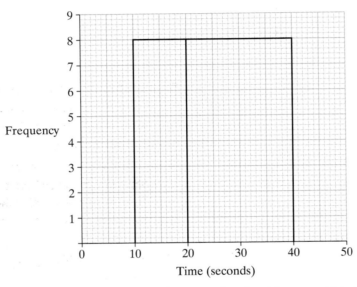

This is not a histogram because the area for the $20 < t \leqslant 40$ interval is larger than for the $10 < t \leqslant 20$ interval. It makes it look as if there are twice as many in the interval $20 < t \leqslant 40$ as in $10 < t \leqslant 20$.

> To draw a histogram for **unequal class intervals**, you need to adjust the heights of the bars so the **area is proportional to the frequency**.
>
> The height of the bar, called the **frequency density**, is found by **dividing the frequency by the class width**.

Example 7

This frequency table shows the time taken for each of 470 people to climb up four flights of stairs.

Draw a histogram of these data.

Time, t (s)	Frequency
$40 < t \leqslant 60$	100
$60 < t \leqslant 70$	60
$70 < t \leqslant 80$	90
$80 < t \leqslant 85$	70
$85 < t \leqslant 90$	60
$90 < t \leqslant 120$	90

Time, t (s)	Frequency	Class width	Frequency density
$40 < t \leqslant 60$	100	20	$\frac{100}{20} = 5$
$60 < t \leqslant 70$	60	10	$\frac{60}{10} = 6$
$70 < t \leqslant 80$	90	10	$\frac{90}{10} = 9$
$80 < t \leqslant 85$	70	5	$\frac{70}{5} = 14$
$85 < t \leqslant 90$	60	5	$\frac{60}{5} = 12$
$90 < t \leqslant 120$	90	30	$\frac{90}{30} = 3$

Add two columns to the table: class width and frequency density.

Calculate the class widths. For example:

$60 - 40 = 20$

Calculate each frequency density. For example:

$\frac{70}{5} = 14$

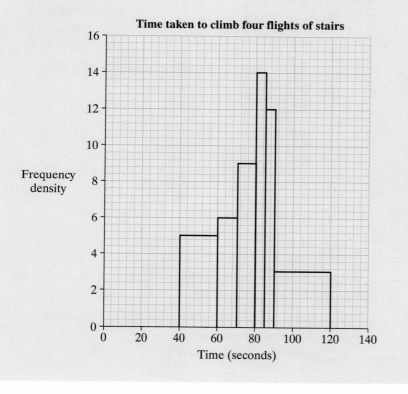

Time taken to climb four flights of stairs

Draw the axes and add scales.

Label the vertical axis 'Frequency density' and label the horizontal axis 'Time (seconds)'.

Now draw the bars for each interval. Use the class boundaries and the frequency density: the first bar goes from 40 to 60 and has a height of 5.

Add a title to the histogram.

3.7 Interpreting histograms with unequal class intervals

The frequency density and the width of the class interval are used to work out the frequency of each class interval in a histogram.

Frequency = frequency density × class width

Example 8

This histogram gives information about the times some students took to complete a puzzle.

Time, t (s)	Frequency
0 < t ≤ 20	12
20 < t ≤ 30	
30 < t ≤ 40	
40 < t ≤ 45	

Use the histogram to complete the frequency table.

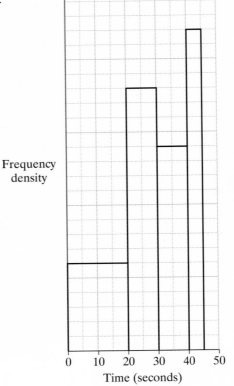

Time to complete a puzzle

Frequency density

Time (seconds)

> There is sometimes more than one way to calculate answers to a question.

Method 1

The area of 0 < t ≤ 20 is 24 squares.

One square has a frequency of $\frac{12}{24} = 0.5$.

Time, t (s)	Frequency
0 < t ≤ 20	12
20 < t ≤ 30	36 × 0.5 = 18
30 < t ≤ 40	28 × 0.5 = 14
40 < t ≤ 45	22 × 0.5 = 11

> Find the area of the 0 < t ≤ 20 bar.
>
> Use the frequency for that interval (12) to find the frequency represented by each square.
>
> Calculate the areas of the other bars and multiply each area by the frequency for one square.

Method 2

The frequency density of the 0 < t ≤ 20 class interval is $\frac{12}{20} = 0.6$.

The frequency of 20 < t ≤ 30 class = 1.8 × 10 = 18.

The frequency of 30 < t ≤ 40 class = 1.4 × 10 = 14.

The frequency of 40 < t ≤ 45 class = 2.2 × 5 = 11.

> Calculate the frequency density for the class interval where the frequency is known $\left(\frac{\text{frequency}}{\text{class width}}\right)$.
>
> Work out the vertical scale. On the vertical axis, frequency density 0.6 is represented by 6 squares. So 1 square represents 0.1.
> Read off the frequency densities of the other class intervals.
>
> Multiply each frequency density by its class width.

Example 9

This table gives information about the times it took some people to read a newspaper article. The histogram shows some of these data.

Time, t (s)	Frequency
0 < t ≤ 20	80
20 < t ≤ 30	70
30 < t ≤ 40	110
40 < t ≤ 45	65
45 < t ≤ 50	70
50 < t ≤ 80	60

Time taken to read article

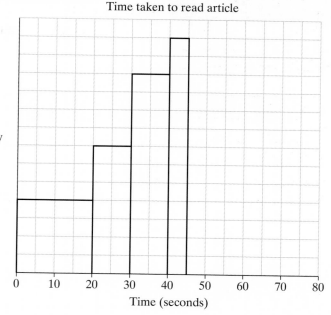

Frequency density

Time (seconds)

a   Use the information given to complete the histogram.

b   Estimate how many people took more than 35 seconds to read the newspaper article.

a   Frequency density for $0 < t ≤ 20 = \frac{80}{20} = 4$.

Frequency density for $45 < t ≤ 50 = \frac{70}{5} = 14$.

Frequency density for $50 < t ≤ 80 = \frac{60}{30} = 2$.

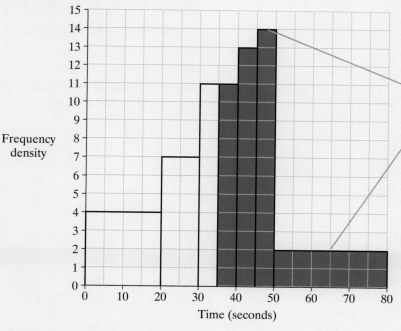

Frequency density

Time (seconds)

Calculate the frequency density of one of the given bars using

$$\text{Frequency density} = \frac{\text{Frequency}}{\text{Class width}}$$

Calculate the frequency density of the two remaining bars.

Label 4 at height of the bar for the class $0 < t ≤ 20$ which you calculated and complete the scale.

Draw the two remaining bars using the frequency density as the height.

The number of people is the area of the histogram between 35 and 80.

b   Number of people = (5 × 11) + (5 × 13) + (5 × 14) + (30 × 2)
          = 250

ResultsPlus
Exam Question Report

8. A shoe manufacturer measured the length (*l* mm) of 200 people's feet.

The results are summarised in the table below.

Length (*l mm*)	Frequency	
230 ≤ *l* < 240	9	
240 ≤ *l* < 250	25	
250 ≤ *l* < 255	26	
255 ≤ *l* < 260	30	
260 ≤ *l* < 265	33	
265 ≤ *l* < 270	30	
270 ≤ *l* < 280	29	
280 ≤ *l* < 295	18	

The incomplete histogram shows information about the data.

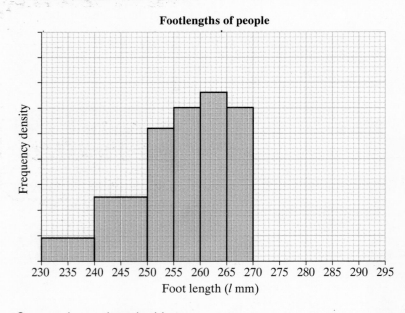

Copy and complete the histogram.

How students answered

Poor   42%

Some students seemed to guess at the heights of the extra bars. They did not show that they had worked out the frequency densities for the different class intervals.

Good   25%

Students calculated the frequency densities correctly and wrote them in the empty column in the table. They didn't use the existing bars in the histogram to work out the scale for the Frequency density axis, so couldn't plot their bars accurately.

Excellent   33%

Students calculated the frequency densities correctly. They used the existing bars to work out the scale for the vertical axis, and drew the extra bars to the correct heights.

Exercise 3E

1 The lengths of 62 songs by Median and the Meanies are represented in the frequency table.

Draw a histogram to display these data.

Song length, L (s)	Frequency
0 < L ⩽ 100	5
100 < L ⩽ 180	8
180 < L ⩽ 210	12
210 < L ⩽ 240	15
240 < L ⩽ 300	12
300 < L ⩽ 500	10

ResultsPlus
Watch out!

■ When asked to draw a histogram for grouped data, many students drew the bars to represent the frequency instead of the frequency density.

● Some students found it difficult to decide on a suitable scale for the vertical axis.

2 The exact ages (in years) of guests at a fancy dress party were recorded. Some information about the ages is shown in the frequency table.

a Draw a histogram to display these data.

b Use your histogram as a guide to draw a frequency polygon for these data.

Age, A (years)	Frequency
15 < A ⩽ 20	5
20 < A ⩽ 23	15
23 < A ⩽ 25	20
25 < A ⩽ 30	20
30 < A ⩽ 40	10

3 Alice measures the heights of 23 students in her class to the nearest centimetre. The data are shown in the table.

Height, h (cm)	Frequency
140.5 < h ⩽ 150.5	5
150.5 < h ⩽ 155.5	3
155.5 < h ⩽ 160.5	6
160.5 < h ⩽ 165.5	3
165.5 < h ⩽ 180.5	6

Examiner's Tip
Remember to use the actual class limits in this question even though they are not whole numbers.

This incomplete histogram shows some information about these data.

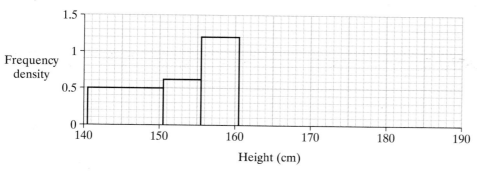

Use the information given to copy and complete the histogram.

4 Justin conducted an experiment to see how far 33 snails would move in 10 minutes. The results are shown in this frequency table.

Distance moved, d (cm)	Frequency
$0 < d \leqslant 5$	3
$5 < d \leqslant 7$	5
$7 < d \leqslant 8$	4
$8 < d \leqslant 9$	6
$9 < d \leqslant 10$	3
$10 < d \leqslant 15$	6
$15 < d \leqslant 25$	6

Construct a histogram to display this information.

5 This histogram gives information about the distances (in metres) thrown in a javelin competition.

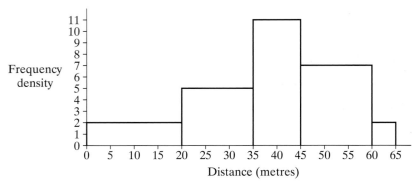

Design and complete a frequency table for the data.

6 A group of children timed how long (in seconds) they could bounce a tennis ball on a racket. This histogram gives information about their times.

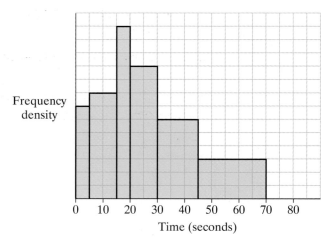

Time, t (s)	Frequency
$0 < t \leqslant 5$	56
$5 < t \leqslant 15$	
$15 < t \leqslant 20$	
$20 < t \leqslant 30$	
$30 < t \leqslant 45$	
$45 < t \leqslant 70$	
$70 < t \leqslant 80$	24

a Copy and complete the table.

b Copy and complete the histogram for the final class.

3.8 Population pyramids

Population pyramids look similar to two back-to-back histograms.
They make it easy to compare aspects of a population, often by gender.

They usually have class sizes, and therefore bars, the same width.

Example 10

This table shows the percentage of the population of a country in each age group.

Age, a (years)	Males (%)	Females (%)
$0 < a \leqslant 10$	13.1	12.3
$10 < a \leqslant 20$	16.4	15.6
$20 < a \leqslant 30$	15.8	15.8
$30 < a \leqslant 40$	17.3	17.2
$40 < a \leqslant 50$	15.1	15.3
$50 < a \leqslant 60$	11.6	12.2
$60 < a \leqslant 70$	6.5	6.6
$70 < a \leqslant 80$	3.4	3.1
$a > 80$	1	1.9

Display this information as a population pyramid.

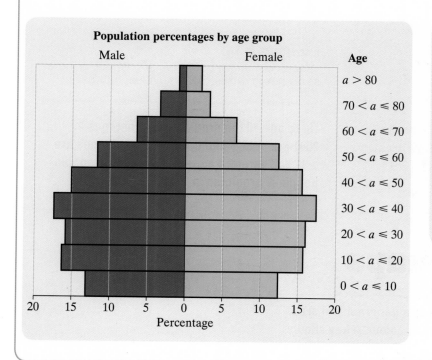

Draw a horizontal axis.

Label the centre 0 and label the percentages to the left and to the right.

Put the age groups on the right.

Draw a vertical line at 0.

Label Males on the left and Females on the right.

Now draw bars for each age group. Make the bars the same width.

Give the chart a title.

Example 11

The first population pyramid shows the percentage of males and females in each age group in the UK in 1991. The second population pyramid shows the predicted population in 2041.

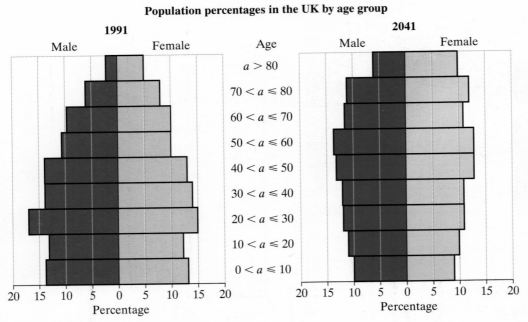

Population percentages in the UK by age group

Source: Office for National Statistics

a Which age group had the greatest percentage of females in 1991?

b Ten per cent of males in 2041 are predicted to be in one age group. Which age group?

c Compare the population in 1991 with the predicted population in 2041.

a $20 < a \leqslant 30$

b $0 < a \leqslant 10$

c There is a greater percentage of younger people in 1991 than in 2041, so conversely there is a greater percentage of older people in 2041 than in 1991.

Examiner's Tip
When comparing two populations, use general statements rather than individual figures.

Up to age 40 the predicted percentages in 2041 are <12% and the 1991 percentages are >12%. Therefore a greater percentage of old people is predicted in 2041.

3.9 Choropleth maps

A **choropleth map** is used to classify regions of a geographical area. Regions are shaded with an increasing depth of colour. A **key** shows what each shade represents.

Example 12

This **choropleth map** shows the east of England split into 14 regions. Each region is shaded to show the percentage of people in it who are four years old or younger.

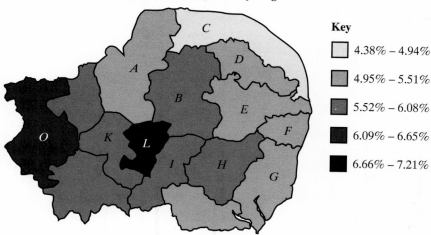

Percentage of population aged 4 or younger

Key

☐	4.38% – 4.94%
☐	4.95% – 5.51%
☐	5.52% – 6.08%
☐	6.09% – 6.65%
☐	6.66% – 7.21%

Source: Office for National Statistics

a Which region has between 4.38% and 4.94% of people four years old or younger?
b Which region has the highest percentage of people four years old or younger?
c What percentage of people are four years old or younger in region O?

a Region C — The shading for 4.38% – 4.94% is pale. Look for the region shaded in this colour.

b Region L — The highest percentage listed in the key is 6.66% – 7.21%. Look for the region shaded in the colour shown for those percentages

c 6.09% – 6.65% — Look at the shading used for region O. Find the shading in the key and read off the percentage.

Example 13

A field is divided up into 25 equal-sized squares.
The number of daisy plants in each square is counted.
The results are shown in the diagram.

5	6	8	1	7
6	3	10	4	2
8	2	0	1	7
6	9	4	3	2
1	3	11	0	0

Key: ☐ 5 means 5 daisy plants in the square

Use the information in the diagram to complete this choropleth map.

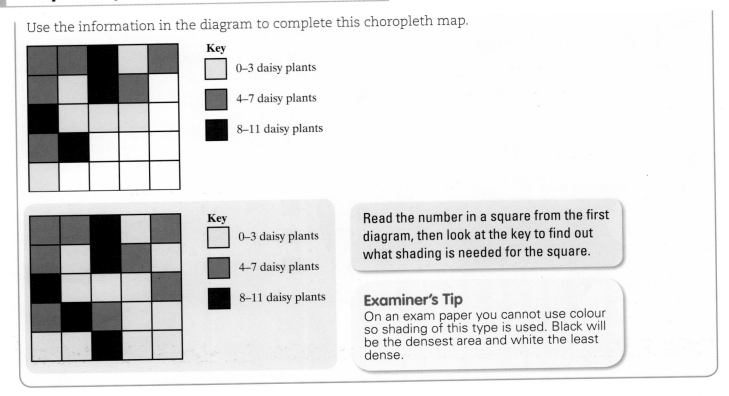

Read the number in a square from the first diagram, then look at the key to find out what shading is needed for the square.

Examiner's Tip

On an exam paper you cannot use colour so shading of this type is used. Black will be the densest area and white the least dense.

ResultsPlus

Exam Question Report

The diagrams show information about the sea temperatures for Peru in 1981 and in 1983.

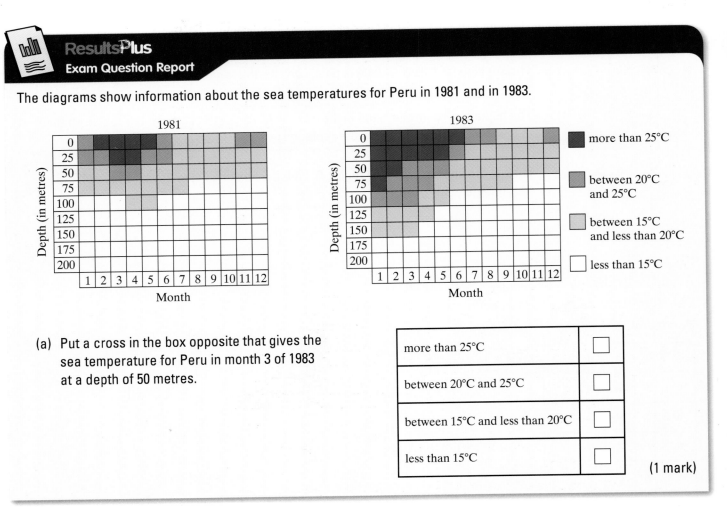

(a) Put a cross in the box opposite that gives the sea temperature for Peru in month 3 of 1983 at a depth of 50 metres.

more than 25°C	☐
between 20°C and 25°C	☐
between 15°C and less than 20°C	☐
less than 15°C	☐

(1 mark)

For two months in 1981 the sea temperature for Peru was between 15°C and less than 20°C at a depth of 100 metres.

(b) Write down these two months   (1 mark)

(c) Use the diagrams to compare the sea temperatures for Peru in 1981 and in 1983.   (2 marks)

(Total 4 marks)

How students answered

Poor   23%

Interpreted the diagrams as showing a change in depth of sea rather than temperature. Compared temperatures in individual cells of the diagram, rather than overall changes over the period.

Good   62%

Interpreted the diagrams accurately in parts a) and b). Compared overall change in temperature between 1981 and 1983 in part c). A typical answer was 'hotter' in 1983.

Excellent   15%

Also mentioned the temperature at different depths, or the length of time a temperature was recorded.

Exercise 3F

1 This table shows the percentage of a country's population in each age group.

Display this information on a population pyramid.

Age, a (years)	Male (%)	Female (%)
$0 < a \leqslant 10$	24	21
$10 < a \leqslant 20$	20	17
$20 < a \leqslant 30$	17	15
$30 < a \leqslant 40$	14	12
$40 < a \leqslant 50$	11	10
$50 < a \leqslant 60$	8	10
$60 < a \leqslant 70$	3	8
$70 < a \leqslant 80$	2	5
$a \geqslant 90$	1	2

2

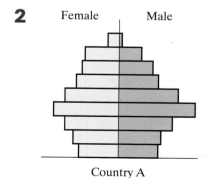
Country A

Age

$a > 80$
$70 < a \leqslant 80$
$60 < a \leqslant 70$
$50 < a \leqslant 60$
$40 < a \leqslant 50$
$30 < a \leqslant 40$
$20 < a \leqslant 30$
$10 < a \leqslant 20$
$0 < a \leqslant 10$

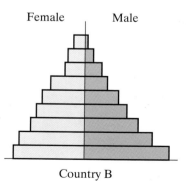
Country B

The population pyramids give information about the ages of the populations in two countries.

a Which country has had a high birth rate in recent years?

   b Are males or females more likely to live longer?

   c A person is chosen at random from Country A. What age are
       they most likely to be?

   d In which country are people more likely to live beyond
       50 years old?

   e Explain why Country B is likely to be a poor country.

3 This choropleth map shows the percentage of adults in each
    region of a town who had studied in higher education.

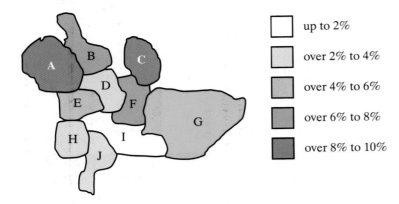

   up to 2%

   over 2% to 4%

   over 4% to 6%

   over 6% to 8%

   over 8% to 10%

   a Which regions have the highest percentage of adults who
       have studied in higher education?

   b Which region has the lowest percentage of adults who have
       studied in higher education?

   c Which region has a similar percentage of adults with
       experience of higher education to region F?

   d Which regions have between 2% and 4% of adults with
       experience of higher education?

4 This table shows data for 10 equally-sized regions.
    It shows the percentage of houses for sale in each
    region.

Region		Percentage of houses for sale
A	Sabury	12%
B	Ellerton	53%
C	Rochwood	5%
D	Barford	38%
E	Marlmore	42%
F	Radford	13%
G	Hatherton	37%
H	Birwich	22%
I	Fishton	18%
J	Carraford	32%

Use the key provided (or similar) to copy and
shade in the choropleth map.

A	B	C	D	E
F	G	H	I	J

Up to 15%

Over 15% to 30%

Over 30% to 45

Over 45% to 60

3.10 Misleading diagrams

Discrete and continuous

Sometimes data are presented as a graph or chart in a way that misleads the viewer. These graphs are said to be poorly presented data. Sometimes graphs are drawn deliberately to mislead, others are unintentionally misleading.

There are many ways in which a graph can be misleading:

- Scales that do not start at zero or have parts of them missed out give a misleading impression of the heights of bars, etc.
- Scales that do not increase uniformly distort the shape of anything plotted on them.
- Lines on a graph that are drawn too thick make it difficult to read information.
- Axes may not be labelled properly.
- Three-dimensional diagrams make comparisons difficult. Often things at the front of the diagram can appear larger than those at the back (e.g. angled pie charts because the angles are distorted). Parts at the back may be hidden behind those at the front and appear smaller than they should.
- Sections of the diagram separated from other parts make comparisons difficult (e. g. pie charts with slices pulled out).
- Using colour so some parts stand out more than others. Generally dark colours stand out more than light colours and make these sections look bigger.
- Using different width bars/pictures. To make charts more interesting, the bars can be made up of pictures of the thing they represent. For example, bags of money to illustrate wages: if the bags are different sizes it is unclear whether it is the height of the money bags that should be compared or the area.
- Some data may be excluded.

This chart shows that the amount of waste recycled over two years has more than doubled. Or does it?

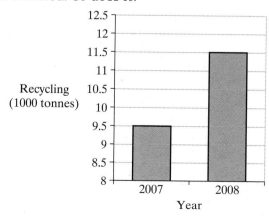

The chart is misleading because the vertical axis scale does not start at zero.

The graph should look like this:

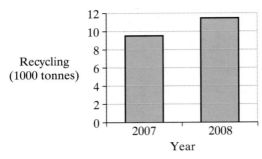

This clearly shows that recycling has not more than doubled.

Example 14

Give **three** reasons why the following graph may be misleading.

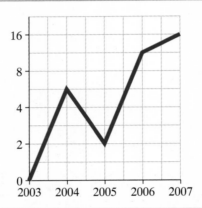

The scale does not go up in steps of equal size.

The line is thick.

The axes are not labelled.

The *y*-axis goes 0, 2, 4, 8, 16 rather than 0, 2, 4, 6, 8, etc.

The line is thick so it is difficult to read values from the vertical axis.

We have no idea what the graph is about as the axes are not labelled (and it has no title).

Example 15

A group of students were asked to choose their favourite soft drink. This pie chart shows the results.

Describe **four** reasons why the diagram could be misleading.

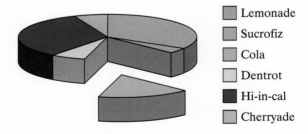

Lemonade
Sucrofiz
Cola
Dentrot
Hi-in-cal
Cherryade

The pie chart is 3-D. Areas at the front look larger. Angles are distorted.

A piece of the pie chart has been pulled out, making comparisons hard. The slices next to the pulled out section look larger because their sides can also be seen.

The dark colours stand out more, so these sections look bigger.

There is no section for 'other soft drinks'.

Exercise 3G

1 This graph shows that the number of mp3s downloaded from a website over two years has more than doubled.

Describe the reasons this graph could be misleading.

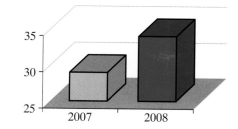

2 The graph appears to shows that the price of a product increased fast between 2003 and 2009.

Why might this graph be misleading?

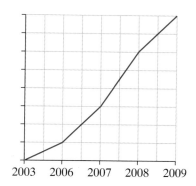

3 A teacher draws a pie chart showing what pets each child in her class has.

Give **two** reasons why this might be misleading.

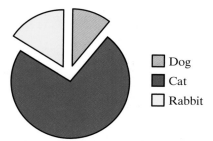

Chapter 3 review

1 Members of a youth club recorded the length of time that each member could balance a dictionary on their head. The frequency polygon shows the results for boys.

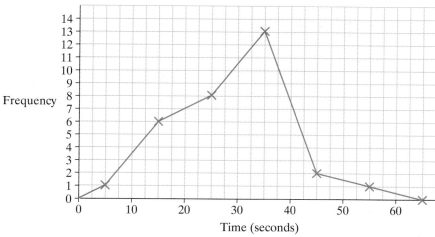

a John said, 'The frequency polygon shows that 13 boys balanced the dictionary for exactly 35 seconds.' Explain why he is wrong.

b Design a frequency table to show these data, using the intervals $0 < T \leqslant 10$ etc.

c This table shows similar results for the girls at the youth club.

Copy the frequency polygon and then draw the polygon for girls on the same axes.

d Were the boys or the girls better at balancing the dictionary? Explain your answer.

Time taken, T (s)	Frequency
$0 < T \leqslant 10$	3
$10 < T \leqslant 20$	10
$20 < T \leqslant 30$	14
$30 < T \leqslant 40$	5
$40 < T \leqslant 50$	2
$50 < T \leqslant 60$	1

2 Map A shows the distribution of the Golden Plover in the UK. Maps B, C and D show three possible factors that may positively influence the distribution of Golden Plovers. Heavier shading implies greater density on maps A and C, higher altitudes on map B and higher rainfall on map D.

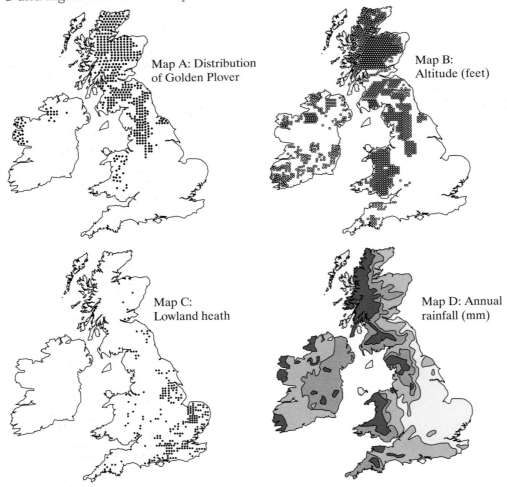

Map A: Distribution of Golden Plover

Map B: Altitude (feet)

Map C: Lowland heath

Map D: Annual rainfall (mm)

Compare maps B, C and D with map A. Decide which of the three factors are most likely to influence the distribution of Golden Plovers. Give a reason for your answer.

edexcel ::: *past paper question*

3 The diagram shows information about the population of Camden in 2001.

  a What is the name of this type of diagram?

Use the diagram to answer the following questions about the population of Camden.

  b Which age group that has the largest population?

  c Estimate the percentage of the female population under 20 years old.

ResultsPlus
Watch out!

⬛ In this type of question, many students lose marks by talking about 'numbers of people', when the population pyramid shows percentages, not numbers

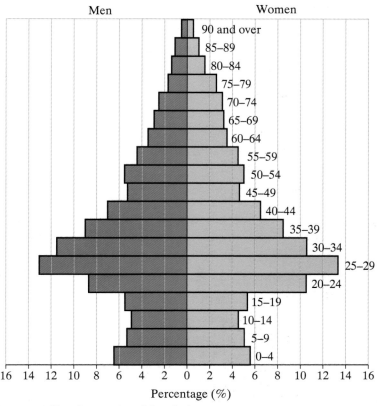

Source: *Office for National Statistics*

The following diagram shows information about the population of Northern Ireland in 2001.

  d Give **one** similarity and **one** difference between the population of Camden and the population of Northern Ireland.

                        edexcel ⠿ *past paper question*

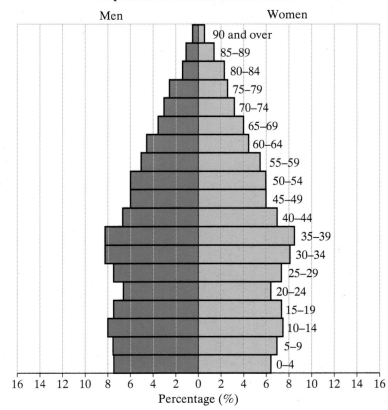

Source: *Office for National Statistics*

4   Thirty-six snails took part in a race. The distances in centimetres travelled by each snail were measured after half an hour. The results are shown below.

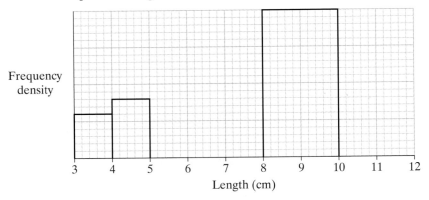

1.3	2.6	3.7	2.2	0.1	4.5	2	0.2	1.1
0.5	1.3	2.7	3.4	3	3	0	3.5	3.2
1.7	2.2	1.1	2.1	1.8	4.2	2.1	4.7	3.8
0.6	1.3	5.2	4.3	2.8	1.9	3.3	3.6	4.2

  a   Construct a stem and leaf diagram from the data.

  b   Construct a frequency table using equal class intervals (e.g. $0 < d \leqslant 1, 1 < d \leqslant 2, \ldots$).

5   Each day Peter recorded the lateness of his train, in minutes. He did this for 23 days. The table shows this information.

Time, t (mins)	Frequency
$0 < t \leqslant 2$	2
$2 < t \leqslant 4$	3
$4 < t \leqslant 6$	4
$6 < t \leqslant 8$	6
$8 < t \leqslant 10$	8

  a   Draw a cumulative frequency diagram to represent these data.

  b   Estimate the number of times the train is more than 6.5 minutes late.

  c   Estimate the number of trains that are between 5 and 7.5 minutes late.

  d   Draw a histogram to represent these data.

6   David randomly selects 50 pebbles from a beach. He measures the length of each one to the nearest centimetre.

The data are shown in this table.

Length, l (cm)	Frequency
$3 \leqslant l < 4$	
$4 \leqslant l < 5$	
$5 \leqslant l < 8$	14
$8 \leqslant l < 10$	20
$10 \leqslant l < 11$	9

This incomplete histogram shows information about these data

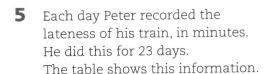

Use the information given to copy and complete the histogram and the frequency table.

Chapter 3 summary

Frequency tables, pie charts and stem and leaf diagrams

1   **Continuous data** can be sorted into a **frequency table**. The **class intervals** must not have gaps in between them or overlap each other.

2   All possible values that round to the same number must fit into the same class interval.

3   **Intervals** do not need to be of equal width. Use narrower intervals where the data are grouped together and wider intervals where the data are spread out.

4   Pie charts can be used to show continuous data. The area of each sector represents the frequency for that class interval. Other than this the chart is drawn in the same way as for discrete data.

Cumulative frequency diagrams

5   If there are not too many observations they can be treated as individual numbers and shown as a stem and leaf diagram. The stem shows the first digit(s) of the numbers and the leaves show the last digit.

6   **Cumulative frequency** is a running total of frequencies.

7   **Cumulative frequency diagrams** can be used to estimate or predict other values.

Histograms and frequency polygons

8   A **histogram** is similar to a bar chart but, because the data are continuous, there are no gaps between the bars.

9   A **frequency polygon** joins the mid-points of the top of the bars with straight lines.

10   A **histogram** shows how the data are distributed across the class intervals.

11   A **distribution** can be **symmetrical**, have **positive skew** or **negative skew**.

12   To draw a histogram for **unequal class intervals** you need to adjust the heights of the bars so the **area is proportional to the frequency**.

13   The height of the bar, called the **frequency density**, is found by **dividing the frequency by the class width**.

14   The frequency density and the width of the class interval are used to work out the frequency of each class interval in a histogram.

$$\text{Frequency} = \text{frequency density} \times \text{class width}$$

Population pyramids

15   **Population pyramids** look similar to two back-to-back histograms. They make it easy to compare aspects of a population, often by gender.

16   They usually have class sizes, and therefore bars, the same width.

Choropleth maps

17   A **choropleth map** is used to classify regions of a geographical area. Regions are shaded with an increasing depth of colour. A **key** shows what each shade represents.

Higher Statistics

Test yourself

1 These two population pyramids show the percentages of males and females in each age group in the United Kingdom and in Northern Ireland.

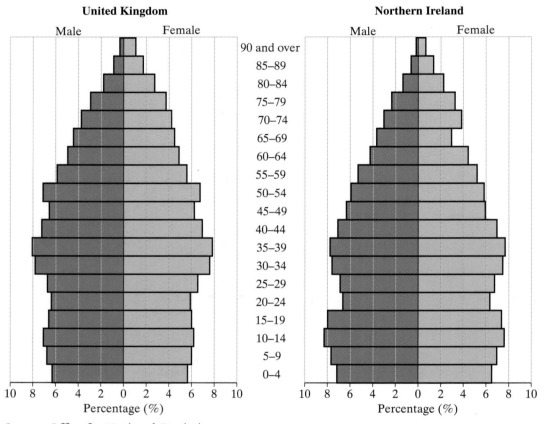

The ages of Males and Females in each age group in 2001

Source: Office for National Statistics

a Which age group has the greatest percentage of females in the United Kingdom?

b Six per cent of males in Northern Ireland are in one age group. Which age group?

c Compare the percentage of people up to the age of 19 in the United Kingdom with the percentage of people up to the age of 19 in Northern Ireland.

2 This diagram gives information about the numbers of ladybirds found on roses in a rose bed.

2	2	3	6	6	6	7	8
0	1	2	4	4	10	9	10

The partially completed choropleth map represents these data.

Copy and complete the map.

Key

0–2   6–8

3–5   9–11

3 Write down **six** different ways in which a diagram representing data might be misleading.

4 Here is a cumulative frequency diagram showing the time taken for students to solve a puzzle.

Use it to estimate the number of students who took between 25 and 45 minutes.

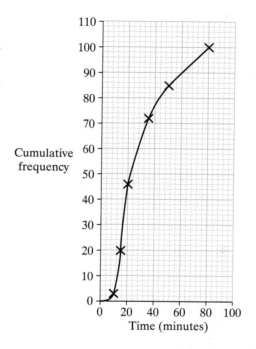

5 A histogram uses a bar of width 2 cm and height 5 cm to represent a frequency of 30. A second bar is 4 cm wide. What height would it need to be to represent a frequency of 50?

Chapter 4:
Summarising data: measures of central tendency and dispersion

After completing this chapter you should be able to

- work out measures of central tendency
  - find the mean, mode and median of raw data given as a list or as a frequency distribution
  - estimate the mean and median of a grouped frequency distribution and identify the modal class
  - understand the effects of transformations of the data on these averages
  - understand when each of these averages is appropriate
  - calculate and use a weighted mean

- work out measures of dispersion
  - work out the range for data given as a list or frequency distribution
  - work out quartiles and percentiles
  - draw box plots
  - identify outliers
  - calculate and use variance and standard deviation
  - compare distributions of data
- use simple index numbers, chain base index numbers and weighted index numbers
  - know about the retail prices index.

School children can be measured in other ways

When you have completed this chapter you will be able to use a measure of central tendency and a measure of spread to summarise data about children.

For example their mean age might be 10.5 years old and range from 5 to 18 years old.

You will also be able to compare sets of data summarised in this way.

4.1 Summarising data

It is possible to measure the heights of 100 men and 100 women, but how can these data be compared?

To make comparisons between sets of numbers it is convenient to have a single figure that represents all the numbers in any given set. This is called an **average** or a **measure of central tendency**.

> An **average** is a single value used to describe a set of data.

How the data values are spread about the average can also be compared: are they closely clustered about the average or are they spread over a wide range of values?

Again, it is useful to have a single figure that represents this property. This is called a **measure of spread** or **dispersion**.

4.2 Averages or measures of central tendency: mode, median and mean

The mode

> The **mode** is the value that occurs most often.

Example 1

There are 17 young people on a school bus. Their ages are:

| 12 | 15 | 13 | 17 | 10 | 14 | 16 | 15 | 16 |
| 12 | 15 | 12 | 13 | 14 | 16 | 15 | 11 | |

Find the mode of their ages.

| 10 | 11 | 12 | 12 | 12 | 13 | 13 | 14 | 14 |
| **15** | **15** | **15** | **15** | 16 | 16 | 16 | 17 | |

The mode is 15.

First arrange the numbers in order.

Look for the number that has the greatest frequency. (15 appears four times.)

The median

> The **median** is the middle number in a list after they have been put in order.

Example 2

Find the median age of the 17 people on the school bus in Example 1.

10	11	12	12	12	13	13	14	**14**
15	15	15	15	16	16	16	17	

The median is 14.

> Arrange the data in order.

> Work in from both ends until you find the middle number.
>
> There are 17 pieces of data, so the middle is the 9th one.

Example 3

Six students took a test and their marks were recorded:

13   13   13   14   14   15

Find the median mark.

13   13   **13**   **14**   14   15

$$\text{Median} = \frac{13 + 14}{2} = 13\tfrac{1}{2}$$

> There are an even number of items in the list. The marks are already ordered, so count in from both ends.
>
> The median is half way between the 3rd and 4th values (which are 13 and 14).

It is not always easy to find the middle number if it is a long ordered list.

There is an easy way to tell which is the middle value that means you only need count from one end of the list.

In a set of n numbers, the median value is the $\frac{1}{2}(n + 1)$th value.

In Example 2, where $n = 17$, the formula gives $\frac{1}{2}(17 + 1) = $ 9th value.

In Example 3, where $n = 6$, the formula gives $\frac{1}{2}(6 + 1) = 3\frac{1}{2}$th value (i.e. half way between the third and fourth values).

> **Examiner's Tip**
>
> In an exam you can either count in from both ends or use the formula to find the middle. You will still get the marks.

> When the number of data values, n, is odd the **median** is the value of the $\frac{1}{2}(n + 1)$**th** observation.
>
> When n is even the **median** is the **mean** of the $\frac{1}{2}(n)$**th** observation and the $\frac{1}{2}(n + 2)$**th** observation.

The mean

A set of data can have several different means. This section looks at the **arithmetic mean**, often called simply the **mean**.

> The **mean** is worked out by adding up the numbers then dividing by how many numbers there are.

Here, x is an observation and n is the total number of observations:

> **Mean** $= \bar{x} = \dfrac{\sum x}{n}$
> where
> - $\bar{x}$ means the mean of all the x values
> - $\sum x$ means the sum of all the x values.

$\sum$ is the Greek letter sigma used to represent the word 'sum'.

Example 4

Find the mean age of the 17 children on the bus in Example 1.

$$12 + 15 + 13 + 17 + 10 + 14 + 16 + 15 + 16 + 12$$
$$+ 15 + 12 + 13 + 14 + 16 + 15 + 11 = 236$$

Add up the ages.

$$\text{Mean} = \frac{\sum x}{n} = \frac{236}{17} = 13.88$$

Divide the result by how many children there are.

Exercise 4A

1 Seven workers recorded the number of minutes it took them to get to their workplace. Their times were:

| 7 | 12 | 12 | 15 | 17 | 18 | 24 |

 a What is the modal time?

 b What is the median time?

 c Work out the mean time.

2 Some council employees were asked how many times they had visited the gym in the past month. These are their answers:

| 5 | 7 | 9 | 9 | 8 | 7 | 10 |
| 9 | 11 | 12 | 5 | 9 | 9 |

 a Order the numbers.

 b What is the modal number of visits?

 c What is the median number of visits?

 d Work out the mean number of visits, to two decimal places.

ResultsPlus

Watch out!

⬤ When given a set of seven numbers and asked to calculate the mean, median and mode, some students confused the three different averages.

When given a set of eight numbers and asked to calculate the median, some students gave the two middle numbers, rather than the mean of these two numbers.

3 Ten applicants for a job did a reaction time test. The times that they took to react were recorded, in seconds.

2   7   1   4   19   11   2   8   5   6

a What is the median time?

b What is the modal time?

c Work out the mean time.

4 The number of cars using a car park during a particular period is summarised in the stem and leaf diagram.

Number of cars

1	2	3				
2	3	4	4	5		
3	2	3	3	3	6	7	9
4	0	1	3	5		
5	0	2				

Key
2 │ 3 means 23

Work out the mode, median and mean for these data.

5 The mean weight of a group of ten students is 52 kg.

a What is the total weight of these students?

One other student joins the group of ten. The student weighs 30 kg.

b What is the mean weight of all 11 students?

4.3 Mode, median and mean of discrete data in a frequency table

Mode

The mode is simple to find as it is the class with the highest frequency. This can be read directly from the table.

Example 5

The frequency table gives information about the number of goals scored by 21 teams in the Premier League last week.

Number of goals, x	0	1	2	3	4	5	6
Frequency, f	1	3	6	2	4	3	2

Find the mode.

The mode is 2.

Look for the highest frequency.

Median

The median is the $\frac{1}{2}(n + 1)$th value. To find this middle number it is often sensible to calculate cumulative frequencies and add these to the table.

Example 6

Find the median number of goals for the data in Example 5.

Number of goals, x	0	1	2	3	4	5	6
Frequency, f	1	3	6	2	4	3	2
Cumulative frequency	1	4	10	12	16	19	21

The median is in position $\frac{1}{2}(n + 1) = \frac{1}{2}(21 + 1)$, which is 11th team.
The median number of goals is 3.

> 10 teams scored 2 goals or fewer.
> 12 teams scored 3 goals or fewer.
> So, the 11th team must have scored 3 goals.

Mean

To calculate the mean, extend the frequency table. Then you can calculate the sum of all the frequencies, $\sum fx$, easily.

Then calculate the mean:

$$\text{mean} = \bar{x} = \frac{\sum fx}{\sum f}$$

> Remember $\bar{x}$ means the mean of all the values of x.

Example 7

Find the mean number of goals for the data in Example 5.

Number of goals (x)	Number of teams (f)	f × x
0	1	1 × 0 = 0
1	3	3 × 1 = 3
2	6	6 × 2 = 12
3	2	2 × 3 = 6
4	4	4 × 4 = 16
5	3	3 × 5 = 15
6	2	2 × 6 = 12
Total	$\sum f = 21$	$\sum fx = 64$

> Add a column for $f \times x$ and a row for totals.
>
> Work out $f \times x$ for each row.
>
> Sum the f and $f \times x$ columns.

> 6 teams scoring 2 goals each makes 12 goals in total.

$$\text{Mean} = \frac{\sum fx}{\sum f} = \frac{64}{21} = 3.048 \text{ (correct to three decimal places)}.$$

> Work out $\frac{\sum fx}{\sum f}$.

When **discrete data** are given in a **frequency table**:

Mode is the one with the highest frequency.

Median $= \frac{1}{2}(n + 1)$th value

Mean $= \frac{\sum fx}{\sum f}$

ResultsPlus
Build Better Answers

Question: Lincoln greenhouses grow Shirley tomatoes. A sample of 26 tomatoes was taken. The weights, to the nearest 5 grams of the tomatoes were:

60	60	55	65	60	50	60	65	50
65	70	50	65	65	50	55	55	70
65	60	65	70	50	55	65	60	

(a) Complete the frequency table.

Weight (*x*)	Frequency (*x*)	(*fx*)
50	5	250
55	4	220
60	6	360
65		
70		
Totals		

(1 mark)

(b) Use the information in the table to work out an estimate of the mean weight of these tomatoes.   (2 marks)

Supermarkets want each tomato they sell to be about the same weight.

(c) What other statistical information might the supermarkets need before deciding whether or not Shirley tomatoes will meet their requirements?   (1 mark)

(Total 4 marks)

■ **Basic   1 mark answers**
Complete the frequency table accurately by adding the frequencies for 65 g and 70 g tomatoes, then calculating their *fx* values, plus the two column totals $\sum f$ and $\sum fx$.

● **Good   2–3 mark answers**
Complete the frequency table in part a), then calculate an estimate for the mean using $\frac{\sum fx}{\sum f}$.

▲ **Excellent   4 mark answers**
Complete the frequency table in part (a) and calculate an estimate for the mean in part (b). For part (c) suggest another **statistical** measure the supermarket might need, for example the range, median or mode. Some students lost marks here for talking about non-statistical qualities such as cost or colour.

Exercise 4B

1   This frequency table summarises the number of cartons of yoghurt sold by a shopkeeper during January.

Number of cartons	10	11	12	13	14	15	16	17
Number of days	1	3	5	10	6	3	2	1

For these numbers of cartons of yoghurt, find

a   the mode

b   the median

c   the mean.

2   This frequency table provides information about the number of times students in a Year 11 class are late during a term.

Times late	2	3	4	5	6	7	8
Number of students	3	12	6	4	7	2	1

For these data, work out

a   the mode

b   the median

c   the mean.

3   During two consecutive months a gardener recorded the temperature, in degrees Celsius, at the same time each day. The results he obtained are shown in the table.

Temperature (°C)	18	19	20	21	22	23
Number of days	5	8	19	14	12	3

For these temperatures, find

a   the mode

b   the median

c   the mean.

4   This frequency table shows the ratings of a disco by a random sample of 40 students.

Rating A means they enjoyed it very much. Rating E means they did not enjoy it at all.

Rating	A	B	C	D	E
Number of students	6	13	10	7	4

a Work out

  i) the mode

  ii) the median rating.

b Explain why you cannot calculate the mean rating.

4.4 Mode, median and mean of grouped data

The table below shows information about the amount of time that some girls spent watching television in one week.

Number of hours, x	Number of girls, f
$0 \leqslant x \leqslant 5$	3
$5 < x \leqslant 10$	7
$10 < x \leqslant 15$	10
$15 < x \leqslant 20$	4

> Each group is called a class interval. $5 < x \leqslant 10$ is a class interval. The class boundaries are 5 and 10. The class width is $10 - 5 = 5$.

> Total number of girls $= 3 + 7 + 10 + 4 = 24$.

We cannot tell exactly how many hours each of the 24 girls spent watching television. The three who spent between 0 and ⩽5 hours could have watched any amount between 0 and 5 hours.

Therefore the mean, mode and median can only be estimated for these data.

Mode or modal class of grouped data

The mode cannot be given for the data above but the modal class is $10 < x \leqslant 15$ as this is the class with the greatest number of girls in it.

> When information is presented in a table of grouped data, the **modal class** is the class with the largest frequency.

Median of grouped data

The median of grouped data cannot be given but it can be estimated.

For the time spent watching television data on page 124, the data are continuous so the variable can take any value. Therefore the $\frac{1}{2}n$th value is the median. There is no need to round up or use $(n + 1)$ unless the data are discrete.

It is also assumed that if there are, say, 10 girls in the class $10 < x \leqslant 15$ hours then their times are equally spread between 10 hours and 15 hours.

> For grouped continuous data the median is the value of the $\frac{1}{2}n$th observation.
>
> There are two ways of estimating the median:
>
> - Using the mid-point of a cumulative frequency diagram.
> - Using interpolation.

Example 8

This table shows information about the time that the group of girls spent watching television in one week.

Find the median time the girls spent watching television.

Number of hours, x	Number of girls, f
$0 \leqslant x \leqslant 5$	3
$5 < x \leqslant 10$	7
$10 < x \leqslant 15$	10
$15 < x \leqslant 20$	4

Answer 1 Using a cumulative frequency diagram

Number of hours, x	Number of girls, f	Cumulative frequency
$0 \leqslant x \leqslant 5$	3	3
$5 < x \leqslant 10$	7	10
$10 < x \leqslant 15$	10	20
$15 < x \leqslant 20$	4	24

Add a cumulative frequency column to the table.

Then draw a cumulative frequency curve or polygon. A polygon is shown here. Unless told otherwise either will be accepted in an exam.

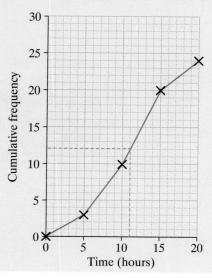

There are 24 observations (girls), so the median will be the $\frac{1}{2}(24)^{th} = 12^{th}$ observation (girl).

Drawing a line to the polygon at Cumulative frequency 12 and then down to the Time axis gives a median of approximately 11 hours.

Answer 2 Using interpolation

Number of hours, x	Number of girls, f	Cumulative frequency
$0 \leqslant x \leqslant 5$	3	3
$5 < x \leqslant 10$	7	10
$10 < x \leqslant 15$	10	20
$15 < x \leqslant 20$	5	24

There are 24 girls, so the median will be the $\frac{1}{2}(24)^{th} = 12^{th}$ girl.

10 girls watched for 10 hours or less.

20 girls watched for 15 hours or less.

The 12th girl is in the $10 < x \leqslant 15$ class interval.

This class interval is for the 10th to 20th girls, so the 12th girl is 2 girls in. There are $20 - 10 = 10$ girls in the class interval.

There are $15 - 10 = 5$ hours between the class limits.

So, we need to go $\frac{2}{10}$ of 5 hours = 1 hour into this class.

The estimated median is $10 + 1 = 11$ hours.

Examiner's Tip
The interpolated answer may differ slightly from that found by using the frequency polygon. This is because of the difficulty of reading the graph accurately.

Mean of grouped data

To estimate the mean of grouped data, assume that in any class the data are evenly spaced about the mid-point of the class limits. The mid-point is found by adding the class limits together and dividing by 2.

Then use the mid-point in the same calculations as for data in an ungrouped frequency table.

An **estimated mean** can be found from a grouped set of data using the formula:

$$\text{Mean} = \frac{\Sigma(f \times \text{mid-point})}{\Sigma f}, \text{ where } \Sigma \text{ means 'the sum of' and } f \text{ is}$$

frequency.

This is an estimate because the actual values of the data are unknown.

Example 9

The table shows information about the time that a group of girls spent watching television in one week.

Find the mean number of hours the girls spent watching television.

Number of hours, x	Number of girls, f
$0 \leqslant x \leqslant 5$	3
$5 < x \leqslant 10$	7
$10 < x \leqslant 15$	10
$15 < x \leqslant 20$	4

Number of hours, x	Mid-point	Number of girls, f	f × mid-point
$0 \leqslant x \leqslant 5$	2.5	3	7.5
$5 < x \leqslant 10$	7.5	7	52.5
$10 < x \leqslant 15$	12.5	10	125.0
$15 < x \leqslant 20$	17.5	4	70.0
		24	255.0

Add a mid-point column and a f × mid-point column.

Sum the columns.

$$\text{Mean} = \frac{\Sigma(f \times \text{mid-point})}{\Sigma f} = \frac{255}{24} = 10.625 \text{ hours}$$

Calculate the mean using the formula.

Exercise 4C

1 A random sample of leaves was collected from a tree. The lengths (x cm) of the leaves were recorded. They are shown in the table.

Length of leaves, x (cm)	Number of leaves, f
$4 \leqslant x \leqslant 6$	6
$6 < x \leqslant 8$	12
$8 < x \leqslant 10$	5

a What is the modal class? $6 < x \leqslant 8$ cm

b Work out an estimate for the median length. 7cm

c Work out an estimate of the mean length.

2 Some people were asked to record the amount of time they spent watching television on one particular Saturday. The results are shown in the table.

Number of hours, x	Frequency, f
$0 \leqslant x \leqslant 3$	6
$3 < x \leqslant 6$	24
$6 < x \leqslant 9$	10

a What is the modal class? 24

b Work out an estimate for the median time.

c Work out an estimate of the mean time.

Results**Plus**

Watch out!

■ Many students find it difficult to work out the mid-points of the class intervals. All you need to do is add the two end numbers and divide by 2.

3 The speeds of some cars on a motorway are given in the frequency table.

Speed, x (mph)	Number of cars
$20 < x \leqslant 30$	3
$30 < x \leqslant 40$	10
$40 < x \leqslant 50$	17
$50 < x \leqslant 60$	30
$60 < x \leqslant 70$	35
$70 < x \leqslant 80$	5

Use this information to work out

a the modal group for the speed of the cars used in this survey

b an estimate for the median car speed

b an estimate for the mean car speed.

4 The ages of some people watching a film are given in this frequency table.

Age, x (years)	Number of people
$10 \leqslant x < 20$	4
$20 \leqslant x < 30$	15
$30 \leqslant x < 40$	11
$40 \leqslant x < 50$	10

Work out

a the modal age of the people watching the film

b an estimate for the median age of the people watching the film

c an estimate for the mean age of the people watching the film.

4.5 Transforming data

Sometimes it is easier to calculate the mean by transforming the data first.

For example, subtracting a chosen number from each number in a set gives smaller numbers that are easier to work with.

> Note: There is no point using transformations for finding the median or the mode. It would make the working more difficult.

Example 10

The ages, in years, at their last birthday, of some people in a retirement home are

76   81   73   92   83

Work out their mean age.

6   11   3   22   13

> Change the numbers to easier numbers by subtracting 70 from each. (Any other number could have been used.)

Mean of transformed ages $= \dfrac{6 + 11 + 3 + 22 + 13}{5} = 11$

> Find the mean of the new numbers.

Mean of original ages $= 70 + 11 = 81$

> Add 70 to get the mean of the original data.

With decimal numbers, it may be better to subtract a whole number and then multiply by 10, 100 or 1000 to get rid of the decimal point.

Example 11

Find the mean of these numbers.

1.04   1.09   1.03   1.12   1.10   1.04

0.04   0.09   0.03   0.12   0.10   0.04

> Subtract 1 from each number.

4   9   3   12   10   4

> Multiply the result by 100.

Mean of transformed numbers $= \dfrac{4 + 9 + 3 + 12 + 10 + 4}{6} = \dfrac{42}{6} = 7$

> Find the mean of the new numbers.

Mean of original numbers $= \dfrac{7}{100} + 1 = 1.07$

> Reverse what you did to the original numbers: divide by 100 and add 1.

Exercise 4D

1 A group of workers make special bolts. The number they each made per hour are shown below.

102   110   104   107   107   102   102   111   102   102   101

Transform this set of numbers to find the mean number of bolts made per hour.

2 Find the mean of these seven numbers by transforming them into smaller numbers.

3003   3005   3001   3010   3004   3009   3002

> Hint: subtract 3000.

3 Work out the mean of the set of numbers by first transforming the data to make it easier to use.

2.14   2.11   2.20   2.18   2.12   2.13   2.17   2.18

> Hint: look at Example 11.

4.6 Deciding which average to use

It is important to be clear about the meaning of the word 'average' when it is used. The mode, median and mean are all 'averages', but each is used for different reasons.

This table compares the three 'averages'.

Average	Advantages	Disadvantages
Mode	• Easy to find. • Can be used with any type of data. • Unaffected by open-ended or extreme values. • The mode will be a data value.	• Mathematical properties are not useful (e.g. it cannot be used to calculate other information about the distribution of the data). • There is not always a mode or sometimes there is more than one.
Median	• Easy to calculate. • Unaffected by extreme values.	• Mathematical properties are not useful (e.g. it cannot be used to calculate other information about the distribution of the data).
Mean	• Uses all the data. • Mathematical properties are well known and useful (e.g. it can be used in the calculation of a measure of spread).	• Always affected by extreme values. • Can be distorted by open-ended classes.

Example 12

The salaries of seven people who work for a small company are

£12 000   £18 000   £120 000   £28 000   £32 000   £22 000   £30 000

a What is the mean salary?

b What is the median salary?

c Which of these two averages is most typical of a person's earnings?

d Why is it not possible to work out the mode of these salaries?

a The mean is $\dfrac{12 + 18 + 120 + 28 + 32 + 22 + 30}{7} = 37.428571$
(i.e. £37 429)

To make the calculation easier divide by 1000 and find the mean of the new figures.
Multiply the answer by 1000.

b The median is

12   18   22   **28**   30   32   120

(i.e. £28 000)

Arrange in order and find the middle value.

c The median is more typical because the single salary of £120 000 affects the mean but not the median (only 1 in 7 earn more than £37 429).

Look to see which is distorted by the high value.

d There is no mode because all the salaries are different.

Exercise 4E

1   Seven friends collect stamps for charity. The number of stamps each person has after one week is

    8   10   10   14   16   17   100

  a   What is the median number of stamps?

  b   What is the mean number of stamps?

  c   Which of these two averages better describes the number of stamps collected by each person? Give a reason for your choice.

2   Ten students received the following pay for working on a Saturday afternoon.

    £18   £22   £16   £26   £23
    £27   £25   £19   £16   £22

Explain why the mode is not the best figure for describing their average pay.

3   Last April a garage sold five different types of car. The numbers sold were

	Type of car				
	Prestige	**Sports**	**Ordinary**	**Coupé**	**4 × 4**
Number sold	12	8	23	2	5

  a   Find the modal type of car sold.

  b   Why is the mode appropriate for these data?

4   Twenty students were asked to name their favourite colour. Here are the results.

red	blue	blue	green	red
red	yellow	red	blue	pink
blue	black	red	red	blue
purple	blue	red	red	red

  a   State clearly, with reasons, the best average to use for these data.

  b   Why can you not state the mean of the favourite colours?

5   Explain briefly how you could work out the average price of a car sold by a garage.

4.7 Weighted mean

The formula used to find the mean of data given in a frequency table

is $\bar{x} = \dfrac{\sum fx}{\sum f}$

When calculating a **weighted mean**, the values f are the actual weightings given to the variables, x. The letter w is usually used to represent weightings since they are not always frequencies. The mean is still calculated in the same way.

> The **weighted mean** of a set of data is given by $\bar{x}_w = \dfrac{\sum wx}{\sum w}$, where w is the weight given to each variable, x.

Example 13

In an exam a candidate's final percentage is worked out using weighted averages.

　Paper 1 has a weight of 40.

　Paper 2 has a weight of 40.

　Paper 3 has a weight of 10.

　Paper 4 has a weight of 10.

A candidate scored the following marks.

　Paper 1: 62%　　Paper 2: 38%　　Paper 3: 58%　　Paper 4: 39%

Work out the candidate's final mark.

Weighted mean, $\bar{x}_w = \dfrac{\sum wx}{\sum w}$

$= \dfrac{(40 \times 62) + (40 \times 38) + (10 \times 58) + (10 \times 39)}{40 + 40 + 10 + 10}$

$= \dfrac{2480 + 1520 + 580 + 390}{100}$

$= \dfrac{4970}{100}$

$= 49.7\%$

> Put the weights and marks in the formula.
>
> Work out the weighted mean.

Example 14

A sample of three children has a mean height of 0.97 m. A second sample of seven children has a mean of 1.06 m. A final sample of five children has a mean height of 1.12 m.

Work out the mean height of all 15 children.

$$\text{Mean height} = \frac{(3 \times 0.97) + (7 \times 1.06) + (5 \times 1.12)}{3 + 7 + 5}$$

$$= \frac{15.93}{15}$$

$$= 1.062 \text{ m}$$

Use the number of children as weightings.

Work out the weighted mean using the formula.

Exercise 4F

1 In an examination
   - Paper 1 is worth 0.35 of the final mark
   - Paper 2 is worth 0.4 of the final mark
   - Controlled assessment is worth 0.25 of the final mark.

Jimmy's and Sumreen's marks are shown in the table.

	Paper 1	Paper 2	Controlled assessment
Jimmy	46	62	55
Sumreen	56	56	65

Work out each student's final mark.

2 In a factory during a typical week
   - 10% of the workers earn £250
   - 35% of the workers earn £290
   - 55% of the workers earn £350.

Work out the average earnings.

3 The weightings used to work out a final average mark in an examination were 0.3, 0.4, 0.2 and 0.1.
Owen scored marks of 62, x, 44 and 58.
The overall pass mark was 55.
Work out the value of x, given that Owen just passed the examination.

4.8 Measures of spread

As well as an average, such as the mode, median or mean, a measure of the spread of the data about the average is needed to describe the data more fully. This section covers a number of different measures for the spread of data.

The range

Range = largest value − smallest value

The **range** is a very crude measure of spread because it compares only the largest and smallest values of the data.

Example 15

For four weeks a headmaster recorded the number of students who arrived late for the start of the day. The minimum number of late arrivals was 8. The maximum number was 25.

Work out the range.

Range = maximum number − minimum number
   = 25 − 8
   = 17 students

Use the formula to find the range.

Example 16

The speeds, v, (to the nearest mile per hour), of cars on a motorway were recorded by the police. This frequency table shows their results.

Speed, v (mph)	Frequency
$20 < v \leqslant 30$	2
$30 < v \leqslant 40$	14
$40 < v \leqslant 50$	29
$50 < v \leqslant 60$	22
$60 < v \leqslant 70$	13

Estimate the range of speeds.

Range = largest value − smallest value
   = 70.5 − 20.5
   = 50 mph

$v > 20$ mph, so the minimum speed is 20.5 mph.
$v \leqslant 70$ mph, so the maximum speed is 70.5 mph.

Quartiles and inter-quartile range

The **lower quartile** is the value such that one quarter (25%) of the values are less than or equal to it.

The **upper quartile** is the value such that three quarters (75%) of the values are less than or equal to it.

Hint: The median and quartiles split the data into four equal parts. This is why they are called quartiles.

The median is the second or middle quartile because it is the value that two quarters (one half) of the values are less than.

Median = $\frac{1}{2}(n + 1)^{\text{th}}$ value.

There are several different formulae for finding quartiles and not all of them give exactly the same answer. The ones below are recommended as the formula they use is similar to that used for the median.

> For discrete data, when n values are written in ascending order
> - the lower quartile, Q_1, is the value of the $\frac{1}{4}(n + 1)$th observation
> - the median, Q_2, is the value of the $\frac{1}{2}(n + 1)$th observation
> - the upper quartile, Q_3, is the value of the $\frac{3}{4}(n + 1)$th observation.

Upper and lower quartiles should each be one of the data set, so if the result for $\frac{1}{4}(n + 1)$ or $\frac{3}{4}(n + 1)$ is a non-integer, round up.

If the result for $\frac{1}{2}(n + 1)$ is a non-integer, take the mean of the two integers either side of this value.

A frequently used measure of spread is the **inter-quartile range**.

> **Inter-quartile range** (IQR) = upper quartile − lower quartile

Example 17

a Find the upper and lower quartiles of this set of data.

7   9   13   5   6   12   3

b Find the inter-quartile range for these data.

a 3   5   6   7   9   12   13 | Order the data.

$n = 7$ | Count the data to find the value of n.

Q_1 is the $\frac{1}{4}(7 + 1) = 2^{nd}$ value, which is 5. | Work out Q_1.

Q_3 is the $\frac{3}{4}(7 + 1) = 6^{th}$ value, which is 12. | Work out Q_3.

b   The inter-quartile range, IQR $= Q_3 - Q_1$ | Subtract Q_1 from Q_3.

$\qquad = 12 - 5$

$\qquad = 7$

Finding quartiles for data presented in a frequency table

The quartiles of discrete data given in a frequency table can be found using the formula or by drawing a step polygon.

Example 18

The table gives information about the number of defective items produced per day by a machine over a known period.

Number of defective items	Frequency
0	17
1	12
2	7
3	6
4	4
5	2
6	1
7	1
8	0

Find the inter-quartile range for these data.

Method 1 – using the formula

Q_1 is the $\frac{1}{4}(50 + 1) = 12.75 = 13^{th}$ value which is 0

Q_3 is the $\frac{3}{4}(50 + 1) = 38.25 = 39^{th}$ value which is 3

$IQR = 3 - 0 = 3$

Use the formulae to find which values give the quartiles.

Look to see what the 13th value is. 17 values are 0 so the 13th must be zero.

There are 36 values that are 2 or less, and 42 that are 3 or less. The 39th value must be 3.

Subtract Q_1 from Q_3.

Method 2 – drawing a step polygon

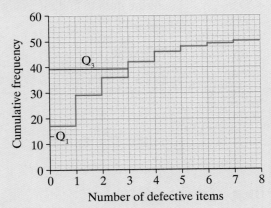

This is discrete data so a step polygon can be drawn.

Draw lines across from 13 and 39 to where they cut the cumulative step polygon. Read off the numbers of defective items.

Q_1 is the 13th value = 0

Q_3 is the 39th value = 3

$IQR = 3 - 0 = 3$

Subtract Q_1 from Q_3.

Finding quartiles for data presented as a grouped frequency table

Quartiles for continuous data can take any value and do not have to be integers.

> For continuous data Q_1 is the value of the $\frac{1}{4}n$th observation, Q_2 is the value of the $\frac{1}{2}n$th observation and Q_3 is the value of the $\frac{3}{4}n^{th}$ observation.

Quartiles for grouped data may be found using a cumulative frequency diagram or by interpolation.

Example 19

The table gives information about the number of women who married for the first time in a town in 2008. The age recorded for each woman is her age last birthday.

Age	16–20	21–25	26–30	31–35	36–45	46–55	56–70	71+
Frequency	8	10	42	25	12	9	4	0

Estimate the three quartiles.

Method 1 – using a cumulative frequency diagram

Age	16–20	21–25	26–30	31–35	36–45	46–55	56–70	71+
Frequency	8	10	42	25	12	9	4	0
Cumulative frequency	8	18	60	85	97	106	110	110

Work out the cumulative frequencies and draw a cumulative frequency diagram.

Age of marriage

$Q_1 = \dfrac{110}{4} = 27.5^{th}$ value

$Q_2 = \dfrac{110}{2} = 55^{th}$ value

$Q_3 = \dfrac{3 \times 110}{4} = 82.5^{th}$ value

Work out the values that give Q_1, Q_2 and Q_3 using $\dfrac{n}{4}$, $\dfrac{n}{2}$ and $\dfrac{3n}{4}$

$Q_1 = 27.5$ years old

$Q_2 = 31$ years old

$Q_3 = 35$ years old

Draw lines across from the Cumulative frequency axis to the cumulative frequency diagram and down to the Age axis.

Read off the ages.

Method 2 – using interpolation

Q_1 is the $\frac{110}{4} = 27.5^{th}$ value.

There are 18 values less than 26, so the 27.5^{th} value is $27.5 - 18 = 9.5$ into the 26–30 range.

$\frac{9.5}{42} \times 5 = 1.13$

So $Q_1 = 26 + 1.13 = 27.13$ years old

Q_2 is the $\frac{110}{2} = 55^{th}$ value.

This is $55 - 18 = 37$ into the 26–31 range.

$\frac{37}{42} \times 5 = 4.40$

So $Q_2 = 26 + 4.40 = 30.4$ years old.

Q_3 is the $\frac{3 \times 110}{4} = 82.5^{th}$ value.

There are 60 values less than 31, so the 82.5^{th} value is $82.5 - 60 = 22.5$ into the 31–35 range.

$\frac{22.5}{25} \times 5 = 4.5$

So $Q_3 = 31 + 4.5 = 35.5$ years old.

To find the 27.5^{th} value using interpolation, look at the table and see in which interval the value falls (26–30).

There are five years in the class 26–30 and the frequency is 42, so the years increase at the rate of $\frac{5}{42}$. If the frequency increases by 9.5 this is $\frac{5}{42} \times 9.5 = 1.13$. Add this to the bottom of the range.

Repeat for Q_2 and Q_3.

Examiner's Tip

These answers are estimates which is why they don't come out exactly the same as the answers using Method 1. The cumulative frequency method is easiest.

Percentiles and deciles

Percentiles are used to divide data into 100 groups. **Deciles** are used to divide data into 10 groups. For example, people taking an exam could be divided into 10 groups and those students in the top decile (top 10%) could be given an A*, those in the next decile (next 10%) an A and so on. (This is not the way exam boards do it!)

Note: foundation students need only to understand percentiles.

A set of data is divided into 100 parts to form **percentiles**.

When the set is divided into 10 equal parts, these are called **deciles**.

Example 20

This cumulative frequency diagram gives information about the speeds of 100 cars.

Use the cumulative frequency diagram to find

a the 35th percentile

b the 80th percentile

c the 80th – 35th percentile range

d the 6th decile.

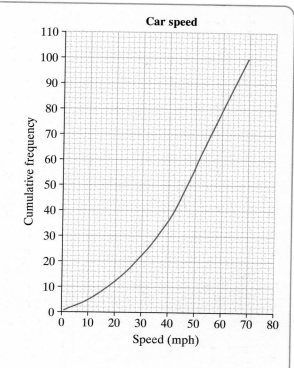

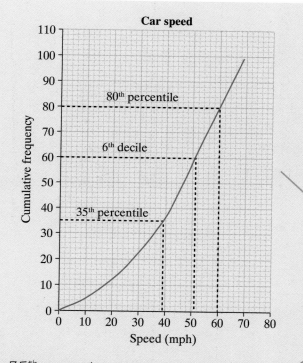

Draw the required percentiles and decile on the graph.

Read the speed value for where the percentile line crosses the curve.

Find the difference between the two percentile speed values.

a The 35th percentile is approximately 39 mph.

b The 80th percentile is approximately 60 mph.

c The 80th – 35th percentile range is 60 − 39 = 21 mph.

d The 6th decile is approximately 51 mph.

Read the speed value of where the 6th decile line crosses the curve.

Exercise 4G

1 The number of computer games owned by 11 teenagers was recorded. Here are the data.

   4   4   6   7   7   9   10   10   10   12   12

For these data find

a the median   b the lower quartile

c the upper quartile   d the inter-quartile range.

2   The maximum running speeds (kilometres per hour) of some mammals was recorded. The data are shown below.

   110   70   98   50   40   78   45   80   69   60   95

For these data find

a   the median speed

b   the lower quartile

c   the upper quartile

d   the inter-quartile range.

3   Find the range, lower quartile, upper quartile and inter-quartile range of the values

a   6, 3, 8, 2, 9, 5, 10

b   21, 16, 72, 40, 67, 65, 55, 34, 17, 48, 32, 19, 44, 61, 73

c   8, 2, 9, 6, 7, 10, 12, 13, 5, 12, 10, 8, 10, 4

4   The ages of a group of 60 people are shown in the table.

Age, a (years)	$10 < a \leqslant 20$	$20 < a \leqslant 30$	$30 < a \leqslant 40$	$40 < a \leqslant 50$	$50 < a \leqslant 60$
Number	4	12	22	19	3

a   Draw a cumulative frequency curve for these data.

b   Find the lower quartile, the upper quartile and the inter-quartile range.

5   The prices of 200 second-hand cars are shown in the frequency table.

Price, x (£1000s)	$1 < x \leqslant 2$	$2 < x \leqslant 3$	$3 < x \leqslant 4$	$4 < x \leqslant 5$	$5 < x \leqslant 6$
Frequency	10	32	95	51	12

a   Draw a cumulative frequency curve for the data.

b   Find

   i)   the median price of a car

   ii)   the upper quartile price of a car

   iii)   the lower quartile price of a car

   iv)   the inter-quartile range in the prices of cars.

c   Work out the approximate value of

   i)   the 85th percentile of these prices

   ii)   the 35th percentile of these prices

   iii)   the 85th – 35th percentile range

   iv)   the 4th decile of these prices

   v)   the 8th decile of these prices.

6 Angela is studying the price of some ladies' dresses. She presents the results of her survey as a stem and leaf diagram.

```
1 | 3  5  8
2 | 1  2  2  2  3  5  6  6  6  7  8  9  9  9
3 | 2  2  2  2  2  3  7  7  9  9
4 | 0  2  3  4  4  7  8  8  8
5 | 1  2  2  6  7  8  8  9
6 | 2  3  8  8  8
7 | 5  5  9
```

Key

$2 \mid 1 = £21$

a Write down the median of these prices.

b Draw a cumulative frequency diagram to show these prices.

c Use your cumulative frequency diagram to work out estimates of

   i) the lower and upper quartiles

   ii) the 6th decile of the prices

   iii) the 15th percentile of the prices.

> Remember the median is the middle value.

> First draw a cumulative frequency table using class intervals 10–19, 20–29, etc. There are 52 prices so the vertical axis should go from 0 to at least 52.

7 The Year 11 students at Mayfield High School completed an IQ test. Here are the results.

IQ	Number of students
60–69	3
70–79	8
80–89	14
90–99	43
100–109	47
110–119	28
120–129	4
130–139	3

a Draw a cumulative frequency polygon.

b Use the graph to work out estimates for

   i) the median IQ

   ii) the lower quartile of these IQs

   iii) the upper quartile of these IQs.

c Use your graph to estimate

   i) the 60th percentile

   ii) the difference between the 2nd and 8th deciles.

4.9 Box plots

A **box plot** represents important features of the data: the maximum and minimum values, the median, and the upper and lower quartiles.

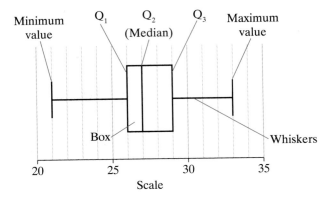

A **box plot** shows whether the distribution is **symmetrical**, **positively skewed** or **negatively skewed**.

If the median is closer to the upper quartile than it is to the lower quartile the data are **negatively skewed**.

If the median is closer to the lower quartile than it is to the upper quartile it are **positively skewed**.

If the median is equidistant from both the upper and lower quartiles it are **symmetrical**.

Example 21

The heights in centimetres of 15 students are given below.

163	170	182	164	155	172	177	184
190	148	193	185	176	158	166	

a Find the median height and the upper and lower quartiles.
b Draw a box plot to represent these data.
c Describe the skewness of this distribution.

a 148   155   158   163   164   166   170   172
   176   177   182   184   185   190   193

$Q_1 = 163$
$Q_2 = 172$
$Q_3 = 184$

b

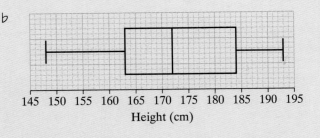

Height (cm)

c The distribution is positively skewed.

> Put the data in ascending order.
> There are 15 pieces of data.
> Q_1 is the $\frac{1}{4}(15 + 1) = 4^{th}$ piece of data.
> Q_2 is the $\frac{1}{2}(15 + 1) = 8^{th}$ piece of data.
> Q_3 is the $\frac{3}{4}(15 + 1) = 12^{th}$ piece of data.

> Draw a box plot. Remember the scale.

> Look to see if Q_2 is closer to Q_1 or Q_3 or if it is half way between.

Outliers

> An **outlier** is any value that is more than 1.5 times the inter-quartile range below the lower quartile or moe than 1.5 times the inter-quartile range above the upper quartile.

Any **outlier** should be marked on the box plot, but instead of the 'whiskers' showing the whole range they should show only

- down to the lowest value that is not an outlier
- up to the highest value that is not an outlier.

Example 22

Represent these numbers using a box plot.

22	30	34	35	35	35	36	37	37	38
39	39	40	40	41	42	42	48	50	

There are 19 pieces of data.

Median = 38

Lower quartile = 35

Upper quartile = 41

IQR = 41 − 35 = 6

1.5 × 6 = 9

35 − 9 = 26 and 41 + 9 = 50.

Anything <26 or >50 is an outlier.

The only outlier is 22.

Find the median and quartiles.

Find the IQR.

Find 1.5 × IQR.

Find lowest and highest values that are not outliers.

Find the outliers.

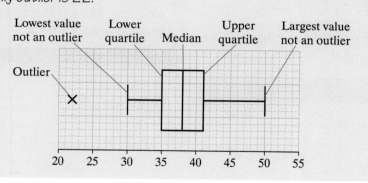

Draw the diagram and add the outliers as crosses.

Exercise 4H

1   The length of 11 mature whales was recorded to the nearest metre.

The lengths were as follows.

21   22   23   23   24   25   25   25   26   27   27

a   For these data work out

  i) the lower quartile

  ii) the median length

  iii) the upper quartile.

b   Draw a box plot to represent these data.

2   The number of eggs in 15 pheasant nests was recorded as follows.

9	8	7	12	15	12	14	9
15	14	11	8	7	14	13	

a   For these data work out
   i) the lower quartile
   ii) the median number of eggs
  iii) the upper quartile.

b   Draw a box plot to represent these data.

3   The marks gained by a group of college students taking a statistics test are shown below.

Boys	62	47	72	50	68	81	45	52	77	46	45	57	80	41	63
Girls	52	89	46	19	22	34	56	97	44	38	47	99	23	20	65

a   Work out the inter-quartile range for
   i) the boys' marks
  ii) the girls' marks.

b   Work out the range of
   i) the boys' marks
  ii) the girls' marks.

c   Draw box plots for both the girls' marks and the boys' marks on the same axes.

d   Comment on the skewness of the two distributions.

ResultsPlus
Watch out!

■   When asked to describe the skewness in a box plot, most students did not describe it correctly. Skewness is negative, positive or symmetrical.

4   The table shows the maximum, minimum and quartiles for the weight, w (grams) of a flock of herring gulls.

Minimum	Q_1	Q_2	Q_3	Maximum
750	810	940	1100	1250

a   Draw a box plot for these data.

b   Describe the skewness of these data.

5   Here are the ages, in years, of a group of 27 people.

14	20	23	25	26	27	27	27	28
28	30	31	21	32	34	34	36	37
40	41	41	42	43	43	45	52	75

a   For these data work out
   i) the median age
  ii) the lower quartile
  iii) the upper quartile

b   Find any outliers.

c   Draw a box plot for these data.

6 Here is a set of scores gained by a group of people playing a computer game.

58	12	34	42	28	33	67	63	51
47	24	32	43	30	21	25	32	44
34	32	76	45	55	43	52	38	44
36	28	32	45					

a For these data find
   i) the median score
   ii) the lower quartile
   iii) the upper quartile
   iv) the inter-quartile range.

b Identify any outliers.

c Draw a box plot to represent these data.

7 Here is a box plot.

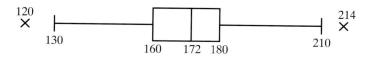

a Write down the upper and lower quartiles, and the median.

b Write down the value of the outliers.

c Describe the skewness of this distribution.

4.10 Variance and standard deviation

The deviation or dispersion of an observation, x, from the mean, $\bar{x}$, is given by $x - \bar{x}$.

One way of measuring the total dispersion (spread) of n observations is to use the **variance**.

The **variance** is a measure of dispersion or spread.

It is given by

$$\text{Var}(x) = \frac{(x_1 - \bar{x})^2 + (x_2 - \bar{x})^2 + \ldots + (x_n - \bar{x})^2}{n}$$

$$= \frac{\sum(x - \bar{x})^2}{n}$$

A friendlier formula for variance is

$$\text{Var}(x) = \frac{x_1^2 + x_2^2 + \ldots + x_n^2}{n - \bar{x}^2}$$

$$= \frac{\sum x^2}{n} - \bar{x}^2$$

Examiner's Tip
The formulae for variance and standard deviationare given on the first page of the exam paper.

The standard deviation is the square root of the variance.

Standard deviation (SD) = $\sqrt{\dfrac{\sum(x - \bar{x})^2}{n}}$

or the simpler version $\sqrt{\dfrac{\sum x^2}{n} - \bar{x}^2}$.

Example 23

Work out the standard deviation of these numbers.

| 32 | 34 | 35 | 35 | 37 | 37 | 37 | 38 | 39 |

Method 1

$$\bar{x} = \frac{32 + 34 + 35 + 35 + 37 + 37 + 37 + 38 + 39}{9}$$

$$= 36$$

Find the mean.

The deviations are $-4, -2, -1, -1, +1, +1, +1, +2, +3$

$$\text{Var}(x) = \frac{(-4)^2 + (-2)^2 + (-1)^2 + (-1)^2 + 1^2 + 1^2 + 1^2 + 2^2 + 3^2}{9}$$

$$= \frac{38}{9}$$

$$= 4.222$$

Work out the deviations from the mean.

Apply the formula for variance.

$$\text{SD} = \sqrt{4.222}$$

$$= 2.055 \ (\text{correct to 3 decimal places})$$

Take the square root of the variance to get the standard deviation.

Method 2 – using the simpler formula

$$\bar{x} = 36$$

$$\sum x^2 = 32^2 + 34^2 + 35^2 + 35^2 + 37^2 + 37^2 + 37^2 + 38^2 + 39^2$$

$$= 11\,702$$

Find the mean (as in Method 1 above).

Find the sum of the squares.

$$\text{Var}(x) = \frac{\sum x^2}{n} - \bar{x}^2$$

$$= \frac{11\,702}{9} - 36^2$$

$$= 4.222$$

Apply the formula for variance.

$$\text{SD} = \sqrt{4.222}$$

$$= 2.055 \ (\text{correct to 3 decimal places})$$

Take the square root of the variance to get the standard deviation.

Finding the variance and standard deviation for a frequency table and a grouped frequency distribution

Use these formulae for data in frequency tables and grouped frequency distributions.

If f stands for the frequency, then $n = \Sigma f$, and the variance formulae become

$$\text{Variance} = \frac{\Sigma f(x - \bar{x})^2}{\Sigma f}$$

or $\dfrac{\Sigma fx^2}{\Sigma f} - \left(\dfrac{\Sigma fx}{\Sigma f}\right)^2$

and standard deviation (SD) $= \sqrt{\dfrac{\Sigma f(x - \bar{x})^2}{\Sigma f}}$

or $\sqrt{\dfrac{\Sigma fx^2}{\Sigma f} - \left(\dfrac{\Sigma fx}{\Sigma f}\right)^2}$

Examiner's Tip

$\dfrac{\Sigma fx^2}{\Sigma f} - \left(\dfrac{\Sigma fx}{\Sigma f}\right)^2$ is the easier formula to use. It is given on the formulae page of the exam paper.

Example 24

This table shows information about the results of a spelling test that Mrs Arnold gave to her class of 30 students.

Mark	0	1	2	3	4	5
Frequency	3	3	3	6	12	3

Work out the standard deviation of the marks.

Mark, x	f	fx	fx^2
0	3	0	0
1	3	3	3
2	3	6	12
3	6	18	54
4	12	48	192
5	3	15	75
Totals	30	90	336

Add columns for fx and fx^2 to the table.

Sum the columns.

$$\text{Var}(x) = \frac{\Sigma fx^2}{\Sigma f} - \left(\frac{\Sigma fx}{\Sigma f}\right)^2$$

$$= \frac{336}{30} - \left(\frac{90}{30}\right)^2$$

$$= 2.2$$

Use the simpler formula to get the variance.

$$\text{SD} = \sqrt{2.2}$$

$$= 1.483 \text{ (3 d.p.)}$$

Take the square root of the variance to get the standard deviation.

Example 25

This table gives information about the time, in seconds, that a group of 16 students took to run 100 m.

Time, t (s)	$10 < t \leqslant 11$	$11 < t \leqslant 12$	$12 < t \leqslant 13$	$13 < t \leqslant 14$
Frequency	2	5	6	3

Estimate the mean and standard deviation.

Time, t (s)	f	ft	ft^2
10.5	2	21.0	220.5
11.5	5	57.5	661.25
12.5	6	75.0	937.50
13.5	3	40.5	546.75
Totals	16	194.0	2366.00

Insert the mid-point of each class.

Add columns for ft and ft^2 to the table.

Sum the columns.

$$\bar{t} = \frac{194}{16}$$
$$= 12.125$$

Work out the mean for t, using $\frac{\sum ft}{\sum f}$

$$\text{Var}(t) = \frac{2366}{16} - 12.125^2$$
$$= 0.859375$$

Work out the variance.

$$SD = \sqrt{0.859375}$$
$$= 0.927 \ (3 \ d.p.)$$

Standard deviation is the square root of variance.

Advantages and disadvantages of measures of dispersion

Measure	Advantages	Disadvantages
Range	• A reasonably good indicator.	• Badly affected by extreme values.
Inter-quartile range	• Not affected by extreme values. • Often used with skewed data.	• Does not tell you what happens beyond quartiles.
Variance	• Good measure. • All values used. • Used when data are fairly symmetrical.	• Mathematical properties not useful (use the standard deviation in preference). • Not so good if data are strongly skewed.
Standard deviation	• Good measure. • All values used. • Used when data are fairly symmetrical. • Can be used in mathematical calculations of other statistics.	• Not so good if data are strongly skewed.

Exercise 4I

1 Calculate the mean and standard deviation for the sets of data below using either the formula $SD = \sqrt{\dfrac{\sum(x - \bar{x})^2}{n}}$ or $SD = \sqrt{\dfrac{\sum x^2}{n} - \bar{x}^2}$.

  a 5, 6, 10, 7, 12

  b 8, 3, 12, 10, 7, 8, 5, 2, 5

  c 2.1, 3.4, 6.2, 1.3, 2.9, 4.3, 5.1, 7.1, 4.2

2 Using the formula that you did not use in question 1, work out the standard deviation for

  a 7, 11, 6, 8, 13

  b 4, 9, 11, 13, 6, 8, 9, 6, 3

  c 3.2, 2.5, 7.3, 1.4, 2.8, 4.4, 6.1, 7.3, 5.1

3 Calculate the mean and the standard deviation for the variable x given that

  a $\sum x^2 = 293$, $\sum x = 19.8$, $n = 12$

  b $\sum x^2 = 3.04$, $\sum x = 1.26$, $n = 8$

4 The marks gained by a sample of 100 students in a GCSE Statistics examination are given in the table below.

Mark, m	$20 < m \leqslant 30$	$30 < m \leqslant 40$	$40 < m \leqslant 50$	$50 < m \leqslant 60$	$60 < m \leqslant 70$
Frequency	18	22	38	20	2

  a Work out an estimate of the mean mark for these 100 students.

  b Work out an estimate of the variance of the marks.

  c Work out an estimate of the standard deviation of the marks.

5 For her controlled assessment, Gemma is examining the maximum speed of a collection of cars.

She has taken a sample of 50 cars and put their maximum speeds in a table.

Max speed (mph)	71 to 80	81 to 90	91 to 100	101 to 110	111 to 120
Number of cars	2	5	18	20	5

Work out

  a an estimate of the mean maximum speed

  b an estimate of the standard deviation of these speeds.

6 Carrie is working on a GCSE controlled assessment looking at the number of eggs laid by different birds in their nests.

She has collected the following data.

Number of eggs	0	1	2	3	4	5	6
Number of nests	2	5	12	27	21	19	4

Work out

  a the mean number of eggs per nest

  b the variance of the number of eggs per nest

  c the standard deviation of the number of eggs per nest.

4.11 Comparing data sets

So far this chapter has explained how to represent and summarise data sets. Sometimes it is necessary to compare two (or more) data sets.

> A full comparison needs, at least, both a **measure of central tendency** and a **measure of dispersion or spread**.
>
> Another comparison that could be made is the **skewness** of the distributions.

Comparisons can be made using diagrams, which should be drawn to the same scale.

Example 26

These histograms show the sizes of fossil fish found at two different locations.

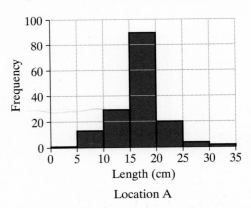

Location A

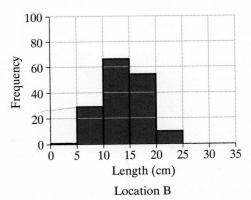

Location B

Compare the two distributions.

The mode at Location A is higher than the mode at Location B.

The range for Location A is greater than the range for Location B.

Location A has slight negative skew. Location B has slight positive skew.

These fossils would seem to come from different populations.

Range A = 35, Range B = 25

A good way of comparing distributions is to use comparative box plots.

Example 27

The box plots give information about the weights of male and female turtles.

a Compare the two distributions.

b It is suggested that the weight of a turtle could be used to indicate its gender. Discuss this suggestion.

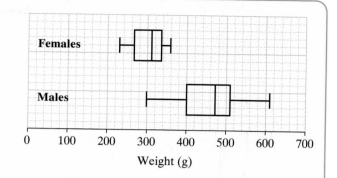

a The median weight for male turtles is higher than the median for females.

  The range and IQR of the males are bigger than those for the females.

  Both distribution s have negative skew.

b Turtles weighing below 300 g are female. Those weighing more than 360 g are male.

  Between 300 g and 360 g they could be either male or female. The nearer to 300 g, the more likely they are to be female. The nearer to 360 g, the more likely they are to be male.

Standardised scores

A way of comparing two individual results is by using **standardised scores**.

$$\text{Standardised score } (z) = \frac{\text{score} - \text{mean}}{\text{standard deviation}}$$

Example 28

Fred and Vicki took an English test and a statistics test. Both tests had a maximum mark of 100. Their results and statistics for each test are given in the table.

	Fred's mark	Vicki's mark	Test mean mark	Test standard deviation
English	46	55	50	12
Statistics	45	42	42	8

a Work out Vicki and Fred's standardised scores in English and statistics.

b Comment on the examination performances of the two students. Who do you think did better overall?

a   Vicki's standardised scores are

English $\dfrac{55 - 50}{12} = \dfrac{5}{12} = 0.4167$

Statistics $\dfrac{42 - 42}{8} = 0$

Fred's standardised scores are

English $\dfrac{46 - 50}{12} = \dfrac{-4}{12} = -0.3333$

Statistics $\dfrac{45 - 42}{8} = \dfrac{3}{8} = 0.375$

> Work out the standardised scores for each student.

b   The lowest standardised score is −0.3333, obtained by Fred in English, and the best is 0.4167, which Vicki scored in English. Fred did better than Vicki in statistics as he got the higher standardised score.

> Use the standardised scores to make the comparison.

Overall the best results appear to be Vicki's, since she did not get any negative standardised scores.

ResultsPlus
Exam Question Report

Tyson took a Statistics test and a Maths test.

Both tests were marked out of 100.

The table gives information about Tyson's marks.

It also shows the mean mark and standard deviation for the group that took the test.

Test	Tyson's mark	Group mean mark	Group standard deviation
Statistics	55	52	15
Maths	48	45	12

(a) i)   Work out Tyson's standardised score for Statistics.

   ii)   Work out Tyson's standardised score for Maths.
(3 marks)

(b) Write down the subject in which Tyson did better. Give a reason for your answer.
(2 marks)

(c) Comment on the group's performance in the two tests.
(2 marks)

(Total 7 marks)

How students answered

Poor   37%

Around one-third of students answered this question poorly. Some candidates wrote down the calculation for (a) i) as $55 - \dfrac{52}{15} (= 51.5)$, instead of $\dfrac{55 - 52}{15} (= 0.2)$.

Good   19%

Some students calculated the standardised scores correctly in part a), but then in part b) stated that Tyson did better in statistics because he got a higher mark, without using the standardised scores that compared Tyson's results with the group.

Excellent   44%

These students answered parts a) and b) correctly. For part c), some commented that statistics was an easier test, but few mentioned the spread or variability of the marks.

Exercise 4J

1   The diagram shows the box plots for the distributions of
heights of male and female adult giraffes.

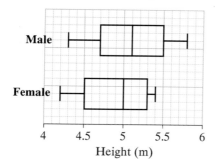

Compare fully the two distributions.

2   The weights of a large number of mallards were recorded. The
summary of the data collected is shown in the table.

	Minimum	Q_1	Q_2	Q_3	Maximum
Mallard drake	850	950	1050	1200	1400
Mallard duck	800	900	1000	1100	1300

  a   Draw comparative box plots of these data.
  b   Compare fully the two distributions.

3   The table shows Raji's waist measurement and his height alongside
the mean and standard deviation of the students in his year group.

	Raji's measurement	Class average	Standard deviation
Waist (cm)	100	89	6
Height (cm)	178	165	4

Calculate the standardised score for Raji's
  a   waist measurement
  b   height.

4   The table gives the history and geography test results for Sean,
Theresa and Victoria along with the mean and standard
deviations of those results of the whole year group.

	Sean	Theresa	Victoria	Year group mean	Year group standard deviation
History	72	58	78	61	6
Geography	34	43	51	44	5

  a   Calculate the six standardised scores.
  b   Comment on the performances of Sean, Theresa and Victoria.

ResultsPlus
Watch out!

▪ When asked to compare
two distributions given in box
plots, most candidates only
compared highest and lowest
values or the quartiles. To
compare distributions you need
to compare an average and a
measure of spread.

Higher
Statistics

4.12 Index numbers

Simple index numbers

Simple **index numbers** are used to see how the price of something varies over a period of time. For example, the relative change in the price of a new house to the price of a new house in the starting year.

To form an index number, pick the year the index should start. This is known as the **base year**.

$$\text{Index number} = \frac{\text{current price}}{\text{base year price}} \times 100$$

Results **Plus**

Watch out!

● When asked to calculate the index number for house prices in different years, some students divided the prices but forgot to multiply by 100, and some divided the prices by the base year.

The **current price** is simply the price for the current period.

Example 29

The table below gives information about the average price of a house in quarter 2 of the years 2005 to 2008.

Year	2005	2006	2007	2008
Price (£)	157 000	165 000	182 000	175 000

Work out the index number for each year using 2005 as the base year.

The index number for 2005 = 100 as this is the base year.

The index number for 2006

$$= \frac{165\,000}{157\,000} \times 100$$

$$= 105.1$$

> Divide the price in the current year by the price in the base year and multiply by 100.

The index number for 2007

$$= \frac{182\,000}{157\,000} \times 100$$

$$= 115.9$$

> Repeat for each year.

The index number for 2008

$$= \frac{175\,000}{157\,000} \times 100$$

$$= 111.5$$

> **Examiner's Tip**
> Remember that the value of the index number for the base year is always 100.

Example 30

This table gives information about a camera's price index for the years 2006 to 2008. The base year is 2005.

Year	2006	2007	2008
Index	110	112	120

a   Work out the price of the camera in 2008 given that its price in 2005 was £300.

The base for the index is to be changed to 2007.

b   Find the new index number for 2008.

a   index number $= \dfrac{current\ price}{base\ year\ price} \times 100$,

so the current price $= \dfrac{index\ number \times base\ year\ price}{100}$.

The price in 2008 $= \dfrac{120 \times £300}{100}$

$= £360$

Transpose the formula to make the price in 2008 the subject.

Enter the figures from the question and calculate the answer.

b   The price in 2007 $= \dfrac{112 \times £300}{100}$

$= £336$

First work out the price in 2007.

So the price index number for 2008 with 2007 as base

$= \dfrac{360}{336} \times 100$

$= 107.1$

Use the formula to find the price index for 2008.

Chain base index numbers

A **chain base index number** is used to see how the price each year changes by comparison with the previous year. For example, the price of a house this year compared with the price of a house last year.

Chain base index number $= \dfrac{current\ price}{previous\ year's\ price} \times 100$

Example 31

This table gives information about the sales price of a certain model of car.

Year	2005	2006	2007	2008
Price (£)	10 000	10 260	10 465	10 780

a   Find the chain base index numbers for 2007 and 2008.

b   By what percentage did the price of this model increase between 2006 and 2007?

a   Chain base index number 2007 $= \dfrac{\text{current price}}{\text{previous year's price}} \times 100$

Use the formula.

$$= \dfrac{10\,465}{10\,260} \times 100$$

$$= 102$$

Chain base index number 2008 $= \dfrac{10\,780}{10\,465} \times 100$

Work out the chain base index number.

$$= 103$$

b $102 - 100 = 2\%$

Subtract 100 from the chain base index number to get the yearly percentage rise (or fall).

Weighted index numbers

When shopping at a supermarket there are some goods that a person buys each visit and some that they only buy every other visit and some they buy only once in a while. To work out how the price of groceries has changed the prices of the different items must be weighted by how frequently that person buys them. The overall price will then be a weighted mean of all the prices.

Weighted means are compared by using **weighted index numbers**.

The calculation of the index number is the same as for simple index numbers except that weighted means are used for the prices.

Weighted index number $= \dfrac{\text{current weighted mean price}}{\text{base year weighted mean price}} \times 100$

Example 32

This table gives information about the price of some groceries in July 2007 and the same items in July 2008. The weightings, out of 100, are the percentages spent on them in an average week.

a Work out the weighted means for 2007 and 2008.

b Work out the weighted index number for the price of the foods in 2008, taking 2007 as the base year.

c By what percentage did the prices of these items increase between 2007 and 2008?

Item	Price 2007 (pence)	Price 2008 (pence)	Weight
Bread	110	130	20
Meat	275	320	26
Fish	950	1000	6
Milk	35	40	18
Fruit	145	160	11
Vegetables	75	85	19

a Weighted mean for

$$2007 = \frac{(110 \times 20) + (275 \times 26) + (950 \times 6) + (35 \times 18) + (145 \times 11) + (75 \times 19)}{20 + 26 + 6 + 18 + 11 + 19}$$

$$= 187p$$

Weighted mean for

$$2008 = \frac{(130 \times 20) + (320 \times 26) + (1000 \times 6) + (40 \times 18) + (160 \times 11) + (85 \times 19)}{20 + 26 + 6 + 18 + 11 + 19}$$

$$= 210p$$

b Index number for $2008 = \frac{210}{187} \times 100$

$$= 112.3$$

c Percentage increase in price $= 112.3 - 100$

$$= 12.3\%$$

> Use the formula for weighted mean
>
> $$\bar{x} = \frac{\sum wx}{\sum w}$$

> Use weighted index number formula.

> Percentage increase = index − 100

Retail prices index

The retail prices index (RPI) is the most familiar general purpose domestic measure of inflation in the United Kingdom. It is commonly used in for updating old age pensions. It is also used for wage bargaining.

It is a weighted index showing how the price of a basket of goods changes from year to year.

The index includes the following categories.

- Food and catering
- Alcohol and tobacco
- Housing and household expenditure
- Personal expenditure
- Travel and leisure

The relative weightings (out of 1000) of each of these are shown in the pie chart.

Each category is made up of a number of different items.

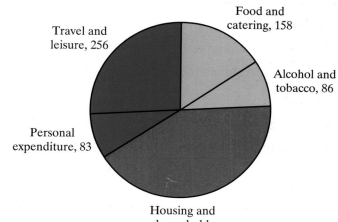

Travel and leisure, 256

Food and catering, 158

Alcohol and tobacco, 86

Personal expenditure, 83

Housing and household expenditure, 417

Source: Office for National Statistics

Exercise 4K

1 In 2007 the price of a box of chocolates was £2.50.

In 2008 the price of the same box of chocolates was £3.00.

Work out the index number for the price of the chocolates in 2008 using 2007 as the base year.

2 The index number for the price of potatoes in May 2008 was 112 based on May 2006. Write down the percentage increase in the price between 2006 and 2008.

3 The simple index numbers for the price of fuel is shown in the table. The base year was 2005.

Year	2005	2006	2007	2008
Index number	100	99	117	167

Describe what has happened to the price of the fuel in the years 2005 to 2008.

4 In 2008 the price of a litre of unleaded petrol was 116p.

In 2006 the price of a litre of unleaded petrol was 92p.

Using 2006 as the base year, work out the simple index number for a litre of unleaded petrol.

5 Jane bought a flat in 2005.

The table shows the value of Jane's flat for the years between 2005 and 2009.

Year	2005	2006	2007	2008	2009
Value (£)	178 000	180 000	184 000	200 000	190 000

a Work out the chain base index numbers for the value of Jane's flat for these five years.

Jane also bought a new car in 2005.

The table shows the value of Jane's car in each of the years 2005 to 2009.

Year	2005	2006	2007	2008	2009
Value (£)	12 000	8500	7300	5900	3900

b Work out the chain base index numbers for the value of Jane's car for these five years.

c Comment on the results of your answers to parts a and b.

6 A particular recipe uses flour and sugar in the ratio 80% to 20%.

In 2005 the price of flour was 100p per kilogram and the price of sugar was 120p per kilogram.

In 2008 the price of flour was 150p per kilogram and the price of sugar was 190p per kilogram.

a Work out the weighted means for this recipe for 2005 and 2008.

b Work out the weighted index number for the price of these items in 2008 using 2005 as the base year.

c By what percentage did the price of these items rise between 2005 and 2007.

Chapter 4 review

1 These are the number of dairy cows owned by nine farms in Cumbria.

72   45   69   72   65   64   71   80   41

For these data work out

a the modal number of dairy cows

b the median number of dairy cows

c the mean number of dairy cows.

2 The mid-day temperatures, in degrees Celsius, in Malta during one week in July were

Sun   Mon   Tue   Wed   Thu   Fri   Sat
27   32   31   28   24   30   29

a Work out the mean mid-day temperature.

b State why you cannot give a mode of the mid-day temperatures.

3 The marks obtained by 30 students in a test out of 50 were

32, 26, 31, 45, 28, 32, 33, 17, 24, 17
42, 32, 22, 32, 47, 24, 32, 32, 42, 48
17, 25, 21, 37, 32, 23, 40, 30, 32, 18

Work out

a the mode   **b** the median   **c** the mean mark.

4 Here are the ages, in years, of seven people.

90   69   69   70   80   83   71

a For these data

   i) what is the mode?

   ii) what is the median?

   iii) work out the mean of these data.

A person aged 73 joins the group.

b Find the median age of the eight people.

edexcel ⠿ *past paper question*

5 The frequency table gives information about the number of GCSE  *challenge*
subjects each student is taking.

Number of subjects	6	7	8	9	10	11
Number of students	12	15	20	29	19	7

For these data work out

a the mode

b the median

c the mean.

6 The ages of some people in a café are given below.  *challenge*

Age	Number of people
$10 \leqslant$ age < 20	7
$20 \leqslant$ age < 30	26
$30 \leqslant$ age < 40	22
$40 \leqslant$ age < 50	10

Find

a the modal class interval

b an estimate for the median age

c an estimate for the mean age.

7 The following table shows the population of Hambleton in 1960 ★★★★★ *challenge*
and in 1980.

Hambleton

Year	1960	1980
Population	6400	7040

a Taking 1960 as the base year, work out the index number for
the population of Hambleton in 1980.

Thorpe is a town near to Hambleton. The population of Thorpe
decreased by 2% between 1970 and 1990.

b Taking 1970 as the base year, complete the table below.

Thorpe

Year	1970	1990
Index number		

edexcel ⠿ *past paper question*

8 The box plots give information about the times, in minutes, some students take to travel from home to school (A), and from school to home (B).

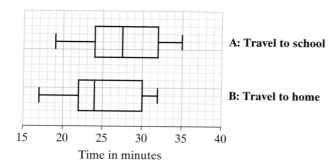

A: Travel to school

B: Travel to home

Time in minutes

a Work out the range for the times taken to travel to school.

b What is the median time taken to travel home?

c What is the shortest time taken to travel home from school?

The students say that it takes longer to travel to school in the morning than it does to travel home in the evening.

d Give one way that the box plots support this claim.

e Which of the box plots shows the most skewness? Describe this skewness.

edexcel ⠿ *past paper question*

9 In a traffic survey the police record the speeds, in miles per hour (mph), of 400 cars on a particular motorway in Britain.

The results were used to work out the information in the table.

Measure	Speed (mph)
Minimum	45
Lower quartile	55
Median	67
Upper quartile	70
Maximum	95

a From this information, work out
   i) the range
   ii) the inter-quartile range.

b Give one disadvantage of using the range as a measure of spread.

c On a grid, draw a box plot to represent the information in the table.

In Britain it is illegal to travel at a speed greater than 70 miles per hour. Speed cameras help to reduce the speed of traffic.

d Use the information to justify the use of speed cameras on this motorway.

edexcel ⠿ *past paper question*

10 John just passed his examination, scoring 52%.

He sat three papers which were weighted 40% : 40% : 20%.

His marks were 48, 50 and x.

Work out x.

11 The price of a camera in 2008 was £320 and in 2009 it was £350.

   a Find the simple index number for the value of the camera in 2009 based on the year 2008.

In 2009 the simple index number for a particular laptop was 96. The base year was 2008

   b Work out the percentage change in the price of the laptop between 2008 and 2009.

12 The value of a motorbike is given in this table.

Age in years	0	1	2	3	4
Value	£12 000	£10 050	£8400	£6800	£5100

Find the chain base index numbers for each year.

13 Calculate the mean and standard deviation for the variable x given that

$$\sum x^2 = 3000, \sum x = 240, n = 20.$$

14 The prices of 160 second-hand cars are given below.

Price, x (£1000s)	$2 < x \leqslant 3$	$3 < x \leqslant 4$	$4 < x \leqslant 5$	$5 < x \leqslant 6$	$6 < x \leqslant 7$
Frequency	13	24	48	70	5

   a Draw a cumulative frequency diagram for these data.

   b On your diagram mark

      i) the median

      ii) the lower quartile

      iii) the upper quartile.

   c Work out the inter-quartile range.

   d Use your diagram to find

      i) the 35th percentile

      ii) the 8th decile.

   e Draw a box plot to represent these data.

15 The table shows Jennifer and Samuel's marks in the Year 10 test, along with the mean marks and standard deviations.

	Jennifer	Samuel	Mean of 4th year	Standard deviation
English	66	52	51	12
Science	43	51	40	6

a Calculate the four standardised scores.

b Comment on Jennifer's performance.

c Comment on Samuel's performance.

16 The table gives the mean and standard deviation of the times, in seconds, for each of two races in the 2004 Olympic Games.

Race	Mean (seconds)	Standard deviation (seconds)
100 m	10.95	0.238
400 m	49.62	1.290

Source: www.athens2004.com

Roman Šebrle ran a time of 10.85 s in the 100 m race and a time of 48.36 s in the 400 m race.

a Calculate Roman Šebrle's standardised score for his time in each of these two races.

b Interpret your answers to part **a**.

edexcel ⠿ past paper question

17 Working days were lost in the Manufacturing Industries and in the Public Administration and Defence Industry between August 2000 and May 2001 as a result of strikes.

The table shows the number of working days, in thousands, lost each month.

Year		2000					2001				
Month		Aug	Sep	Oct	Nov	Dec	Jan	Feb	Mar	Apr	May
Manufacturing		14	4	2	6	8	2	6	9	2	4
Public Administration and Defence		14	13	0	15	5	6	5	7	2	0

Source: Government statistics

The mean value of the number of working days lost each month in the Manufacturing Industries is 5.7 thousand.

The standard deviation of the number of working days lost each month in the Public Administration and Defence Industry is 5.3 thousand, to one decimal place.

a   **i)** Calculate the mean value of the number of working days lost each month in the Public Administration and Defence Industry.

      ii) Calculate the standard deviation of the number of working days lost each month in the Manufacturing Industries. Give your answer to one decimal place.

You may assume that, between August 2000 and May 2001, there were approximately the same number of people employed in the Manufacturing Industries as there were employed in the Public Administration and Defence Industry.

b   Using the given summary statistics and your answers to part a, compare the numbers of working days lost each month due to strikes in the two industries between August 2000 and May 2001.

Chapter 4 summary

Averages

1   An **average** is a single value used to describe a set of data.

2   The **mode** is the value that occurs most often.

3   The **median** is the middle number in a list after the numbers have been put in order.

4   When the number of data values, n, is odd the **median** is the value of the $\frac{1}{2}(n + 1)$th observation.
When n is even the **median** is the **mean** of the $\frac{1}{2}(n)$**th** observation and the $\frac{1}{2}(n + 2)$**th** observation.

5   The **mean** is worked out by adding up the numbers then dividing by how many numbers there are.

6   **Mean** $= \bar{x} = \frac{\sum x}{n}$ where
  - $\bar{x}$ means the mean of all the x values
  - $\sum x$ means the sum of the x values

Mode, median and mean of discrete data in a frequency table

7   When **discrete data** are given in a **frequency table**:

Mode is the one with the highest frequency.

Median $= \frac{1}{2}(n + 1)$th value

Mean $= \bar{x} = \frac{\sum fx}{\sum f}$

Mode, median and mean of grouped data

8 When information is presented in a table of grouped data, the **modal class** is the class with the largest frequency.

9 For grouped, continuous data the median is the value of the $\frac{1}{2}n$th observation.

There are two ways of estimating the median.
- Using the mid-point of a cumulative frequency diagram.
- Using interpolation.

10 An **estimated mean** can be found from a grouped set of data using the formula

$$\text{mean} = \frac{\Sigma(f \times \text{mid-point})}{\Sigma f}, \text{ where } \Sigma \text{ means 'the sum of' and } f \text{ is frequency.}$$

Weighted mean

11 The **weighted mean** of a set of data is given by $\bar{x}_w = \dfrac{\Sigma wx}{\Sigma w}$, where w is the weight given to each variable, x.

Measures of spread

12 **Range** = largest value − smallest value

13 The **lower quartile** is the value such that one quarter (25%) of the values are less than or equal to it.

14 The **upper quartile** is the value such that three quarters (75%) of the values are less than or equal to it.

15 For discrete data, when n values are written in ascending order
- the lower quartile, Q_1, is the value of the $\frac{1}{4}(n + 1)$th observation
- the median, Q_2, is the value of the $\frac{1}{2}(n + 1)$th observation
- the upper quartile, Q_3, is the value of the $\frac{3}{4}(n + 1)$th observation.

16 **Inter-quartile range** (IQR) = upper quartile − lower quartile

17 For continuous data Q_1 is the value of the $\frac{1}{4}n$th observation, Q_2 is the value of the $\frac{1}{2}n$th observation and Q_3 is the value of the $\frac{3}{4}n$th obserrvation.

18 A set of data is divided into 100 parts to form **percentiles**.

19 When the set is divided into 10 equal parts, these are called **deciles**.

Box plots

20 A **box plot** represents important features of the data: the maximum and minimum values, the median, and the upper and lower quartiles.

21 If the median is closer to the upper quartile than it is to the lower quartile the data are **negatively skewed**.

22 If the median is closer to the lower quartile than it is to the upper quartile they are **positively skewed**.

23 If the median is equidistant from both the upper and lower quartiles they are **symmetrical**.

24 An **outlier** is any value that is more than 1.5 times the inter-quartile range below the lower quartile or more than 1.5 times the inter-quartile range above the upper quartile.

Variance and standard deviation

25 The **variance** is a measure of dispersion or spread.

It is given by

$$\text{Var}(x) = \frac{(x_1 - \bar{x})^2 + (x_2 - \bar{x})^2 + \ldots + (x_n - \bar{x})^2}{n}$$

$$= \frac{\Sigma(x - \bar{x})^2}{n}$$

A friendlier formula for variance is

$$\text{Var}(x) = \frac{x_1^2 + x_2^2 + \ldots + x_n^2}{n} - \bar{x}^2$$

$$= \frac{\Sigma x^2}{n} - \bar{x}^2$$

26 The **standard deviation** is the square root of the variance.

Standard deviation (SD) $= \text{R}\Sigma(x - \bar{x})^2 \backslash n$

or the simpler version $\sqrt{\dfrac{\Sigma x^2}{n} - \bar{x}^2}$.

Comparing data sets

27 A full comparison needs, at least, both a **measure of central tendency** and a **measure of dispersion or spread**.

28 Another comparison that could be made is the **skewness** of the distributions.

29 **Standardised score (z)** $= \dfrac{\text{score} - \text{mean}}{\text{standard deviation}}$

Higher Statistics

Index numbers

30 Index number $= \dfrac{\text{current price}}{\text{base year price}} \times 100$

31 Chain base index number $= \dfrac{\text{current price}}{\text{previous year's price}} \times 100$

32 Weighted index number $= \dfrac{\text{current weighted mean price}}{\text{base year weighted mean price}} \times 100$

Test yourself

1 Find the mean, mode and median for the following set of numbers.

    3   9   5   8   7   3   1   7   4   7

2 Find the mean mode and median for the data given in the frequency table.

Number, x	1	2	3	4	5	6
Frequency	2	5	10	12	16	5

3 Estimate the mean and median for the data given in the grouped frequency table.

Class	$10 < x \leqslant 15$	$15 < x \leqslant 20$	$20 < x \leqslant 25$	$25 < x \leqslant 30$
Frequency	2	3	3	1

4 Find the range, inter-quartile range and standard deviation for the following set of data.

    2   5   6   8   9   9   10   10   11   13   14

5 Estimate the median and quartiles of the following set of data.

Class	$1 < x \leqslant 3$	$3 < x \leqslant 5$	$5 < x \leqslant 7$	$7 < x \leqslant 9$
Frequency	4	6	4	6

6 Julius got 35 marks out of 100 in an examination. The mean mark for the examination was 48 and the standard deviation was 6.5 marks. Work out Julius's standardised mark for the exam.

Chapter 5:
Scatter diagrams and correlation

After completing this chapter you should be able to

- draw scatter diagrams
- recognise correlation
- know what is meant by a causal relationship
- draw lines of best fit on scatter diagrams
- know what is meant by a mean point
- draw lines of best fit through mean points

- use lines of best fit to make predictions
- find the equation of a line of best fit
- fit a line of best fit to a non-linear model
- calculate and interpret Spearman's rank correlation coefficient
- know what to do with tied ranks.

Does it get colder the further you climb up a mountain?

Is there a relationship (correlation) between how high you go and the temperature?

A scatter diagram can be used to answer questions such as this and, given the temperature at the bottom of a mountain, it is possible to predict the temperature at the top!

5.1 Scatter diagrams

Scatter diagrams are used to show whether two sets of data are **associated**.

 Scatter diagrams should always be drawn on graph paper or by using ICT.

Example 1

A teacher thinks that students do equally well at mathematics as they do at science.

The marks out of 50 gained by nine students in mathematics and science exams are shown in the table.

Candidate	A	B	C	D	E	F	G	H	I
Mathematics mark, x	16	20	20	28	29	31	37	39	45
Science mark, y	15	18	22	26	29	33	38	36	45

a Draw a scatter diagram of these data.

b Explain whether the teacher's belief is correct.

a

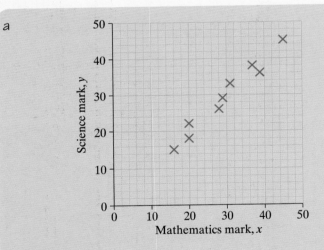

Use each pair of mathematics and related science marks as coordinates to plot a point on the graph. For example, the first point plotted will have the coordinates (16, 15), (16 for mathematics and 15 for science).

Label the axes. Do not join up the points on a scatter diagram.

b The scatter diagram shows that as the mathematics marks increase the science marks also increase. There is an association between science and mathematics marks.

Look to see if an increase in one variable is associated with an increase (or decrease) in the other variable.

If the points on a **scatter diagram** lie approximately on a straight line, there is a linear relationship between them. They are **associated**.

Example 2

Ten £1 coins were weighed and their age, in years, noted. The results are given in the table.

Coin	1	2	3	4	5	6	7	8	9	10
Age, x (years)	16	18	9	11	13	20	21	25	12	14
Mass, y (g)	8.95	9.10	9.18	9.31	9.10	8.95	8.87	8.91	9.20	9.15

a Plot these data on a scatter diagram.

b Is there an association between the age and the mass of the coins? Suggest an explanation for your answer.

a

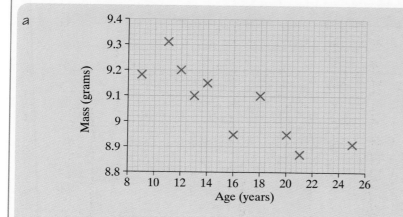

Choose scales to make the plotting of all the pairs of data as easy as possible.

Scales do not have to begin at zero. (A suitable scale for age would be from 8 to 26 years, and for mass from 8.8 to 9.4.)

Plot the points.

b Generally older coins weigh less than newer ones. There is some association between the age and weight of coins. This might be because coins wear away as they are used.

Look to see if an increase in one set of data brings about an increase or decrease in the other (here there is a decrease). The plotted points lie less on a straight line than in Example 1, so the association is not as strong.

Exercise 5A

1 A group of students sat a mock exam in English and later they sat the final exam. The marks obtained by the candidates in both exams are shown in this table.

Student	A	B	C	D	E	F	G	H
Mock mark, M	10	15	23	31	42	46	70	75
Final mark, F	11	16	20	27	38	50	68	70

a Plot the marks on a scatter diagram.

b Are good marks in the mock exam associated with good marks in the final exam?

2 The table shows the marks awarded to six skaters by two judges in an ice skating competition.

Skater	1	2	3	4	5	6
Judge A	6.5	7.0	7.2	8.1	8.6	9.0
Judge B	7.4	8.2	6.4	6.8	8.5	8.5

a Plot a scatter diagram of these marks.

b Comment on how the marks of the two judges compare.

3 The table shows the number of hours of sunshine and the maximum temperature in ten British towns on one particular day.

Town	A	B	C	D	E	F	G	H	I	J
Number of hours of sunshine	11	17	15	13	12	10	10	10	12	14
Maximum temperature (°C)	13	21	20	19	15	16	12	14	14	17

a Plot this information on a scatter diagram. (Use graph paper.)

b What does your diagram tell you about the maximum temperature as the number of hours of sunshine increases?

4 The weight and height of 10 adults are as follows.

Adult	A	B	C	D	E	F	G	H	I	J
Height, H (cm)	155	163	183	198	164	178	205	203	213	208
Weight, W (kg)	58	61	85	93	70	76	84	98	100	101

a Select suitable scales and plot a scatter diagram for these measurements.

b Jacob thinks that the taller a person is the more they weigh. Comment on Jacob's belief.

5 A general knowledge test was given to seven boys of different ages with the following results.

Boy	A	B	C	D	E	F	G
Age (years)	11.5	13	13.5	15	16	17	17
Number of correct answers	18	23	25	24	33	32	40

a Select suitable scales and plot a scatter diagram for these data.

b Is there an association between age and the number of correct answers?

5.2 Correlation

> **Correlation** is a measure of the strength of the linear association between two variables.

The scatter diagram below shows the price of six second-hand bicycles in relation to their age.

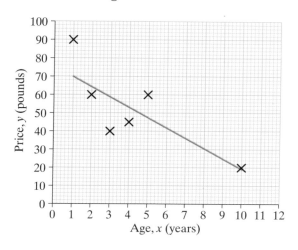

As the age of a bicycle increases the second-hand value of a bicycle decreases. This is called **negative correlation**.

Because the points are very scattered about a straight line this is called a **weak linear correlation**.

The scatter diagram below shows the petrol consumption of five cars with different engine sizes, over a measured distance.

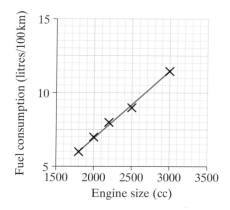

As the engine size increases the petrol consumption increases. This is called **positive correlation**.

There is a **strong linear correlation** because the points lie almost in a straight line.

It is possible to have **strong** or **weak positive linear correlations**, and **strong** or **weak negative linear correlations**.

ResultsPlus
Watch out!

■ When asked to describe the correlation shown on a scatter diagram, most students can identify it as positive or negative, but not many explain the correlation in terms of the context, for example 'as you go deeper in the water the temperature gets lower'.

Positive correlation is when one variable increases as the other increases.

Negative correlation is when one variable decreases as the other increases.

The scatter diagrams below show the possibilities.

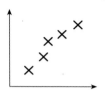    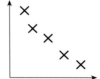

Strong positive linear correlation (e.g. the weight on a wire and how much it stretches).

Weak positive linear correlation (e.g. the height and weight of people).

No correlation (e.g. a student's height and maths test mark).

Weak negative linear correlation (e.g. the speed of a car and its petrol consumption).

Strong negative linear correlation (e.g. the weight on top of a spring and its length).

The middle scatter diagram shows two variables that are **associated** – the points lie on a circle. The linear correlation between them is however zero, since they do not lie close to a straight line.

Association does not necessarily mean there is a linear correlation.

Exercise 5B

1 State the type of correlation shown in each of these scatter diagrams.

a
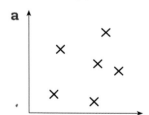

b
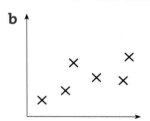

c

2 The exam grades of a sample of students in both their mock and final exams are shown on the scatter diagram.

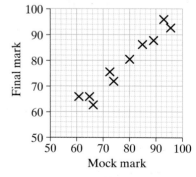

a Describe the correlation shown in the diagram.

b What conclusion can you draw regarding the relationship between the mock and the final marks of the students?

3 The number of goals scored by football teams and their positions in the league were recorded for the top twelve teams.

Team	A	B	C	D	E	F	G	H	I	J	K	L
Goals scored	49	44	43	36	40	39	29	21	28	30	33	26
League position	1	2	3	4	5	6	7	8	9	10	11	12

  a Plot a scatter diagram of these data.

  b Describe and interpret, in context, the type of correlation.

4 It is thought that there is a correlation between the numbers of times a computer game is played and the score gained at the next attempt. Seven students were asked how many times they had played the game and their next score was noted with the following results.

Student	A	B	C	D	E	F	G
Number of times played, x	6	8	4	5	3	7	9
Score, y	33	37	28	30	22	37	40

  a Draw a scatter diagram for these two variables on graph paper.

  b What correlation does this scatter diagram suggest?

  c What conclusions can you draw regarding the score at the next attempt?

5 The heights and the weights of six Labrador dogs were recorded as follows.

Dog	1	2	3	4	5	6
Height, x (cm)	61	45	51	48	53	56
Weight, y (kg)	37	30	32.5	32	34	36

  a Draw a scatter diagram of the heights and the weights of the dogs.

  b Describe the correlation between the height and the weight of the dogs.

  c What conclusion can you draw about the correlation between the heights and the weights of Labrador dogs?

5.3 Causal relationships

The amount of petrol a car uses depends on the size of its engine since bigger engines use more petrol. The size of the engine **causes** the car to use more petrol. There is a **causal relationship** between the amount of petrol used and the engine size.

When a change in one variable directly causes a change in another variable, there is a **causal relationship** between them.

Causal relationships and correlation

Example 3

The scatter diagram gives information about the value of a second-hand bicycle. The age of the bicycle and its value are negatively correlated: the older it is, the less its second-hand value.

a Do you think there is a causal relationship between the age of the bicycle and its value? Give a reason for your answer.

b How strong is the correlation? Give a possible reason for this.

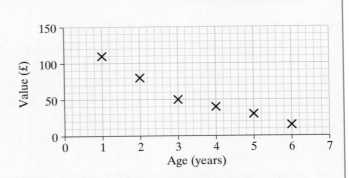

a There is a causal relationship between a bicycle's age and its second-hand price. The age of the bicycle causes its value to decrease.

> Think if there is a reason why an increase in age could cause a decrease in value. If there is, it is a causal relationship.

b The condition of the bicycle also plays a part in its second-hand price, therefore the correlation is not very strong.

> Look to see how big the departure from a straight line actually is. Think of anything else that could cause a change in value.

Example 4

The scatter diagram shows the number of television sets and the number of calculators sold by an electrical shop over a period of seven years.

There is a positive linear correlation between the number of televisions sold and the number of calculators sold. Is there a causal relationship between these two variables?

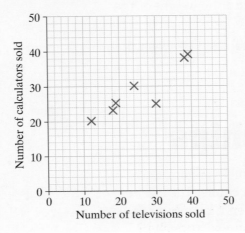

Buying a television does not cause you to buy a calculator.

Think if there is any reason why one should affect the other.

Both variables may depend on another factor (e.g. an advance in technology or an increase in wages).

Think if there are other factors which cause the sales of both to go up.

Correlation does not necessarily imply a causal relationship.

Exercise 5C

1 Which of the following pairs of variables are likely to have a causal relationship?

   a A car's weight and its petrol consumption.

   b Sales of chocolates and sales of clothes.

   c Low temperature and snowfall.

   d Sales of computers and sales of software.

2 Choice Cars have eight cars for sale. The scatter diagram shows the ages and prices of the eight cars.

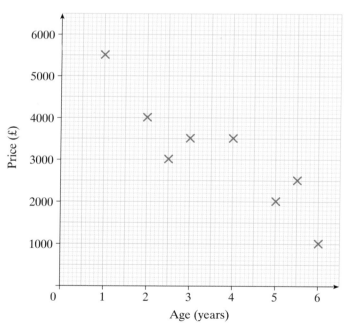

   a What is the price of the car that is 4 years old?

   b How old is the car that costs £2000?

   c Describe the correlation between a car's age and its price.

   d Is the correlation between age and price a causal relationship?

3   The times some young people took to reach a required standard of proficiency on a training scheme and their ages are given in the table.

Trainee	A	B	C	D	E	F	G	H	I	J
Age of trainee (years)	16	17	18	19	19	21	20	21	18	20
Training time (months)	8	6	9	8	12	9	10	12	7	11

a   Draw a grid like this on graph paper.

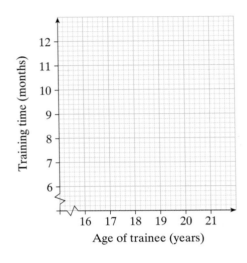

b   Plot a scatter diagram for these data.

c   State whether or not the diagram gives evidence of correlation, and, if it does, whether the correlation is negative or positive and if it is strong or weak.

d   Would you say there was a causal relationship between the two variables?

4   Twelve students sat two biology tests, one theoretical and one practical. Their marks are shown below.

Student	A	B	C	D	E	F	G	H	I	J	K	L
Marks in theory test, x	5	9	7	11	20	4	6	17	12	10	15	18
Marks in practical test, y	6	8	9	14	21	8	7	16	15	8	18	18

a   Draw a scatter diagram to represent these data.

b   Describe the correlation between the theory and the practical tests.

c   Is there a causal relationship between the two variables?

5 The table shows the number of cars per 100 of the population and the number of road deaths per 100 000 of the population for ten different countries.

Country	Cars per 100 of population, x	Road deaths per 100 000 of population, y
A	31	15
B	32	35
C	55	23
D	64	20
E	28	26
F	19	19
G	34	22
H	38	22
I	48	31
J	60	36

  a Draw a scatter diagram to represent these data.
  b Describe the correlation between the number of cars and the number of deaths.
  c Is there a causal relationship between these variables?

5.4 Line of best fit

> Points on a scatter diagram are strongly correlated if they lie along a straight line.

This scatter diagram shows the height and shoe sizes of ten pupils from a class of thirty.

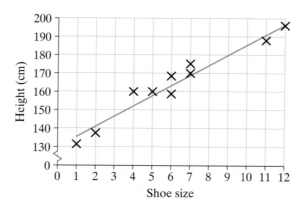

 A straight line can be drawn that passes as close to (or through) as many points as possible because height and shoe size are strongly correlated. This line is called the **line of best fit**.

> A **line of best fit** is a straight line drawn so that the plotted points on a scatter diagram are evenly scattered either side of the line.

Example 5

This table shows the height above sea level, x (m), and the temperature, y (°C), on the same day at nine different places.

Place	A	B	C	D	E	F	G	H
Height, x (m)	1400	400	280	800	920	560	1220	680
Temperature, y (°C)	7	15	19	9	10	12	8	14

a Draw a scatter diagram and add a line of best fit.

b Use your line of best fit to estimate the air temperature at 600 m above sea level.

a

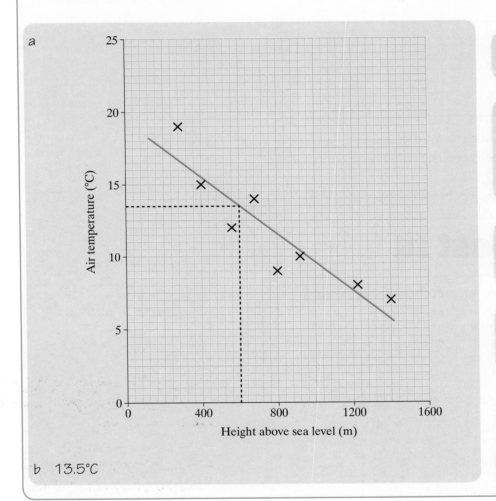

Draw the scatter diagram as normal.

Draw a line so that the points are evenly spaced either side of it. The line does not necessarily have to pass through any of the points on the scatter diagram.

Draw a vertical line from 600 m on the horizontal axis (x-axis) until it meets the line of best fit.

Draw a horizontal line from the meeting point on the line of best fit to the vertical axis (y-axis).

Read off the air temperature.

b 13.5°C

In Example 5, the air temperature at 600 m above sea level was estimated using the line of best fit. The line of best fit is a **model** for how air temperature changes with height above sea level.

> A line of best fit is a **model** for the association between the two variables.

The mean point

The line of best fit can be drawn by eye but is best if it passes through the **mean point**.

> A line of best fit should pass through the **mean point** $(\bar{x}, \bar{y})$.

For the data in Example 5:

$$\text{Mean height} = \frac{1400 + 400 + 280 + 800 + 920 + 600 + 560 + 1220 + 680}{9}$$
$$= 762.2\,\text{m}$$

$$\text{Mean temperature} = \frac{7 + 15 + 19 + 9 + 10 + 14 + 12 + 8 + 14}{9}$$
$$= 12$$

So the mean point $(\bar{x}, \bar{y}) = (762.2, 12)$.

Exercise 5D

1 An investigation was undertaken into the length of main road, x (in 1 000 000 miles), and the number of road accident injuries per year, y (in 100 000), for seven industrialised countries. The results are shown in the table.

Country	A	B	C	D	E	F	G
Main road length, x	12.4	28.5	31.2	45.8	18.4	44.4	18.4
Injuries, y	3.1	2.4	4.5	2.2	1.7	1.1	2.7

a Draw a scatter diagram for these data.

b Would you draw a line of best fit for these data? Give reasons for your answer.

2 The following table shows the exam marks of eight students in English and French.

Student	A	B	C	D	E	F	G	H
English, x	10	20	30	32	49	52	61	74
French, y	20	24	35	30	48	59	72	80

a Find the mean of each set of marks.

b Draw a scatter diagram for your data. Both axes can start at 0.

c Plot the mean point found in **a** and draw a line of best fit through it.

ResultsPlus
Watch out!

When asked to plot the point $(\bar{x}, \bar{y})$ and then draw a line of best fit, some students didn't draw the line through $(\bar{x}, \bar{y})$.

3 **a** Draw a horizontal axis from 20 to 40 and a vertical axis from 0 to 12.

b Plot the following points on your diagram: (22, 3), (24, 4), (25, 4), (29, 7), (32, 8), (36, 10).

c Find the mean point and mark it on your diagram.

d Draw a line of best fit.

4 The following table shows the heights and the weights of ten boys.

Boy	A	B	C	D	E	F	G	H	I	J
Height, x (cm)	130	129	133	135	136	140	142	145	150	160
Weight, y (kg)	30	33	33	38	37	40	44	52	61	72

a Draw a scatter diagram of these data.
b Find the mean height and the mean weight. Mark the mean point on your diagram.
c Draw a line of best fit.
d What sort of correlation does your diagram show?

5.5 Using scatter diagrams and lines of best fit

Lines of best fit can be used to estimate other values from the graph.

In Example 5, the line of best fit was used to estimate the air temperature at 600 m above sea level. The readings were for heights from 400 m to 1400 m, so 600 m fell within this range.

This process is known as **interpolation**.

If you wanted to estimate the temperature at 1600 m you would have to extend the line of best fit, as shown in the diagram below, and then read off the temperature. This is known as **extrapolation**.

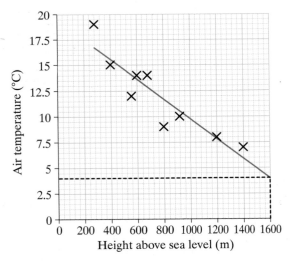

The estimate is 4°C.

> **Interpolation** is when the data point you need lies within the range of the given values.
>
> **Extrapolation** is when the data point you need lies outside the range of the given values.

Example 6

The table shows the results of an experiment on the effect of water temperature on the number of heartbeats per minute of *Daphnia* (a water flea).

Observation	1	2	3	4	5	6
Water temperature, x (°C)	5	10	15	20	25	30
Number of heartbeats per minute	110	200	240	300	380	400

a   Draw a scatter diagram of these data. Add a line of best fit that passes through the mean point.

b   Using the line of best fit

   i)   interpolate to estimate the number of heartbeats per minute when the water temperature is 22°C.

   ii)   extrapolate to estimate the number of heartbeats per minute when the water temperature is 35°C.

a

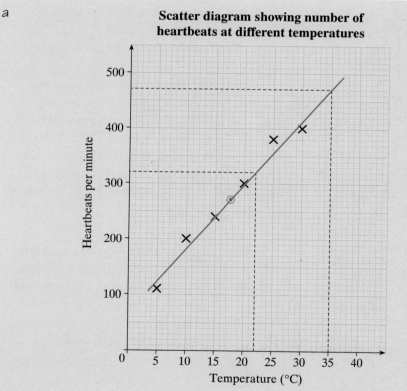

Scatter diagram showing number of heartbeats at different temperatures

> Work out the mean point.
>
> Plot the points, including the mean point.
>
> Draw a line of best fit through the mean point.

Mean temperature $= \dfrac{5 + 10 + 15 + 20 + 25 + 30}{6}$

$\qquad\qquad\qquad = 17.5°C$

Mean heartbeat $\quad = \dfrac{110 + 200 + 240 + 300 + 380 + 400}{6}$

$\qquad\qquad\qquad = 271.7$ beats per minute

The mean point will be (17.5, 271.7).

b   From the diagram

   i   Number of heartbeats when the temperature is 22°C = 320 per minute.

> Draw a line from 22 on the horizontal axis (*x*-axis) to the line of best fit and then across to the vertical axis (*y*-axis).

   ii   Number of heartbeats when the temperature is 35°C = 470 per minute.

> Extend the line of best fit.
>
> Draw a line from 35 on the horizontal axis (*x*-axis) to the line of best fit and then across to the vertical axis (*y*-axis).

Build Better Answers

Question: On a particular day, a scientist recorded the air temperature at 8 different heights above sea level.
The scatter diagram shows the air temperature, *y*°C, at each of these heights, *x* km above sea level.

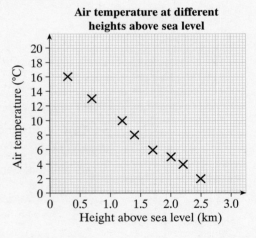

Air temperature at different heights above sea level

(a) Using the scatter diagram, write down the air temperature recorded at a height of 2.5 km above sea level.   **(1 mark)**

(b) Describe the correlation between the air temperature and the height above sea level.   **(1 mark)**

The mean point of the data $(\bar{x}, \bar{y})$ is (1.5, 8)

(c) On the scatter diagram,

   (i)   plot the point (1.5, 8),

   (ii) draw a line of best fit through (1.5, 8).   **(2 marks)**

(d) Using your line of best fit, find an estimate of the height above sea level when the air temperature is 0°C.   **(1 mark)**

(Total 5 marks)

■ **Basic   1–2 mark answers**
In part a), read the air temperature correctly from the graph. In part b), describe the correlation as 'negative'.

● **Good   3–4 mark answers**
Complete parts a) and b) correctly. In part c) plot the mean point correctly and draw the line of best fit through it, with equal numbers of points either side. Some students lost a mark here for drawing the line of best fit through the origin.

▲ **Excellent   5 mark answers**
Complete parts a), b) and c) correctly. For part d) extend the line of best fit to insersect the horizontal axis and read off the value for the height above sea level. Be careful reading the scale – some students lost marks by reading the value incorrectly.

Interpolated values are reasonably accurate.

The dangers of extrapolation are shown by Example 6. The estimate for the number of heartbeats at 35°C was 470, but the actual value was observed to be 389. An estimate at 40°C would be 520, but the correct value is 0 since the water flea would be dead in water at this temperature.

Be careful with extrapolated values – the further the extrapolation is outside the range of given values the less reliable the estimated value.

Exercise 5E

1 A measure of personal fitness is the time taken for a person's pulse rate to reach normal after strenuous exercise. Gordon recorded his pulse rate y at time x minutes after finishing some strenuous exercise. The results are shown in the table.

Time, x (min)	0.5	1.0	1.5	2.0	2.5	3.0	3.5
Pulse rate, y (beats/min)	125	113	110	94	81	83	71

   a Draw a scatter diagram and add a line of best fit.

   b Estimate Gordon's pulse rate 5 minutes after stopping exercise.

   c Is your estimate a reliable one?
      Give a reason.

2 A bar was supported at its ends in a horizontal position and various weights, x (kg), were hung from the mid-point of the bar. The deflection (how much the middle of the bar sags), y (cm), was recorded each time. The results are shown in the table.

Mass, x (kg)	20	25	30	35	40	45	50
Deflection, y (cm)	0.20	0.32	0.34	0.40	0.49	0.59	0.65

   a Draw a scatter diagram and add a line of best fit drawn by eye.

   b Estimate the deflection under a weight of 28 kg.

   c Estimate the deflection under weights of 15 kg and 55 kg.

   d Which of your three estimates is likely to be the most accurate? Explain your answer.

3 Ten students were selected at random from those visiting the tuck shop at mid-morning break. The students were asked their age and how much pocket money they got each week. The results are shown in the table.

Student	1	2	3	4	5	6	7	8	9	10
Age, x (years)	17	16	18	13	10	$11\frac{1}{2}$	14	11	15	12
Pocket money, y (£)	12	15	20	10	2	2.25	10.5	2.5	11	13

a Draw a scatter diagram for these data.

b Draw a line of best fit by eye and use this to predict how much a child of $13\frac{1}{2}$ is likely to get.

c Explain why you would not bother to extrapolate to find how much a 25-year-old would get for pocket money.

Examiner's Tip
When plotting the graph remember $11\frac{1}{2}$ is the same as 11.5.

4 The sales in units of £1000 of a certain company for the years 2002 to 2008 are given in the table.

Year, x	2002	2003	2004	2005	2006	2007	2008
Sales, y (£1000s)	65	70	73	78	83	83	88

a Draw a scatter diagram for these data.

b Add a line of best fit drawn by eye and use your line to predict the sales for 2010. Comment on the validity of this estimate.

5 The length, y mm, of a metal rod was measured at various temperatures, x°C, giving the following results.

Temperature, x (°C)	60	65	70	75	80	85
Length, y (mm)	100.2	100.8	101.8	102	103.4	104.5

a Draw a scatter diagram for these data.

b Add a line of best fit that passes through the mean point.

c Use your line to predict the length of the rod when the temperature was 68°C and what it would be at 100°C. Comment on the validity of these estimates.

5.6 Finding the equation of a line of best fit

You may be familiar with the equation for a straight line in the form $y = mx + c$. In statistics it is used in the form $y = ax + b$.

> The **equation of a line $y = ax + b$** has a **gradient a**, and its **intercept** on the y-axis is $(\mathbf{0}, \mathbf{b})$.

Example 7

The graph shows the length of a spring, y cm, under different loads, x kg. Find the equation of the line.

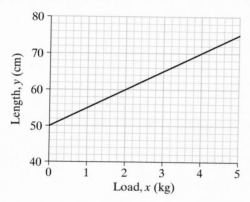

$b = 50$

> Look to see where the line intercepts with the vertical line $x = 0$ (the y-axis).

As x increases from 1 to 4, y increases from 55 to 70.
So, as x increases by 3, y increases by 15.

> Pick any two points on the line and work out the gradient of the line.

For every increase of 1 in x, y increases by $\frac{15}{3} = 5$.

> The gradient, $a = \dfrac{\text{increase in } y}{\text{increase in } x}$

So the gradient is 5, giving $a = 5$.
Therefore the equation of the line is $y = 5x + 50$.

> Remember to give the equation of the line.

If the graph had not crossed the y-axis, then you would have to find b another way.

Since every point on the straight line fits the equation $y = ax + b$ and $(1, 55)$ is a point on the line,

then $55 = (5 \times 1) + b$,

giving $b = 55 - (5 \times 1) = 50$.

Finding the values of *a* and *b* in general

Calculating *a*, the gradient

Select two points on the graph. These should be well apart and at points where the values of x and y are whole numbers, if possible. Let the point with the lower x-value be (x_1, y_1) and let the point with the higher x-value be (x_2, y_2).

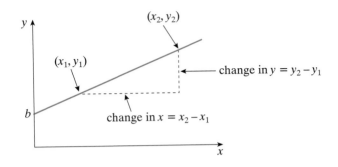

Then the gradient, $a = \dfrac{\text{change in y}}{\text{change in x}}$

$\qquad\qquad = \dfrac{y_2 - y_1}{x_2 - x_1}$

Finding or calculating *b*, the intercept with the *y*-axis

If the x-values on the scatter diagram go down to 0, then the value of b can be read directly from the graph.

If the x-values do not go down to 0, the formula $y = ax + b$ can be rearranged to give $b = y - ax$. The value of b is then calculated by substituting the values from one of the points on the line and the value for a in this formula, for example $b = y_1 - ax_1$ or $b = y_2 - ax_2$.

> The value of the constants in the equation of a line of best fit are calculated using $a = \dfrac{y_2 - y_1}{x_2 - x_1}$ and $b = y_1 - ax_1$ or $b = y_2 - ax_2$.

Example 8

The scatter diagram gives information about the temperature at different heights above sea level. Find the equation of the line of best fit in the form $y = ax + b$. Explain the values of a and of b in context.

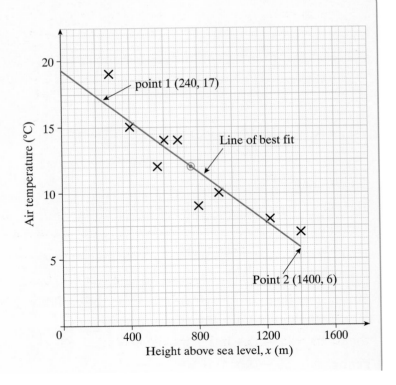

Two points on the line are $(x_1, y_1) = (240, 17)$ and $(x_2, y_2) = (1400, 6)$ as shown on the diagram.

$$a = \frac{y_2 - y_1}{x_2 - x_1}$$

$$= \frac{6 - 17}{1400 - 240}$$

$$= -0.009\,482...$$

$b = 19$

Alternatively,

$b = y_2 - ax_2$
$= 6 - (-0.009482... \times 1400)$
$= 6 + 13.27$
$= 19.27$
$= 19$ (to nearest whole number)

The equation of the line of best fit is $y = -0.0095x + 19$.

b is the temperature at zero metres above sea level. In this example, the temperature is approximately 19°C at zero metres above sea level.

a is the rate of change in temperature as the height above sea level increases. In this example, the temperature goes down by 0.0095°C for every metre further above sea level.

> First select two points on the line of best fit.

> Calculate a.

> Find b.
> The value of b can be read from where the extended line would cut the vertical axis (y-axis) on the scatter diagram. You cannot read the value exactly but you can make a good estimate.

> Alternatively, calculate b using the formula $b = y - ax$ with a known coordinate and the gradient a calculated above.

> State in full the equation of the line of best fit.

> Comment on the equation in the context of the question.

 You can use ICT to produce a scatter diagram and the line of best fit, and calculate and display the equation of the line.

Exercise 5F

1 The scatter diagram shows the results of a survey on the thickness of the soles of trainers, x, and the number of years that they had been worn, y.

For the line of best fit shown in the diagram, find:
  a the gradient of the line
  b the intercept of the line with the y-axis
  c the equation of the line of best fit.

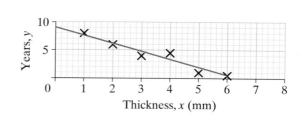

2   Work out the equation of the line of best fit for the scatter diagram shown below.

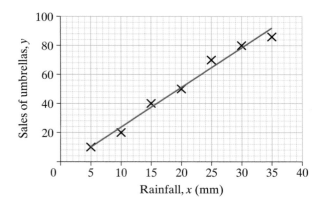

3   This scatter diagram shows the controlled assessment mark and the exam mark for seven students. Copy the diagram.

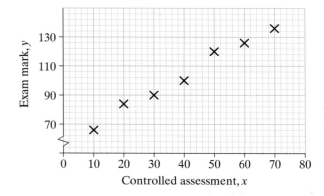

a   Draw a line of best fit by eye.

b   Work out the equation of the line of best fit.

4   Ten students were selected at random from those visiting the tuck shop at mid-morning break. The students were asked their age and how many hours they watched television each week. The results are shown in the table.

Student	1	2	3	4	5	6	7	8	9	10
Age, x (years)	17	16	18	13	10	$11\frac{1}{2}$	14	11	15	12
Hours of TV watching, y	12	15	20	10	2	2.25	10.5	2.5	11	13

a   Draw a scatter diagram of these data.

b   Add a line of best fit that passes through the mean point.

c   Calculate the equation of the line of best fit and use it to predict how many hours a student aged $16\frac{1}{2}$ watches television.

d   Explain why you would not bother to extrapolate to find how many hours a 40-year-old would watch television.

5 **a** For the data in Exercise 5E question 4, find the equation of the line of best fit.

   b Use the equation to predict the sales for the year 2010. Comment on the validity of this estimate.

6 For the data in Exercise 5E question 5:

   a Calculate the equation for your line of best fit in the form $y = ax + b$, and use it to predict the length of the rod at:

   i) 68°C

   ii) 100°C

   Comment on the validity of these estimates.

   b What do the constants a and b represent in this case?

7 The manager of a factory decided to give the workers an incentive by introducing a bonus scheme. After the scheme was introduced the manager thought that the workers might be making more faulty products because they were rushing to make articles quickly. A study of the number of articles rejected, y, and the amount of bonus earned, £x, gave the figures shown in the table.

Employee	A	B	C	D	E	F	G	H
Bonus, x (£)	14	23	17	32	16	19	18	22
Number of rejects, y	6	14	5	16	7	12	10	14

   a Draw a scatter diagram for these data and add a line of best fit that passes through the mean point.

   b What sort of correlation is there between the two variables? What does this mean in terms of the manager's belief?

   c Calculate the equation of the line of best fit in the form $y = ax + b$.

   d The maximum number of rejects acceptable is 9. At what level should the maximum bonus be set?

8 The height of a seedling, y millimetres, x weeks after it is planted is given in the table.

x	5	6	7	8	9	10
y	102	111	123	135	148	153

   a Plot a scatter diagram for these data.

   b Draw a line of best fit that passes through the mean point.

   c Calculate the equation of the line of best fit in the form $y = ax + b$.

   d What do the constants a and b represent in this case?

   e Use the equation found in **c** to estimate the height of the seedling after 20 weeks. How accurate will this estimate be?

5.7 Fitting a line of best fit to a non-linear model of the form $y = ax^n + b$ and $y = ka^x$

Fitting a line of best fit to a non-linear model of the form $y = ax^n + b$

In an investigation the lengths and areas of nine privet leaves were recorded.

Length, x (mm)	8	12	18	20	26	33	38	40	45
Area, y (mm²)	25	48	124	162	220	315	420	515	650

The scatter diagram for these two variables is shown opposite.

A straight line is not the best model for the association between these two variables. A curve is better, so these data fit a **non-linear model**.

This curve is of the form $y = ax^n + b$.

The look of curves of the form $y = ax^n + b$ vary according to the value of n. It is important to be able to recognise and use values for n of 2, −1 or $\frac{1}{2}$. These values give the following equations.

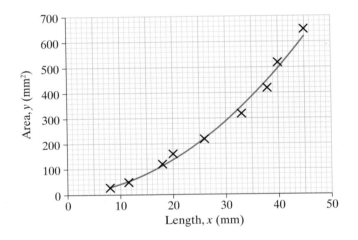

n = 2

$y = ax^2 + b$

n = −1

$y = ax^{-1} + b$

$\qquad = \frac{a}{x} + b$

n = $\frac{1}{2}$

$y = ax^{\frac{1}{2}} + b$

$\qquad = a\sqrt{x} + b$

The graphs of non-linear models of the form $y = ax^n + b$ look like these.

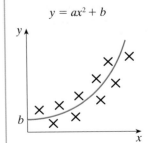

$y = ax^2 + b$

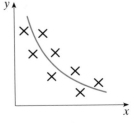

$y = \frac{a}{x} + b$

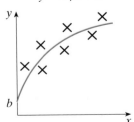

$y = a\sqrt{x} + b$

Examiner's Tip

In an exam you will need to recognise these models.

Example 9

The data in the table are thought to follow the equation $y = \frac{a}{x} + b$.

x	1	2	3	4	5
y	5	4.1	3.6	3.5	3.2

a Plot a scatter diagram for these data.

b Draw a straight line of best fit by eye, and a curve of the form $y = \frac{a}{x} + b$.

c Comment on the fit of these lines.

a and b

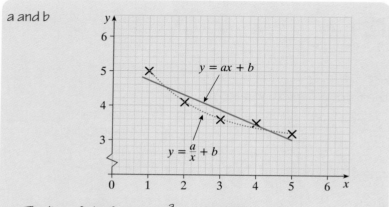

Plot the points on the scatter diagram.

Draw a straight line through the points so that the points are equally spread each side of the line.

Draw a curve of the form $y = \frac{a}{x} + b$ so that the points are equally spread about the curve.

c The line of the form $y = \frac{a}{x} + b$ fits better.

Look to see which is the better fit to the points.

Fitting a line of best fit to a non-linear model of the form $y = ka^x$

Sometimes a curve of the form $y = ka^x$ may be a better shape than the type shown above.

The function $y = ka^x$, where a is a positive number and x is a variable, is called an exponential function. An exponential function models the way things grow and the way things decay.

The graph below shows the value, v, of an investment after x years. The initial investment is £1000 and the interest rate is 10%.

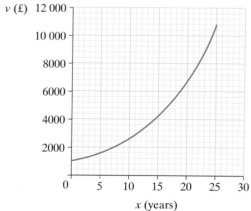

Examiner's Tip
You will be told which model, $y = ax^n + b$ or $y = ka^x$, you should consider.

This curve shows exponential growth and is of the form $y = ka^x$ ($a > 1$).

The graph below shows the value of an investment of £1000 if 10% of
the remaining sum is spent every year.

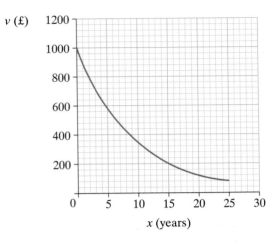

This curve shows exponential decay and is of the form $y = ka^x$ ($0 < a < 1$).

Exercise 5G

1   Suggest an equation for each of the following diagrams.

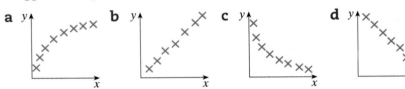

2   A scatter diagram of the following data suggests that an equation
of the form $y = ax^2 + b$ would be a suitable model.

Maximum diameter of apples, x (cm)	1.1	1.7	2.2	2.4	3.3	4.0	5.2	6.4
Surface area, y (cm²)	5	10	17	20	32	53	85	136

a   Plot the data on a scatter diagram.

b   Draw a line of best fit and a curve of the appropriate shape.

c   Does the line of best fit or the curve fit better?

3   The cost per component, y, of setting up tooling for the components'
manufacture, and the number of components produced, x, are
shown in the table below.

Number of components, x	1	2	3	4	5
Cost/component, y (£)	5	4.1	3.6	3.5	3.2

The data in the table are thought to follow the equation $y = \frac{a}{x} + b$.

a   Draw a scatter diagram.

b   Draw a line of best fit and a curve of the suggested shape.

c   Which of the two models is more suitable?

Examiner's Tip
Questions won't always say
whether to draw the line of
best fit through the mean point
or by eye. Make a sensible
choice.

4   In an experiment with a simple pendulum, a large weight was fixed at the end of a thin piece of wire. The length, x (cm), of the wire and the time for one oscillation of the pendulum, y (seconds), were recorded. The results are given in the table.

Length, x (cm)	10	20	30	40	50
Time, y (seconds)	0.6	0.9	1.1	1.3	1.4

   a   Draw a scatter diagram for these data.

   b   Which of the following equations might be a suitable model?

   i) $y = a\sqrt{x} + b$ **ii)** $y = ax^2 + b$

5   The current, y, flowing through two parallel resistors is measured as the resistance, x, of one of the resistors is varied. The result is shown in the table.

Resistance, x (ohm)	1	2	3	4	5
Current, y (amp)	5	4.1	3.5	3.5	3.4

   a   Plot these points on a scatter diagram.

   b   Decide if a curve or a straight line would be the better model.

   c   Give the general equations of two possible curves that might model the association between x and y.

5.8 Spearman's rank correlation coefficient

Ranking

Correlation is a measure of the strength of the linear association between two variables.

> **Spearman's rank correlation coefficient** is a numerical measure of the correlation between two sets of data. It tells us how close the agreement is.

The basic idea behind this coefficient is **ranking**.

Example 10

Two judges were judging a gardening competition. There were six entries in the sweet pea section. The marks, out of 12, awarded by the two judges, x and y, are shown in the table. Rank these data.

Entry	A	B	C	D	E	F
Mark (*x*-value)	1	2	5	8	6	3
Mark (*y*-value)	6	3	8	11	7	4

x-value	*y*-value	*x*-rank	*y*-rank
1	6	6	4
2	3	5	6
5	8	3	2
8	11	1	1
6	7	2	3
3	4	4	5

To rank the observations, assign

- the largest value the rank 1
- the next highest the rank 2
- the third highest the rank 3
- and so on until all observations are ranked.

Alternatively, the smallest value could be ranked 1, the next smallest ranked 2, etc. This makes no difference to the answer provided the same is done for both variables.

Spearman's rank correlation coefficient

To use **Spearman's rank correlation coefficient** both variables must be ranked.

Spearman's rank correlation coefficient is concerned with the difference, *d*, in the ranking of the two sets of data.

d = rank of the *x*-observation − rank of the *y*-observation

Spearman's rank correlation coefficient is given by

$$r_s = 1 - \frac{6 \sum d^2}{n(n^2 - 1)}$$

where *d* is the difference in ranks and *n* is the number of observations.

Examiner's Tip

Remember the square of a negative number is always positive.

This formula is given on the formula sheet in the exam.

Remember that $\sum d^2$ means the sum of all the differences squared.

 A spreadsheet can be used to calculate Spearman's rank correlation coefficient.

Example 11

Work out Spearman's rank correlation coefficient for the two sets of marks in Example 10.

x-value	y-value	x-rank	y-rank	d	d^2
1	6	6	4	+2	4
2	3	5	6	−1	1
5	8	3	2	+1	1
8	11	1	1	0	0
6	7	2	3	−1	1
3	4	4	5	−1	1
					$\sum d^2 = 8$

First rank the two sets of observations, x and y.

Find the difference, d, between each pair of values.

Square the d-values to get d^2.

Sum the d^2-values.

$n = 6$

$r_s = 1 - \dfrac{6\sum d^2}{n(n^2 - 1)}$

$= 1 - \dfrac{6 \times 8}{6(36 - 1)}$

$= 1 - 0.2286$

$= 0.7714$

Count the number of observations, n.

Put these values in the formula and work out the value for r_s.

Spearman's rank correlation coefficient for these marks is 0.7714.

Tied ranks

There may be two or more observations that are equal in value. These are called **tied** values. When two or more values are tied they are each given the mean value of the ranks that they would have had if they were not tied.

Consider the numbers 93, 87, 74, 74, 72, 60, 45.

These two values are tied

The two numbers 74 would have been ranked 3 and 4, so they are both given the rank $\dfrac{3 + 4}{2} = 3.5$.

The numbers and their ranks are shown in the table.

Number	93	87	74	74	72	60	45
Rank	1	2	3.5	3.5	5	6	7

Examiner's Tip

Exam questions will not contain tied ranks but they may occur in the controlled assessment.

The value given by the Spearman formula will not be very reliable if there are many sets of tied ranks.

Exercise 5H

1 The following marks were given to 11 candidates taking a driving test exam. Rank these data.

49   44   43   36   40   39   29   28   30   33   26

2 A factory gave eight of their apprentices a practical test before and after a short course. The marks gained before the course, y, and those gained after the course, x, are given in the table. Separately rank each set of marks.

Marks after, x	12	22	40	33	18	25	14	4
Marks before, y	10	30	45	12	28	18	19	4

3 In an investigation into smoking, eight patients were assessed for the amount of lung damage they had suffered, x, and were asked how many years they had been smoking, y. Copy the table and replace each of the numbers by their separate rank.

Lung damage, x	42	56	12	6	83	44	58	23
Years smoked, y	16	18	14	19	22	13	17	25

4 Calculate Spearman's rank correlation coefficient for each of the following.
  a $n = 12, \sum d^2 = 18$
  b $n = 6, \sum d^2 = 14$
  c $n = 8, \sum d^2 = 100$

5 Apples were ranked for flavour (rank 1) and juiciness (rank 2). The ranks are shown in the table. Copy and complete the table and then calculate r_s.

Rank 1	Rank 2	d	d^2
1	5		
2	4		
3	1		
4	6		
5	3		
6	2		

6 The following table shows the rank orders put on six collie dogs by two judges.

Collie	A	B	C	D	E	F
Judge 1	3	1	6	2	5	4
Judge 2	2	3	6	1	4	5

Work out Spearman's rank correlation coefficient.

ResultsPlus
Watch out!

● When asked to calculate Spearman's rank correlation coefficient for two sets of data, some students thought that $(-4)^2 = 16$. Others used $n(n - 1)$ instead of $n(n^2 - 1)$ in the formula – even though the formula is on the exam formula sheet.

7   The table shows the marks given by two judges to six skaters in an ice skating competition.

Skater	A	B	C	D	E	F
Judge 1	6.5	8.2	9	6	8	7.5
Judge 2	7	8.4	8.6	6	9	6.8

a   Rank the marks of each judge.

b   Work out Spearman's rank correlation coefficient.

5.9 Interpretation of Spearman's rank correlation coefficient

The value of Spearman's rank correlation coefficient, r_s, gives a measure of how well the ranks fit a straight line.

> If r_s is close to 1 there is a strong positive linear correlation.
>
> If r_s is close to -1 there is a strong negative linear correlation.
>
> If r_s is close to zero then there is no linear correlation.

The diagram below shows the scale of values.

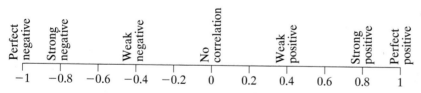

In Example 11, $r_s = 0.7714$ describes a fairly strong positive linear correlation of the ranks.

Example 12

Two experts tried to rank eight antiques in order of value. The results are shown in the table.

Antique	1	2	3	4	5	6	7	8
Rank of expert A	8	2	7	1	3	4	6	5
Rank of expert B	8	2	6	1	4	3	7	5

Calculate Spearman's rank correlation coefficient for these data and comment on the result.

Antique	1	2	3	4	5	6	7	8
Rank of expert A	8	2	7	1	3	4	6	5
Rank of expert B	8	2	6	1	4	3	7	5
d	0	0	1	0	−1	1	−1	0
d^2	0	0	1	0	1	1	1	0

$n = 8$

$\sum d^2 = 4$

$r_s = 1 - \dfrac{6\sum d^2}{n(n^2 - 1)}$

$ = 1 - \dfrac{6 \times 4}{8(64 - 1)}$

$ = 1 - 0.0476$

$ = 0.9524$

There is a strong positive correlation between the rankings of the two experts. The two experts are in agreement.

> The ranking has already been done, so add two rows to the table for d and d^2.

> Fill in the two rows.

> Find n and $\sum d^2$.

> Put these into the formula for Spearman's rank correlation coefficient.

> Comment on the correlation. Remember to put it in the context of the question.

Examiner's Tip
Remember that the formula is

$r_s = 1 - \dfrac{6\sum d^2}{n(n^2 - 1)}$

and NOT $r_s = \dfrac{1 - 6\sum d^2}{n(n^2 - 1)}$

Results Plus
Exam Question Report

Question: At one location on a river, the distance from the bank and the corresponding depth of water were measured.
The table shows the results.

Distance from bank (cm)	0	50	150	200	250	300	350	400	450	500
Depth (cm)	0	10	26	44	59	52	74	104	85	96
Rank of distance	1	2	3	4	5	6	7	8	9	10
d										
d^2										

(a) Complete the table.   (2 marks)

(b) Calculate Spearman's rank correlation coefficient for these data.
                                                            (2 marks)

(c) Interpret your answer to (b)   (2 marks)

                                                    (Total 6 marks)

Poor   37%

Some students reversed the ranks in part (a), and some squared and got negative numbers.

Good   24%

Most of these students correctly completed the table in part (a). Some students forgot the 1 in the Spearman's rank correlation coefficient formula, or put it at the top of the fraction. Some made errors in their calculation and got an answer outside the range −1 to +1. They did not seem to realise that this was an incorrect value and went on to try and interpret it in part (c).

Excellent   39%

These students calculated the values in parts (a) and (b). In part (c) most candidates got one mark for identifying positive correlation, but not many interpreted the result in context to gain full marks. Vague answers such as 'there is an association between the two' were fairly common.

Exercise 5I

1   Two teachers were asked to rank 10 students in a fancy dress competition. The ranks they gave are shown in the table.

Child	A	B	C	D	E	F	G	H	I	J
Teacher 1	1	9	3	6	8	2	7	4	10	5
Teacher 2	8	4	7	1	3	5	6	9	10	2

Calculate a value for Spearman's rank correlation coefficient, and comment on its value.

2   The eight photographers who entered a photographic competition were ranked by each of two judges. The Spearman's rank correlation coefficient value for the rankings was −0.92. Comment on the agreement between the two judges.

3   In a song contest six songs were given marks out of 10 by two judges. The marks awarded are shown in the table.

Song	1	2	3	4	5	6
Judge A	3	8	9	2	7	5
Judge B	2	6	9	3	8	7

a   Rank the two sets of marks and calculate Spearman's rank correlation coefficient.

b   Comment on the degree of agreement between the two judges.

4   Students in Year 11 at a school can choose to study from a selection of ten subjects. The numbers of boys and girls in each of the subject classes are shown.

Subject	Art	Bio	Chem	Eco	Eng	French	Geog	Hist	Math	Phy
Girls	17	18	10	21	22	25	23	19	20	2
Boys	15	10	22	20	42	19	16	14	17	12

a   Work out Spearman's rank correlation coefficient for these data.

b   Describe the correlation found in part **a**.

c   Interpret you answer to part **a** in context.

> **Examiner's Tip**
> Remember to rank first.

5   In a cookery competition two tasters gave each of eight dishes a mark out of 50. The results are shown in this table.

Dish	A	B	C	D	E	F	G	H
Taster 1	29	35	40	38	34	47	28	36
Taster 2	29	35	45	38	26	40	28	30

a   Calculate Spearman's rank correlation coefficient.

b   Describe and interpret the correlation

6 Gemma believes that people with long surnames will have been given short first names. She counts the number of letters in the first names and in the surnames of 10 children chosen at random from the register. The table shows the results.

Person	1	2	3	4	5	6	7	8	9	10
Length of first name	5	9	8	7	11	12	6	10	4	3
Length of surname	5	11	12	4	14	7	3	9	6	10

a Calculate Spearman's rank correlation coefficient for these data.

b How true is Gemma's belief? Give a reason for your answer.

7 The table shows the initial weights, x (g), of eight insects just after hatching and their weights, y (g), after feeding for 25 days.

Insect	1	2	3	4	5	6	7	8
Initial weight, x (g)	0.74	0.77	0.79	0.83	0.9	0.96	0.98	1.10
Final weight, y (g)	0.85	0.90	0.95	1.00	0.98	1.10	1.07	1.30

a Calculate Spearman's rank correlation coefficient for these data.

b What do you conclude about the final weight in relation to the original weight?

Chapter 5 review

1 A small electrical shop recorded the yearly sales of radios (y) and televisions (x) over a period of 10 years. The results are shown in the table below.

Year	1	2	3	4	5	6	7	8	9	10
Number of television sold (x)	60	68	73	80	85	88	90	96	105	110
Number of radios sold (y)	80	60	72	65	60	55	52	44	42	36

a Draw a scatter diagram for these data. Use a scale from 50 to 120 for the sales of televisions and 30 to 90 for the sales of radios.

b What sort of correlation does the scatter diagram suggest?

c Is there a causal relationship between the television sales and radio sales?

2 In a woodland, the number of breeding pairs of blackbirds (x) and the mean numbers of young blackbirds raised by each breeding pair (y) were recorded for 12 consecutive years. The table below shows the results.

Year	1	2	3	4	5	6	7	8	9	10	11	12
Number of breeding pairs (x)	45	46	60	60	63	40	42	65	55	48	95	55
Mean number of young (y)	6	7	5.5	4	1.8	4.5	6	5.2	4	4.5	1.3	2.5

It is suggested that there could be a negative correlation between the number of breeding pairs and the average number of young that were raised per breeding pair.

a Draw a scatter diagram for these data.

b How strong is the correlation?

c Draw a line of best fit by eye.

d Use your line of best fit to estimate the average number of young blackbirds per breeding pair if there were 50 breeding pairs.

3 It has been said that when the stock market in America goes up or down the stock market in London goes up and down with it. The table below gives the values of the stock market indices in America (Dow-Jones) and London (FTSE 100) at the end of eight consecutive weeks.

Week	1	2	3	4	5	6	7	8
American index	10 600	10 400	10 100	10 200	9900	9750	10 000	10 010
London index	5800	5650	5500	5700	5600	5620	5720	5730

a Draw a scatter diagram for these data.

b Calculate the mean point and draw a line of best fit through it.

c Describe the correlation shown by the scatter diagram.

d The American stock market index was 10 500. What do you estimate the London stock market index to be?

e The American stock market index went up to 12 500. Would you expect any estimate you make for the London stock market to be totally reliable? Give a reason for your answer.

f Calculate the equation of the line of best fit in the form $y = ax + b$.

Statistics

4 The number of mites found in wheat grain was assessed every two days over a 12-day period. The results are shown in the table.

Day	0	2	4	6	8	10	12
Number of mites	100	180	300	550	900	1400	1900

  a Draw a scatter diagram for these data. ☆☆☆☆☆ *challenge*

  b Suggest a suitable model for the relationship between these data. ☆☆☆☆☆ *challenge*

5 The numbers of people (in thousands) engaged in the manufacture of motor vehicles and in the manufacture of accessories for motor vehicles over a period of nine years are shown in the table below.

Year	1	2	3	4	5	6	7	8	9
Manufacture of motor vehicles (1000s)	280	290	297	300	310	295	311	330	320
Manufacture of accessories (1000s)	90	95	110	125	140	145	155	170	175

  a Rank these data. ☆☆☆☆★ *challenge*

  b Calculate Spearman's correlation coefficient for these data. ☆☆☆★★ *challenge*

  c Comment on the strength of the correlation. Interpret the correlation in context. ☆☆☆☆★ *challenge*

6 Table 1 shows seven countries, selected at random, their Human Development Index (HDI – a measure of their quality of life) and their Gross National Product per person (GNP – a measure of their wealth).

Table 1

Country	Niger	Rwanda	India	Oman	China	Cuba	UK
HDI	0.116	0.304	0.439	0.535	0.716	0.877	0.970
GNP	20	26	25	93	22	66	113

Table 2 shows the countries ranked in descending order for their HDI.

Table 2

Country	Niger	Rwanda	India	Oman	China	Cuba	UK
HDI rank	7	6	5	4	3	2	1
GNP rank							
Difference in ranks (d)							
d^2							

  a Copy and complete Table 2. ☆☆☆★★ *challenge*

  b Use the information in Table 2 to calculate Spearman's rank correlation coefficient for these data. Give your answer to three decimal places. ☆☆☆☆★ *challenge*

  c Interpret your answer to part **b**.  ☆☆☆☆★ *challenge*

Chapter 5 summary

Scatter diagrams

1 If the points on a **scatter diagram** lie approximately on a straight line, there is a linear relationship between them. They are **associated**.

Correlation

2 **Correlation** is a measure of the strength of the linear association between two variables.

3 **Negative correlation** is when one variable decreases as the other increases.

4 **Positive correlation** is when one variable increases as the other increases.

5 **Association** does not necessarily mean there is a linear correlation.

Causal relationships

6 When a change in one variable directly causes a change in another variable, there is a **causal relationship** between them.

7 Correlation does not necessarily imply a causal relationship.

Line of best fit

8 Points on a scatter diagram are strongly correlated if they lie along a straight line.

9 A **line of best fit** is a straight line drawn so that the plotted points on a scatter diagram are evenly scattered either side of the line.

10 A line of best fit is a **model** for the association between the two variables.

11 A line of best fit should pass through the **mean point** $(\bar{x}, \bar{y})$.

12 **Interpolation** is when the data point you need lies within the range of the given values.

13 **Extrapolation** is when the data point you need lies outside the range of the given values.

14 The **equation of a line** $y = ax + b$ has a **gradient** a, and its **intercept** on the y-axis is $(0, b)$.

15 The value of the constants in the equation of a line of best fit are calculated using

$$a = \frac{y_2 - y_1}{x_2 - x_1} \text{ and } b = y_1 - ax_1 \text{ or } b = y_2 - ax_2.$$

Non-linear models

16 The graphs of **non-linear models** of the form $y = ax^n + b$ look like these.

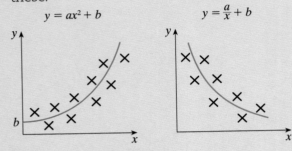

Spearman's rank correlation coefficient

17 **Spearman's rank correlation coefficient** is a numerical measure of the correlation between two sets of data. It tells us how close the agreement is.

18 Spearman's rank correlation coefficient is given by

$$r_s = 1 - \frac{6\sum d^2}{n(n^2 - 1)},$$

where d is the difference in ranks and n is the number of observations.

Interpretation of Spearman's rank correlation coefficient

19 If r_s is close to 1 there is a strong positive linear correlation.

20 If r_s is close to -1 there is a strong negative linear correlation.

21 If r_s is close to zero then there is no linear correlation.

Test yourself

1 The table gives information on the age (years) and price (£) of a
particular electronic item.

Age (years)	1	2	3	4	5	6
Price (£)	4500	4000	3450	3000	2550	2000

Find the coordinates of the mean point.

2 A scatter diagram is drawn for the data in question **1**.

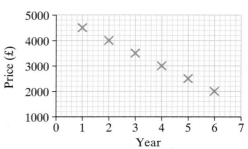

  a Copy the diagram and mark in the mean point.

  b Draw a line of best fit passing through the mean point.

  c Use your diagram to find the value, to the nearest £100,
      of an item 4.5 years old.

  d Use your diagram to find the value, to the nearest £100,
      of an item 7 years old.

  e Is answer **c** or **d** more reliable? Give a reason for your answer.

  f Is there a causal relationship between these two variables?
      Explain your answer.

3 The mass, x, in grams, put on a spring and the length, y, of the
spring, in millimetres, was measured for a number of different
masses. An equation in the form $y = ax + b$ was found for the
relationship between the mass and the length of the spring. The
equation was $y = 40 + 0.2x$.

Write down the contextual meaning of the values 40 and 0.2.

4 In a training scheme for young people the age of six of them and
the time each took to reach a certain level of proficiency was
measured. These data are shown in the table.

Age (years)	16	17	18	19	20	21
Time (hours)	22	18	19	16	15	12

  a Work out Spearman's rank coefficient for these data.

  b Describe and interpret this correlation coefficient.

Higher
Statistics

Chapter 6:
Time series

After completing this chapter you should be able to

- draw line graphs
- plot points as a time series graph
- draw a trend line by eye
- use a trend line to make a prediction
- calculate and plot appropriate moving averages
- identify seasonal variations by eye

- work out mean seasonal variations
- draw a trend line based on moving averages
- recognise seasonal effects at a given point and mean seasonal effect
- interpret time series graphs

The price you pay to a travel agent for a holiday is different at different times of the year.

Is this true? When are the cheapest and most expensive times of the year?

Time series graphs show observations taken at intervals. They make it possible to answer questions like these.

6.1 Line graphs

Graphs that look like this are often shown in magazines and newspapers.

This is an example of a **line graph**.

No reliable information can be found from the graph for values lying between the points plotted. For example, on this graph, there could be a tea break at 11.30 reducing the production to zero for a short time.

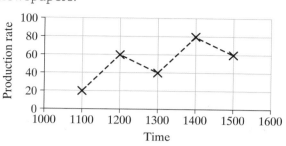

Straight, dotted lines are therefore used to join the points.

> A **line graph** is used to display data when the two variables are not related by an equation and it is uncertain what happens between the plotted points.

It is important to choose scales that fit the graph paper being used when plotting a graph.

Example 1

This table shows the monthly rainfall at a seaside town last year.

Month	Jan	Feb	Mar	Apr	May	Jun	Jul	Aug	Sep	Oct	Nov	Dec
Rainfall (cm)	18	21	12	16	11	9	6	10	7	15	18	23

Show these data on a line graph and comment on it.

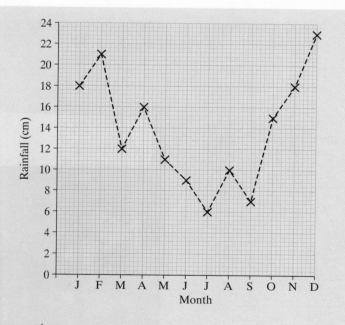

Choose a suitable scale for each axis and label them clearly.

Plot the points given in the table.

Join the points with dotted lines.

It can be seen that

- the highest rainfall was in December
- the lowest rainfall was in July
- rainfall increases in the autumn.

Look to see the highest and lowest values and if there is any pattern.

Plotting a graph

Example 2

This table shows the temperatures in degrees Celsius at different times of a day in March. Draw a line graph of these data.

Time	0900	1000	1100	1200	1300	1400	1500
Temperature (°C)	10	11	13	14	15	12	11

Comment on the graph.

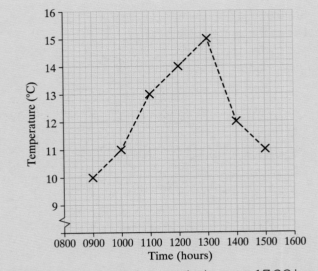

Look at the temperature values and decide on a suitable scale.

The temperature values are in the range 10°C to 15°C. The vertical axis could go from 9°C to 16°C.

This larger scale makes it easier to plot and read the values.

It can be see that the temperature was highest at 1300 hours when it was 15°C.

Comment on what the graph shows.

Exercise 6A

1   The graph shows the number of hours Joan watched television on each day of one week.

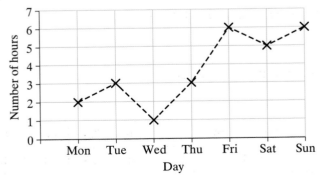

a   For how many hours did Joan watch television on Tuesday?

b   On which day did Joan watch least television?

c   Suggest a reason why Joan watched television more on the last three days of the week.

2   The table shows the amount of money in Wing's bank account at the end of each month during the course of a year.

Month	Jan	Feb	Mar	Apr	May	Jun	Jul	Aug	Sep	Oct	Nov	Dec
Money (£)	50	65	78	84	100	96	84	24	38	43	78	48

   a   Draw a line graph of these data.

   b   There were two months when Wing had large bills to pay. Which months do you think these were?

   c   At the end of which month did Wing have the greatest amount of money in his bank account?

3   The table shows the numbers, in thousands, of people who died from flu over a ten-year period.

Year, x	1999	2000	2001	2002	2003	2004	2005	2006	2007	2008
Number of deaths, y (1000s)	3.8	1.5	4.6	3.4	16.0	3.2	6.8	3.4	4.2	3.0

   a   Draw a line graph of these data. Start the x-axis at 1999, and the y-axis at 0.

   b   What happened between 2003 and 2005?

4   The table shows the wind speed in Eskdale during one day in May.

Time	03 00	06 00	09 00	12 00	15 00	18 00	21 00	24 00
Wind speed (mph)	10	15	20	25	20	25	30	35

   a   Draw a line graph of these data.

   b   Describe, in words, what happened to the wind speed between 03 00 and 24 00.

5   The table below shows the monthly profits made by a market stall selling vegetables.

Month	Jan	Feb	Mar	Apr	May	Jun	Jul	Aug	Sep	Oct	Nov	Dec
Profit (£100s)	10	11	12	14	12	15	20	24	23	16	10	18

   a   Draw a line graph of these data.

   b   At what time of the year are profits at their highest? Suggest a reason for this.

6.2 Time series

In weather stations, instruments measure temperature and rainfall. Readings are usually taken at a set time each day. These readings can be used to produce monthly rainfall figures.

Sets of observations of variables taken over a period of time are called **time series**.

In most cases the observations are made at equal time intervals.

> A **time series** is a set of observations taken over a period of time.
>
> A line graph can be used to show a time series. When plotting a time series, time is plotted on the horizontal axis.

Example 3

The table shows the quarterly sales of ice cream over a period of time.

Year	2005		2006				2007				2008			
Quarter	3	4	1	2	3	4	1	2	3	4	1	2	3	4
Sales (£m)	3.6	1.4	2.0	3.6	4.6	2.6	3.0	4.6	5.6	3.4	3.6	5.2	6.3	4.6

a   Draw a time series graph for these data.

b   Comment on the graph.

a
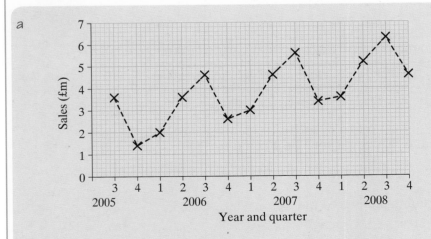

b   You can see that:

- the sales of ice cream are not the same throughout the year

- the sales are lowest in Quarter 4 of a year. Sales then increase in Quarter 1, reaching a peak in Quarter 3 before dropping back in Quarter 4 in each year.

- Sales seem to be generally increasing each year.

Exercise 6B

1 Which of the following could be shown on a time series graph? Explain your answer.

   a Sales of cold drinks.

   b Numbers of pets owned by a group of children.

   c Hours of sunshine.

2 The table shows the quarterly rainfall figures for a town in the centre of Great Britain.

Year	2005		2006				2007				2008			
Quarter	3	4	1	2	3	4	1	2	3	4	1	2	3	4
Rainfall (cm)	8	21	26	14	8	20	24	13	4	19	19	11	3	20

   a Draw a time series graph for these data.

   b Comment on the graph.

> **Examiner's Tip**
> You could comment on highest and lowest rainfall.

3 The quarterly sales of computer games in a shop are shown below.

Year	2005	2006				2007				2008			
Quarter	4	1	2	3	4	1	2	3	4	1	2	3	4
Sales (100s)	14	2	3	3	12	3	4	5	10	4	5	6	9

   a Draw a time series graph for these data.

   b Comment on the graph.

4 The number of nurses at a large hospital who resigned from work was recorded. The result for a three-year period is shown below.

Year	2006				2007				2008			
Quarter	1	2	3	4	1	2	3	4	1	2	3	4
Number resigning	28	20	26	20	34	30	31	26	42	34	32	34

   a Draw a time series graph for these data.

   b In which quarter do most nurses resign?

   c Did nurses tend to resign more in 2008 than in 2006?

5 A factory manager looked at the monthly profits on a particular manufacturing process last year. The figures are shown in the table.

Month	Jan	Feb	Mar	Apr	May	Jun	Jul	Aug	Sep	Oct	Nov	Dec
Profit (£10 000s)	14	16	15	15	17	16	15	18	19	18	19	20

   a Draw a time series graph for these data.

   b The manager thinks profits are falling. Comment on his thoughts.

6.3 Trend lines

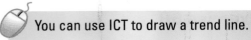

You can use ICT to draw a trend line.

This graph, from Example 3, shows the sales of ice cream over three and a half years. In Example 3 it was commented that the sales of ice cream seemed to be increasing over this period. A **trend line** can be used to show this more clearly.

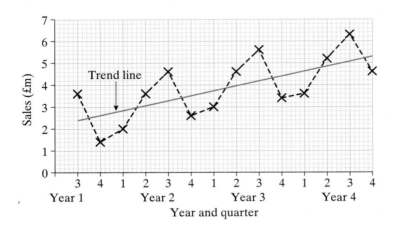

Although the sales seem to increase and decrease in different quarters, the general trend is for sales to rise. The trend line shows clearly that overall sales are rising.

> A **general trend** is the way that the data change over time.
> A **trend line** shows the general trend of the data.

To draw a trend line, put a ruler on the graph and move it until the areas enclosed either side of the ruler are equal. This will be roughly half way between the highest and lowest point for each year.

ResultsPlus
Watch out!

Some students describe trends incorrectly as 'positive' or 'negative'. Don't get confused with correlation. Trends can be 'rising', 'falling' or 'level' (no change).

A trend line may show a tendency to rise, to fall or to stay level.

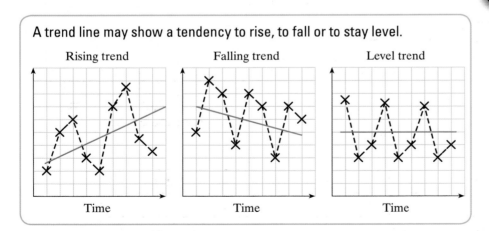

Examiner's Tip
Some students do not recognise that a trend line could be level. They often say that there is no trend.

Exercise 6C

1 Here are three line graphs.

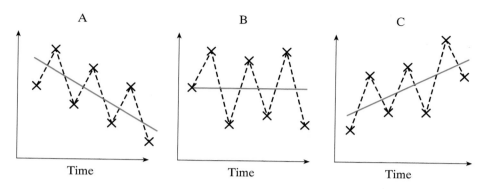

a Which graph shows a rising trend?

b Which graph shows a level trend?

c What trend does graph A show?

2 The table shows the quarterly sales of new cars by a garage over a three-year period.

Year	2006				2007				2008			
Quarter	1	2	3	4	1	2	3	4	1	2	3	4
Sales in 100s	10	3	8	5	14	7	13	9	16	10	18	9

a Copy this time series graph. Complete it by plotting the remaining points. Join them up with a dotted line.

b Draw in a trend line.

c Comment on the trend of the data.

Examiner's Tip
Remember: a trend is either rising, falling or level.

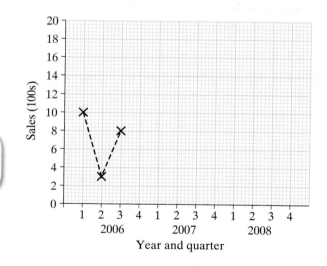

3 A company records its quarterly sales figures. The figures for three years are shown below.

Year	Year 1				Year 2				Year 3			
Quarter	1	2	3	4	1	2	3	4	1	2	3	4
Sales in (£10 000s)	45	48	50	52	40	25	22	45	28	30	25	36

a Draw a time series graph for these data.

b Draw in a trend line.

c Comment on the trend of these data.

4 A tour bus company organises trips to Scotland. They wish to find out if the tours are getting more or less popular so that they can plan for the future. Data for the past two years are shown in the table below.

Year	Year 1						Year 2					
Months	Jan–Feb	Mar–Apr	May–Jun	Jul–Aug	Sep–Oct	Nov–Dec	Jan–Feb	Mar–Apr	May–Jun	Jul–Aug	Sep–Oct	Nov–Dec
Number of people on trip	24	38	40	52	40	22	20	36	30	46	30	20

a Draw a time series graph for these data.

b Draw in a trend line.

c Comment on the trend of the data.

> **Examiner's Tip**
> Are the tours getting more or less popular?

6.4 Variations in a time series

A time series may show variations in its pattern.

> Variations in a time series may be
> - a general trend (as shown by the trend line)
> - seasonal variations (a pattern that repeats).

Here is the ice cream sales graph from Example 3 again.

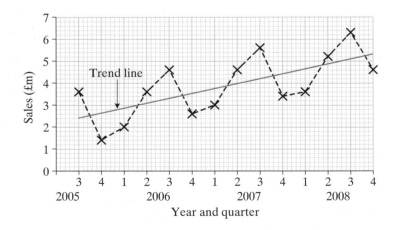

> **ResultsPlus**
> **Watch out!**
>
> ■ When asked to describe this type of variation on a graph, most students didn't identify it as 'seasonal variation'.

Although the general trend for the sale of ice cream is upwards, the sales in the first and fourth quarters of the year are less than the trend would suggest, while those in the second and third quarters are above the trend.

This is because ice cream sells better in the warmer summer months and less well in the colder winter months.

These variations are due to the seasons of the year. This four season cycle repeats itself each year.

The size of a seasonal variation is the difference between its actual value and the trend value. Variations above the trend will be positive. Those below the trend will be negative.

Seasonal variations do not always correspond with the four seasons of the year. They could, for example, be days of the week. The number of hours a person watches television could be related to the days of the week, peak viewing being on Saturdays or Sundays when they are at home all day. This would give a seven season weekly cycle. Each season would be a different day.

Example 4

This table shows quarterly figures over three years for the number of unemployed builders.

Year	2006				2007				2008			
Quarter	1	2	3	4	1	2	3	4	1	2	3	4
Unemployed builders (1000s)	20	13	13	17	19	12	12	14	18	9	9	12

a Draw a time series graph of these data.

b Draw a trend line on the graph.

c Describe the variations shown in the graph. Suggest a reason why these seasonal variations take place.

a and b

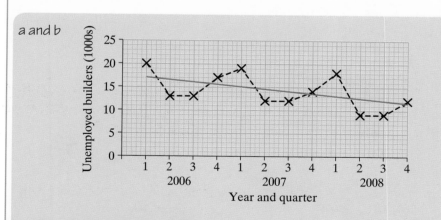

Draw the time series graph in the usual way, and plot the points.

Draw the trend line so that the points are equally distributed either side of it.

c The graph shows a falling trend – the number of unemployed builders tends to fall each year. There is a seasonal variation with unemployment being higher than the trend value in the first quarters and lower than the trend value in the second and third quarters. More builders are unemployed in the winter when weather conditions make building work difficult.

Look to see what the trend line is doing.

Always relate the answer to the context of the problem.

Exercise 6D

1 Which of the following are likely to show seasonal variations? Explain your answer.

    a The number of bank accounts opened.

    b The number of hours of sunshine.

    c The sales of swimsuits.

    d The sales of breakfast cereals.

2 The graph below shows the quarterly sales of hot dogs from a market stall during three consecutive years.

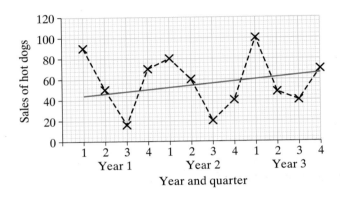

    a The sales seem to go up and down about the trend line. Give a reason for this.

    b In which quarter of the year are the sales highest?

3 The line graph below shows the number of people going on a particular coach tour in each quarter of the years 2005 to 2008.

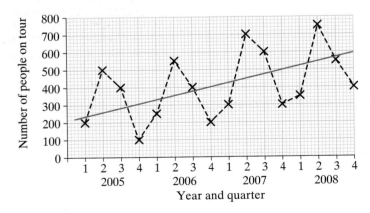

    a How many people went on the tour during the second quarter of 2005?

    b How many people went on the tour during the whole of 2005?

    c Describe the general trend of the data.

    d Comment on the seasonal variations.

4 The table shows the quarterly sales of hot baked potatoes at a stall in a busy market square over 3 years.

Year	2006				2007				2008			
Quarter	1	2	3	4	1	2	3	4	1	2	3	4
Sales (£1000s)	14	10	2	8	9	7	4	12	11	6	3	11

a Draw a time series graph for these data.

b Draw in a trend line.

c Comment on the trend and seasonal variations.

6.5 Moving averages

Time series may show large variations with upward and downward peaks. This can make it difficult to see the trend of the data and to draw a trend line.

A good way of seeing the trend is to use **moving averages**.

Moving averages are the averages of successive observations of a time series.

The number of points in each moving average should cover one complete cycle of seasons. This makes sure that each moving average contains one reading from each season.

The averaging may be done over 3, 4, 5, etc. readings and are known as three-point, four-point, five-point, etc. moving averages.

Examiner's Tip
There will be a maximum of five points in a foundation exam, and a maximum of seven points in a higher exam.

A **moving average** is an average worked out for a given number of successive observations.

The number of points in each moving average should cover one complete cycle of seasons.

Moving averages are plotted on the time series graph to help show the trend.

They are plotted at the mid-point of the time intervals they cover.

The points plotted for moving averages should not be joined up.

Example 5

Quarterly car sales over a two-year period are shown in this table.

Quarter	1	2	3	4	5	6	7	8
Cars sold/quarter	16	26	30	24	28	36	35	42

a Find the four-point moving averages, and plot them on a time-series graph.

b Comment on the trend of the moving averages.

a The first point is $\dfrac{16 + 26 + 30 + 24}{4} = 24$.

The second point is $\dfrac{26 + 30 + 24 + 28}{4} = 27$.

The third point is $\dfrac{30 + 24 + 28 + 36}{4} = 29.5$.

The fourth point is 30.75 and the fifth point is 35.25.

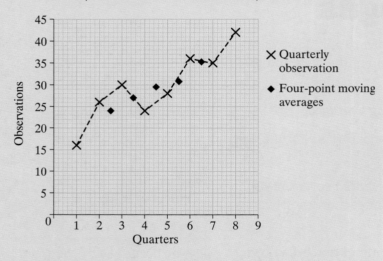

Point 1 is the mean of the values for quarters 1 to 4 inclusive.

Point 2 is the mean of the values for quarters 2 to 5 inclusive.

Point 3 is the mean of the values for quarters 3 to 6 inclusive.

Calculate the next two moving averages in the same way.

The eight points produce five moving averages.

Plot the first moving average point at the mid-point of quarters 1, 2, 3 and 4 (i.e. at 2.5).

Plot the second moving average point at the mid-point of 2, 3, 4 and 5 (i.e. at 3.5).

Plot the third moving average point at the mid-point of 3, 4, 5 and 6 (i.e. at 4.5) and so on.

b The moving averages clearly show an upward trend. Over the two years the sales have risen.

Comment on the trend. Remember to put the trend in the context of the question.

Example 6

The table shows a shopkeeper's takings (in £1000s) in each quarter of three successive years.

a Draw a time series graph to illustrate these data. Show the four-point moving averages on the same graph.

b Give the reason why a four-point moving average is used in this case.

c What conclusion may be drawn from the graph?

Year	Quarter			
	1st	2nd	3rd	4th
1	15	25	53	24
2	15	27	59	26
3	17	28	60	25

a

Year	Quarter	Takings	Four-point moving average
1	1	15	
	2	25	(15 + 25 + 53 + 24)/4 = 29.25
	3	53	(25 + 53 + 24 + 15)/4 = 29.25
	4	24	(53 + 24 + 15 + 27)/4 = 29.75
2	1	15	(24 + 15 + 27 + 59)/4 = 31.25
	2	27	(15 + 27 + 59 + 26)/4 = 31.75
	3	59	(27 + 59 + 26 + 17)/4 = 32.25
	4	26	(59 + 26 + 17 + 28)/4 = 32.50
3	1	17	(26 + 17 + 28 + 60)/4 = 32.75
	2	28	(17 + 28 + 60 + 25)/4 = 32.50
	3	60	
	4	25	

Calculating the moving averages is best done by making a table.

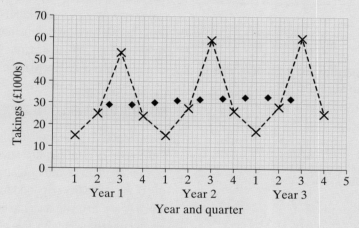

Plot the time series and the four-point moving averages.

✗ Quarterly observations
◆ Four-point moving averages

b A four-point moving average is used because the seasonal changes take place over four quarters.

Explain why.

c The general trend is slightly upwards. A seasonal variation is present. The takings are above the trend in the third quarter of the year and below trend in the first quarter.

Look at the general trend of the moving averages.

Remark on the seasonal changes.

Exam Question Report

Question: The table gives information about the numbers of cars made in the UK for each four-month period from 2003 to 2005, to the nearest ten thousand.

Year	Period		Number of cars made (nearest ten thousand)	Three-point moving average
2003	1	Jan–Apr	57	
	2	May–Aug	53	55.3
	3	Sep–Dec	56	55.7
2004	1	Jan–Apr	58	55.3
	2	May–Aug	52	55.0
	3	Sep–Dec	55	54.7
2005	1	Jan–Apr	57	
	2	May–Aug	50	
	3	Sep–Dec	52	

(a) (i) Complete the table to show the two missing three-point averages.
    (ii) Plot the moving averages on this time series graph.

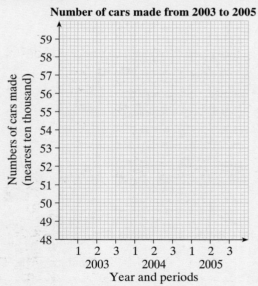

Number of cars made from 2003 to 2005

(4 marks)

(b) What do the moving averages show about the trend in the number of cars made in the UK from 2004 to 2005?   (1 mark)

The time series graph shows there are seasonal variations in the numbers of cars made.

(c) (i) Write down the priod when there are fewer cars made than the general trend.
    (ii) Suggest a reason why.   (2 marks)
                                                           (Total 7 marks)

Poor   49%

Only about a quarter of students calculated the moving averages correctly in the table in part (a). Students who calculated them wrongly, but then plotted three values from the table accurately on the graph, still gained a mark.

Good   13%

Most students realised that moving averages needed to be plotted at the mid-points (at 2, 3, 1, etc.) though some started at 1. In part b) most students were able to describe the trend as 'falling' or 'decreasing'.

Excellent   38%

Some students answered (c) (i) incorrectly by referring to a specific period in the table, rather than the common seasonal period for these years. In (c) (ii) only the best candidates were able to give a sensible reason why fewer cars were made in this period. A good answer was 'workers on holiday, so fewer made'. A poor answer that didn't gain marks was 'summer so people like to walk'.

Drawing a trend line through moving averages

A trend line may be drawn through the moving averages by eye.

It will be more accurate as a measure of trend than one drawn using the original data.

The diagram shows the line graph for Example 6 with a trend line added for the moving averages.

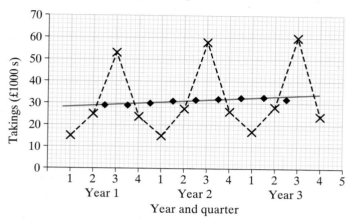

> **Examiner's Tip**
> You will not be asked to draw your trend line through a mean point.

> **Note:** A trend line should not be drawn by joining the points. The trend line is a straight line with roughly equal numbers of points above and below the line.

The trend line helps you to see more easily that the general trend is slightly upwards. A seasonal variation is present. The takings are above the trend in the third quarter of the year and below trend in the first quarter.

Exercise 6E

1 This table gives the monthly values of a company's production, in thousands of items, from January 2007 to February 2008. Calculate the three-monthly moving averages.

Month	Jan	Feb	Mar	Apr	May	Jun	Jul
Production	108	83	100	108	111	115	106

Month	Aug	Sep	Oct	Nov	Dec	Jan	Feb
Production	103	118	119	124	114	120	125

2 a Explain what is meant by a four-point moving average.

This table gives the numbers (in millions of pounds) of exports from Great Britain to a continental country between July 2007 and June 2008.

Month	Jul	Aug	Sep	Oct	Nov	Dec
Exports	9.8	10.2	9.7	11.4	10.5	10.7

Month	Jan	Feb	Mar	Apr	May	Jun
Exports	11.4	10.9	12.6	11.3	13.3	11.8

ResultsPlus
Watch out!

■ Many students could not calculate four-point moving averages for a set of data and plot them on a graph. Very few knew that the moving averages should be plotted at the mid-points of the time intervals they cover.

b Draw a time series graph for these data.

c Calculate the three-point moving averages and add these to your graph.

3 This table shows the number of new houses completed in a city in twelve consecutive quarters.

Year	Quarter	New houses	Moving average
1	1	260	
	2	285	252.50
	3	200	247.50
	4	265	238.75
2	1	240	256.25
	2	250	260.00
	3	270	270.00
	4	280	275.00
3	1	280	
	2	270	
	3	290	
	4	295	

a Draw a time series graph of these data.

b Calculate the last two moving averages and plot all the moving averages on the graph.

c Describe the trend.

4 This table gives the annual circulation of library books (in thousands) for the library in a small town.

a Draw a line graph to illustrate these figures.

b Calculate the last three three-point moving averages and plot the moving averages on the graph.

c Comment on the trend.

Year	Circulation (1000s)	Moving average
1997	14.5	
1998	16.5	15.0
1999	14.0	15.0
2000	14.5	15.5
2001	18.0	16.0
2002	15.5	16.0
2003	14.5	15.0
2004	15.0	15.5
2005	17.0	
2006	16.0	
2007	19.5	
2008	18.5	

5 The half-yearly profits, in £1000s, made by a computer shop are shown in this table.

Year	Months	Profit (£1000s)	Moving average
1	Jan–Jun	18	
	Jul–Dec	22	
2	Jan–Jun	18	
	Jul–Dec	26	
3	Jan–Jun	22	
	Jul–Dec	26	
4	Jan–Jun	24	
	Jul–Dec	28	
5	Jan–Jun	26	
	Jul–Dec	34	

a Plot these data on a time series graph.

b Calculate appropriate moving averages and plot them on the graph.

c Draw a trend line by using the moving averages. Comment on the trend.

6 The average weekly sales (in £1000s) for two competing firms over a three-year period are shown in this table.

Year	2006			2007			2008		
Months	Jan–Apr	May–Aug	Sep–Dec	Jan–Apr	May–Aug	Sep–Dec	Jan–Apr	May–Aug	Sep–Dec
Firms A	280	260	310	480	450	530	730	710	740
B	480	360	420	500	430	480	640	520	550

a Plot both sets of data on the same time series graph.

b Calculate appropriate moving averages for both firms.

c Plot the moving averages on your graph.

d Draw trend lines for the moving averages for both firms.

e Use your trend lines to estimate in which period of time the sales of firm A first equalled the sales of firm B.

7 The table gives the monthly totals (in 100s) of new motorcycles bought during 2008.

Month	Jan	Feb	Mar	Apr	May	Jun
Motorcycles	120	90	105	120	85	85

Month	Jul	Aug	Sep	Oct	Nov	Dec
Motorcycles	90	45	50	55	20	25

a Draw a time series graph to illustrate these data.

b Calculate the three-monthly moving averages and add them to the graph.

c Comment on the trend.

6.6 Estimating seasonal variations

The seasonal variation at a specific point is calculated by subtracting the trend value (on the trend line) from the actual value.

> **seasonal variation at a point = actual value − trend value**

Mean seasonal variations

A particular season's variations differ from year to year. An estimate of a particular season's seasonal variation is found by taking the mean value of all the seasonal variations for that season.

estimated mean seasonal variation for any season = mean of all the
seasonal variations for that season

Example 7

The quarterly sales of a company over a three-year period, (in
millions of pounds), are shown in this table.

Year		1				2				3		
Quarter	1	2	3	4	1	2	3	4	1	2	3	4
Sales (£1 000 000s)	14	9	16	21	18	11	21	23	18	15	25	27

a Draw the time series graph and plot the moving averages.

b Calculate the mean seasonal variations.

a

Year	Quarter	Sales (£m)	Four-point moving average
1	1	14	
	2	9	
	3	16	15.00
	4	21	16.00
2	1	18	16.50
	2	11	17.75
	3	21	18.25
	4	23	18.25
3	1	18	19.25
	2	15	20.25
	3	25	21.25
	4	27	

Calculate the four-point moving averages.

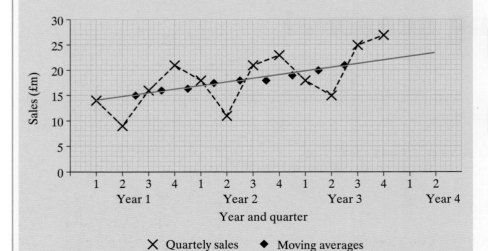

Sales (£m) vs Year and quarter

× Quartely sales   ◆ Moving averages

Draw the time series graph and add the moving averages.

Draw the trend line for the moving averages.

b

Year	Quarter	Actual sales (£m)	Trend (from trend line)	Seasonal variation at a point
1	1	14	14.0	0
	2	9	14.7	−5.7
	3	16	15.5	+0.5
	4	21	16.2	+4.8
2	1	18	16.9	+1.1
	2	11	17.6	−6.6
	3	21	18.4	+2.6
	4	23	19.1	+3.9
3	1	18	19.8	−1.8
	2	15	20.6	−5.6
	3	25	21.3	+3.7
	4	27	22.0	+5.0

> Find the seasonal variations at each plotted point on the trend line, using:
>
> **actual value − trend value**

	Quarter			
Year	1	2	3	4
1	0	−5.7	+0.5	+4.8
2	+1.1	−6.6	+2.6	+3.9
3	−1.8	−5.6	+3.7	+5.0
Totals	−0.7	−17.9	+6.8	+13.7
Estimated mean seasonal variation	$\frac{-0.7}{3} =$ −0.23	$\frac{-17.9}{3} =$ −5.97	$\frac{6.8}{3} =$ +2.27	$\frac{13.7}{3} =$ +4.57

> Find the estimated mean seasonal variations by finding the mean of each season separately.
>
> The estimated mean seasonal variation for the second quarter is −5.97 and it is +2.27 for the third quarter.

6.7 Making predictions

A trend line and the estimated mean seasonal variations can be used to predict the sales figures at some future time.

> **predicted value = trend line value (as read from trend line on graph) + estimated mean seasonal variation**

Examiner's Tip

The trend line often needs to be extended beyond the plotted values in order to make a prediction.

In Example 7 the predicted sales figures for the first quarter of Year 4 would be

predicted value = **trend line value (as read from trend line on graph) + estimated mean seasonal variation**

$$= 22.7 - 0.23$$
$$= 22.47 \text{ million pounds}$$

For the second quarter
Predicted sales $= 23.5 - 5.97$
$\qquad\qquad\quad = 17.53$ million pounds

ResultsPlus
Watch out!

■ Many students forgot that to predict a value they need to add the mean seasonal variation to the trend line value – not to one of the plotted point values. Sometimes the mean seasonal variation is called the 'average seasonal effect'.

ResultsPlus
Build Better Answers

Question: The time series graph shows the values, to the nearest £1000 million of the Total Exports from the United Kingdom between 1997 and 2003.

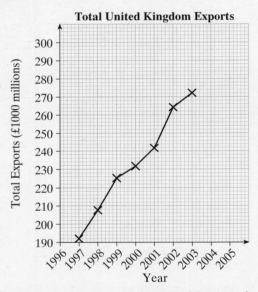

Total United Kingdom Exports

(Data source: *Office for National Statistics*)

(a) Draw a trend line on the time series graph (1 mark)

(b) What does the trend line show about the Total United Kingdom Exports between 1997 and 2003? (1 mark)

(c) Use your trend line to predict the Total United Kingdom Exports in 2004. (1 mark)

(d) Why might this prediction for the Total United Kingdom Exports in 2004 be unreliable? (1 mark)

(e) Does this figure follow the overall trend shown by the trend line on the time series graph?
Give a reason for your answer. (2 marks)

(Total 6 marks)

■ **Basic   1–2 marks**
Draw a trend line on the graph by eye, using a ruler placed so that the areas of the graph enclosed either side of the ruler are equal. Describe the trend as 'rising', and refer back to the context of the problem, for example, 'the trend was for Exports to rise between 1997 and 2003'.

● **Good   3–4 marks**
Draw the trend line and describe what it shows in parts (a) and (b). In part (c) make sure your trend line extends to 2004 and read off the predicted value. In part (e) identify that the figure does not follow the overall trend.

▲ **Excellent   5–6 marks**
Complete parts (a) to (c) accurately. In part (d) comment that predicting into the future is inaccurate, as the past does not necessarily affect the future. You could use the word 'extrapolation', which means extending the trend line outside the plotted values to make predictions. In part (e) identify that the figure does not follow the overall trend and for full marks explain why (is it higher or lower than expected?).

Accuracy of predictions

The accuracy of any prediction will depend upon two things.
- How far into the future the prediction is made. The further into the future the prediction is made the less accurate it will be. (This is extrapolation).
- How good the estimates of the mean seasonal variations are at predicting future seasonal variations. Trends and variations can unexpectedly change.

Exercise 6F

1 Work out the predicted value given a trend line value of £13.39 and a mean seasonal variation of £3.20.

2 This table shows the seasonal variations in the price of lettuce. Copy and complete the table.

Year	Seasonal variation			
	Quarter 1	Quarter 2	Quarter 3	Quarter 4
1	0.0	−6.2	8.1	4.0
2	3.1	−4.0	5.0	3.5
Total				
Mean seasonal variation				

3 This table shows actual and trend values of the sales of mountain bicycles at a small shop. Copy and complete the table.

Year	Quarter	Actual value	Trend	Seasonal variation
1	1	26	22	
	2	38	24	
	3	14	18	
	4	10	14	
2	1	20	19	
	2	32	23	
	3	13	17	
	4	10	10	

4 Use the values in Question **3** to draw up a table and work out the mean seasonal values.

5 This table shows the quarterly profits of a factory (in £1000s) for
the years 2006 to 2008.

Year	Quarter			
	1	2	3	4
2006	30	54	60	40
2007	46	90	92	60
2008	74	122	132	96

The information for the years 2006 and 2007 has been plotted as
a line graph.

The first five four-point moving averages have been plotted on
this graph. They are 46, 50, 59, 67, 72.

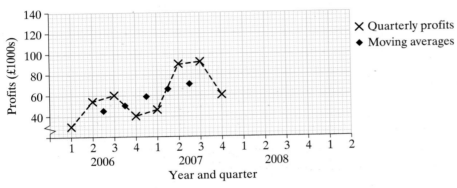

a Copy this graph and complete it by plotting the remaining
points and the moving averages.

b Draw the trend line and extend it to the first quarter of 2009.

c The mean seasonal trend for the first quarter is −14. Estimate
the profit for the first quarter of 2009.

6 This table shows the quarterly electricity costs (in £s) for a one
bedroom flat over a two-year period.

Year	Quarter			
	1	2	3	4
1	180	142	60	110
2	196	146	76	122

a Plot these data on a time series graph.

b Work out the four-point moving averages and plot these on
the time series graph.

c Draw a trend line through the moving averages.

d Work out the mean seasonal variations for the first two
quarters of the year.

e Predict the actual electricity bills for the first two quarters of
Year 3.

7　**a**　Describe what is meant by seasonal change
and give an example of this.
The table shows company A's quarterly sales (in £1000s).

　b　**i)** Plot the sales and moving averages on a time series graph.
　　ii) Add a trend line.

　c　Work out the mean seasonal variations.

　d　Predict the sales for the third and fourth quarters of 2008.

Year	Quarter 1	2	3	4
2004			120	100
2005	86	138	132	92
2006	90	122	128	84
2007	110	122	140	80
2008	118	130		

8　This table shows the number of articles sold over a three-year
period by a manufacturing company.

Year	Quarter 1	2	3	4
2006	686	590	660	720
2007	754	642	732	808
2008	842	738	808	900

The first seven four-point moving averages are 664, 681, 694, 712,
734, 756 and 780.

　a　Work out the last two four-point moving averages.

　b　Represent the data and moving averages on a time series graph.
Use a scale of 1 cm to represent 20 articles on the vertical axis,
and start at 500 on the horizontal axis.

　c　Draw a trend line on the time series graph.

　d　Take readings from your time series graph and draw up a table
to work out the mean seasonal variations.

　e　Use your result in **d** to estimate the first two quarterly figures
for 2009.

6.8 Calculating the equation of the trend line

An equation of the form **y = ax + b** can be fitted to the trend line using
the same method as used in the last chapter (see pp. 187–9).

> The **equation of a trend line** is **y = ax + b**, where **a** is the **gradient** of
> the line and **b** is the **intercept with the y-axis**.

Examiner's Tip
Remember $a = \dfrac{y_2 - y_1}{x_2 - x_1}$ and
$b = y_1 - ax_1$ or $b = y_2 - ax_2$.

Before working out the equation, label the periods on the x-axis 1, 2, 3,
etc. (Measures such as '2002 Quarter 1' are difficult to deal with.)

The constant *a* is the amount by which the trend increases per time period (quarter, two-months, etc).

The constant *b* is the value of the trend at the end of the last period before observations began.

Examiner's Tip
You will be expected to relate the constants *a* and *b* to their meaning in the context of the problem.

Example 8

A factory keeps records of the number of work days lost through illness. The records for two years are summarised in the table. Each figure represents the number of days lost over a two-month period.

Year	Number of days lost					
2007	222	246	168	130	100	184
2008	255	273	198	154	133	187

a i) Draw a time series graph to illustrate these figures.

   ii) Plot the moving averages using an appropriate number of observations.

b i) Draw a trend line on the graph.

   ii) Find the equation of the trend line and give an interpretation of it.

a

Two-monthly period	Days lost	Moving average
1	222	
2	246	
3	168	
		175
4	130	
		180.5
5	100	
		185
6	184	
		190
7	255	
		194
8	273	
		199.5
9	198	
		200
10	154	
11	133	
12	187	

Use a six-point moving average to cover one complete cycle of seasons.

Number the two-monthly periods from 1 to 12.

Calculate the moving averages.

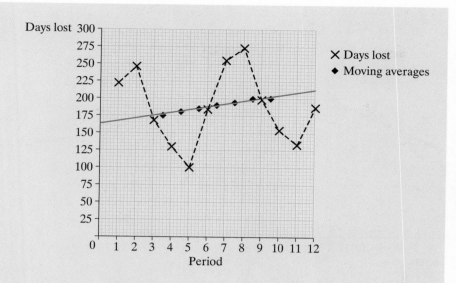

Draw the graph by

- plotting the points given in the table
- joining them up with a dotted line
- plotting the moving averages
- drawing a trend line using the moving averages.

b   If y is the number of days lost and x is the period, the equation will be $y = ax + b$.

From the graph, two points on the trend line are $(x_1, y_1) = (3, 175)$ and $(x_2, y_2) = (9, 200)$.

So $a = \dfrac{y_2 - y_1}{x_2 - x_1}$

$= \dfrac{200 - 175}{9 - 3}$

$= \dfrac{25}{6}$

Examiner's Tip
Equivalent numbers are accepted, e.g. 4.16 instead of $\dfrac{25}{6}$.

and $b = y_2 - \dfrac{25}{6}x_2$

$= 200 - \dfrac{25}{6}x_2$

$= 200 - \dfrac{25}{6} \times 9$

$= 162.5$

The equation is $y = \dfrac{25}{6}x + 162.5$

The trend is for the number of days lost by illness to increase by $\dfrac{25}{6}$ over each two-month period.

The estimated trend value at the last period before recording started was 162.5.

Pick two points on the trend line (one at either end).

Work out a.

Work out b.
Alternatively b can be read off the y-axis at period $= 0$. In this case it would not be very accurate because the scale is small.

Remember: a is the gradient and b is the value when $x = 0$. Relate the answer to the real-life context of the problem.

b has little meaning in this context.

Exercise 6G

1   The diagram shows a trend line.

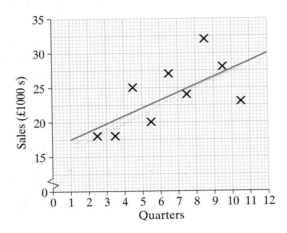

a Work out the gradient of the line.

b Describe in words what this gradient represents.

2   This table shows the turnover (in £1000s) of a department store.

Year	Four-month period	Sales (£1000s)
2005	May–Aug	134
	Sep–Dec	162
2006	Jan–Apr	142
	May–Aug	149
	Sep–Dec	180
2007	Jan–Apr	166
	May–Aug	182
	Sep–Dec	210
2008	Jan–Apr	196

a Plot the sales on a time series graph.

b   **i)** Work out appropriate moving averages and add these to
          your graph.

   ii) Draw a trend line using the moving averages.

c Work out the equation of your trend line.

d Work out the mean seasonal variations.

e Predict the sales for the second and third four-monthly
       periods of 2008.

3 The number of pairs of size 6 trainers sold by a shoe shop is shown.

   a Plot these data on a time series graph.

   b Explain why a three-point moving average should be used for these data.

   c Work out the three-point moving averages and plot them on your graph.

   d Draw a trend line.

   e Work out the gradient of the trend line and give an interpretation of this.

   f Work out the mean seasonal variations.

   g Predict the number of size 6 trainers that will be sold in the three periods of Year 4.

Year	Period	Number of trainers sold
1	Jan–Apr	49
	May–Aug	130
	Sep–Dec	70
2	Jan–Apr	40
	May–Aug	121
	Sep–Dec	55
3	Jan–Apr	31
	May–Aug	112
	Sep–Dec	31

Chapter 6 review

1 The total monthly turnovers of the engineering industry in England over six consecutive months are shown in the table.

⭐⭐⭐⭐⭐ *challenge*

Month	Jul	Aug	Sep	Oct	Nov	Dec
Turnover (£ billion)	6.4	6.0	7.0	6.5	6.7	6.4

Draw a line graph of these data.

2 The quarterly takings of a post office in £1000s for three successive years are shown in this table.

Year	Quarter			
	1	2	3	4
1	26	42	46	74
2	26	40	47	76
3	34	48	56	90

> **Examiner's Tip**
> When commenting on a trend do not describe every up and down, just the general trend.

   a Draw a time series graph for these data.

   b Draw a trend line on your graph.

   c Describe the trend shown by your trend line.

   d Does the graph suggest that there is a seasonal variation? If so, in which quarter are the sales highest? Give a reason for this.

3   A multiplex cinema shows a set of films for three consecutive weeks instead of for one week only. The attendances (in 100s) are shown below.

	Week		
Day	**1**	**2**	**3**
Mon	6	4	4
Tues	6	5	5
Wed	8	7	6
Thu	11	9	6
Fri	14	12	12
Sat	25	20	18
Sun	20	16	15

a   Draw a time series graph for these data.

b   Calculate seven-point moving averages and add them to your graph.

c   Describe the trend and comment on the seasonal variations.

4   The table shows the number of houses sold by Houses Direct in successive four-month periods in the years 2006 to 2008.

Year	Period	Number of houses	Three-point moving average
2006	1	9	
	2	26	$\frac{9 + 26 + 19}{3} = 18$
	3	19	$\frac{26 + 19 + 12}{3} = 19$
2007	1	12	$\frac{19 + 12 + 32}{3} = 21$
	2	32	
	3	22	
2008	1	15	
	2	38	
	3	25	

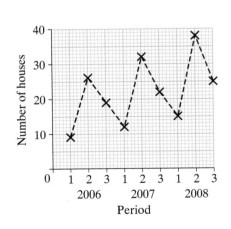

a   **i)** Copy and complete the table.

ii) Copy the graph and plot the moving averages on it.

There was an increase in the number of houses sold in some areas of England in the years 2006 to 2008. It is claimed that Houses Direct is in one of these areas.

b i) Write down one reason why this claim might be true.

   ii) Write down one reason why this claim might be false.

edexcel ::: *past paper question*

5 The table shows information about the quarterly gas bill, in £s, for Samira's house, over a period of two years.

Year	Quarter			
	1	**2**	**3**	**4**
1	£200	£162	£80	£130
2	£216	£166	£96	£142

The data have been plotted as a time series.

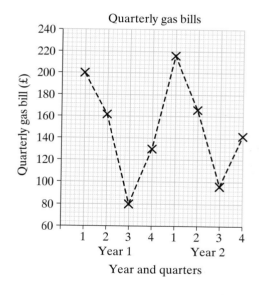

a The first three four-point moving averages are £143, £147 and £148.

   i) Work out the last two four-point moving averages.

   ii) Plot all five of the moving averages on the graph.

b What do the moving averages show about the trend of the quarterly gas bills?

The time series shows that the quarterly gas bills are varying from the general trend.

c i) Write down what these variations are called.

   ii) Write down a reason for these variations.

edexcel ::: *past paper question*

6   The table shows the numbers of new houses (in 100s) completed in southern England over 12 consecutive quarters.

	Quarter			
Year	**1**	**2**	**3**	**4**
1	186	182	220	240
2	170	180	250	280
3	190	200	260	300

a   Draw a time series graph for these data.

b   Calculate the four-point moving averages and plot them on the graph.

c   Draw a trend line through the moving averages.

d   Work out the equation of the trend line.

e   Work out the mean seasonal variations.

f   Using your answers to **d** and **e** predict the number of houses that will be completed in the second quarter of Year 4.

Chapter 6 summary

Line graphs

1   A **line graph** is used to display data when the two variables are not related by an equation and it is uncertain what happens between the plotted points.

2   A **time series** is a set of observations taken over a period of time. A line graph can be used to show a time series. When plotting a time series, time is plotted on the horizontal axis.

Trend lines

3   A **general trend** is the way that the data change over time.

4   A **trend line** shows the general trend of the data.

5   A trend line may show a tendency to rise, to fall or to stay level.

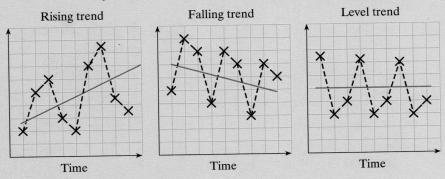

6 Variations in a time series may be
- a general trend (as shown by the trend line)
- seasonal variations (a pattern that repeats).

Moving averages

7 A **moving average** is an average worked out for a given number of successive observations.

8 The number of points in each moving average should cover one complete cycle of seasons.

9 Moving averages are plotted on the time series graph to help show the trend.

10 They are plotted at the mid-point of the time intervals they cover.

11 The points plotted for moving averages should not be joined up.

Estimating seasonal variations and making predictions

11 seasonal variation at a point = actual value − trend value

12 estimated mean seasonal variation for any season = mean of all the seasonal variations for that season

13 predicted value = trend line value (as read from trend line on graph) + estimated mean seasonal variation

Calculating the equation of a trend line

14 The **equation of a trend line is $y = ax + b$**, where a is the **gradient** of the line and b is the **intercept with the y-axis**.

15 The constant a is the amount by which the trend increases per time period (quarter, two-months, etc).

16 The constant b is the value of the trend at the end of the last period before observations began.

Test yourself

This time-series graph shows the mean quarterly temperatures in degrees Centigrade.

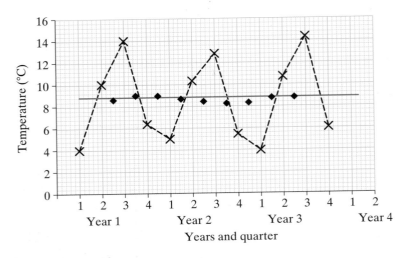

1 What was the mean temperature in Year 2, Quarter 3?

2 The mean temperatures for Quarters 1, 2, 3 and 4 in Year 3 were 4.0°C, 10.8°C, 14.4°C and 6.0°C, respectively. Calculate the exact numerical value of the four-point moving average for Year 3.

3 Describe the trend of the trend line.

4 What is the seasonal variation in the mean temperature for Year 3, Quarter 3?

5 The seasonal variations for the Quarter 1s were −4.8°C, −3.6°C and −4.4°C. Calculate the mean seasonal variation for Quarter 1.

6 What is the trend line value for Year 4, Quarter 1?

7 Predict the mean temperature for Year 4, Quarter 1.

Chapter 7:
Probability

After completing this chapter you should be able to

- understand the meaning of the words 'event' and 'outcome'
- use a likelihood scale
- put outcomes in order in terms of probability
- put probabilities in order on a probability scale
- understand the terms 'random', 'equally likely', 'mutually exclusive' and 'exhaustive'
- understand and use measure of probability from a theoretical perspective and from a limiting frequency or experimental approach

- compare expected frequencies and actual frequencies
- use simulation to estimate more complex probabilities
- use probability to assess risk
- produce, understand and use a sample space
- understand and use Venn diagrams and sample space diagrams
- understand and use the addition and multiplication laws
- use and calculate conditional probabilities
- draw and use probability tree diagrams.

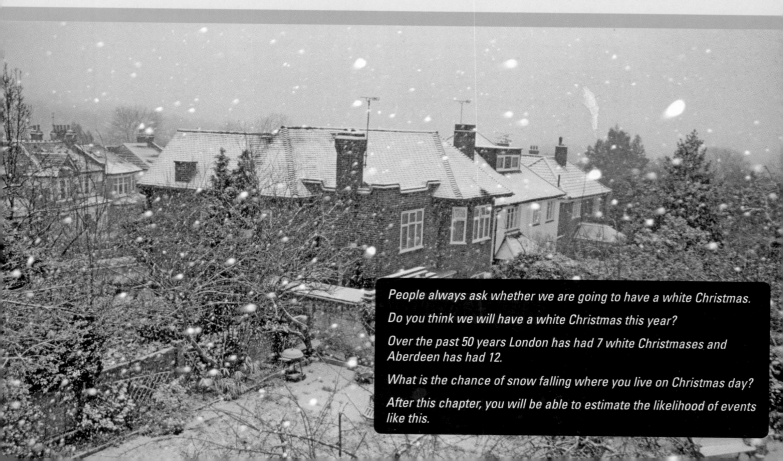

People always ask whether we are going to have a white Christmas.

Do you think we will have a white Christmas this year?

Over the past 50 years London has had 7 white Christmases and Aberdeen has had 12.

What is the chance of snow falling where you live on Christmas day?

After this chapter, you will be able to estimate the likelihood of events like this.

7.1 Events and outcomes

When you flip a coin you get either a head or a tail.

The act of flipping the coin is called a **trial**.

The results of flipping the coin – head or tail – are called the **outcomes**.

> A **trial** is the act of testing/doing something.
> **Outcomes** are the possible results of a trial.

Example 1

In a game a six-sided dice is rolled.

A player needs a 6 to start the game.

a What is the trial?

b What are the possible outcomes of the trial?

a The trial is the act of rolling the dice.

b The possible outcomes are 1, 2, 3, 4, 5 and 6.

> A trial is the act of testing/doing something. What is being done?

> Outcomes are the possible results of a trial. What numbers could show when the dice is rolled?

> An **event** is a set of one or more successful outcomes.

Example 2

A six-sided dice is rolled and the number showing on the top is recorded.

Write down the successful outcomes for these events:

a Rolling a number less than 4.

b Rolling an even number.

a The successful outcomes are 1, 2 and 3.

b The successful outcomes are 2, 4 and 6.

> Which outcomes match the requirements of the question when the dice is rolled?

7.2 The meaning of probability

You may often hear people making comments such as 'my team is certain to win'. Words like 'certain', 'impossible', 'likely', 'unlikely' and 'evens' can be used to describe the chance of an event happening.

Here is a **likelihood scale**.

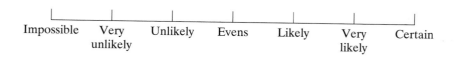

Impossible   Very unlikely   Unlikely   Evens   Likely   Very likely   Certain

Example 3

Write down the most suitable word from the likelihood scale above to describe the probability of these events.

A   A flower will die.

B   A baby born will be a boy.

C   A cow will turn into a horse.

D   There will be a hurricane in Britain in the next week.

A   Certain

B   Evens

C   Impossible

D   Very unlikely

Choose a word that describes how likely you think the event is.

Example 4

Copy the likelihood scale given above. Mark the likelihood of each event with its letter.

A   A baby will be born somewhere in the world today.

B   You will score ten when a normal six-sided dice is rolled.

C   Someone will win the jackpot in the next lottery draw.

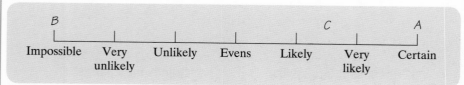

B   C   A
Impossible   Very unlikely   Unlikely   Evens   Likely   Very likely   Certain

Event A is certain and event B is impossible.

Event C is somewhere between evens and certain. It may not be exactly where it is on this diagram.

In statistics a numerical scale is used to describe probability.

> Probability is a numerical measure of the chance of an event happening.
> - A probability of 0 means it is impossible for the event to happen.
> - A probability of 1 means the event is certain to happen.

Here is a **probability scale**.

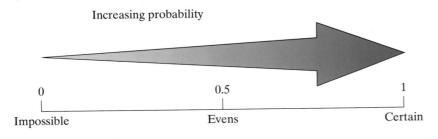

Increasing probability

0   0.5   1
Impossible   Evens   Certain

Example 5

Put the following on a probability scale.

A   Great Britain will win 40 Gold medals in the next Olympics.

B   A person chosen at random was born on a weekday.

C   A dice is rolled and does not show a six.

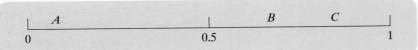

Put *A* near 'impossible'.

Put *B* and *C* so that *B* is lower than *C* and both are greater than 0.5.

Examiner's Tip

On a probability scale you only need to put letters in roughly the right place.

> Probabilities can be written as fractions, decimals or percentages.

Example 6

There is a 20% chance that the Bank of England will change interest rates next month. Write the probability of this change as a percentage, a fraction and a decimal.

$P(C) = 20\% = \dfrac{20}{100} = 0.2$

$P(C)$

Stands for 'probability'   Stands for 'change' (choose any letter)

When something is chosen 'at **random**', it means it is selected without a conscious choice.

When a card is selected at random from a full pack of 52 cards, each card is equally likely to be picked.

On a fair six-sided dice, each number from 1 to 6 is equally likely to be rolled.

When outcomes have the same chance of happening they are called **equally likely outcomes**.

Examiner's Tip
Rolling a **fair** dice gives a random choice from the numbers 1 to 6.

Exercise 7A

1 List the possible outcomes for these trials.
  a The weather tomorrow is studied to see if it rains.
  b A fair eight-sided dice, numbered 1 to 8, is rolled.
  c A drawing pin is dropped to see how it lands.

2 Explain what 'equally likely outcomes' means.

3 Write down the most suitable word from the likelihood scale to describe the probability of these events.
  a A tree will die.
  b A fair coin will fall showing heads.
  c There will be a hurricane in England next May.

4 Draw a probability scale. Mark each event with its letter on the scale.
  a A There will be four aces in a complete pack of 52 cards.
  b B The Queen of England will be US President.
  c C Everyone in Great Britain will make at least one telephone call this week.

5 **a** If an event is impossible what is its numerical probability?
  b If an event is certain what is its numerical probability?
  c Rewrite this statement in words P(W) = 0.5.

6 There is an 80% chance of catching a train. Write down this probability as a decimal.

7 Draw a probability scale and mark on it
  a G the probability that the grass will grow this summer
  b D the probability that a rose bush will die
  c O the probability that a dice will show an odd number when rolled.

ResultsPlus
Watch out!

■ When asked to mark probabilities of $\frac{1}{7}$ and $\frac{2}{7}$ on a probability scale marked 0, 0.5, 1, most students put them in the wrong place. If they had measured the line they would have realised it was 14 cm long, so $\frac{1}{7}$ would be at 2 cm from 0, and $\frac{2}{7}$ would be 4 cm from 0.

7.3 The probability of an event

If *all* possible outcomes are equally likely then it is easy to find the probability of an event happening.

> If all possible outcomes are equally likely,
>
> the **probability of an event** = $\dfrac{\text{number of successful outcomes}}{\text{total number of possible outcomes}}$.

Example 7

A fair six-sided dice is rolled. Work out

a the probability of an even number

b the probability of a number $\leqslant 2$

c the probability of a multiple of 3.

a $P(even) = \dfrac{3}{6} = \dfrac{1}{2}$

> There are six possible outcomes 1, 2, 3, 4, 5 and 6.
> Count how many are even numbers. (There are three: 2, 4, and 6.)
> Write the fraction in its lowest terms.

b $P(\leqslant 2) = \dfrac{2}{6} = \dfrac{1}{3}$

> Count how many of the outcomes are less than or equal to 2. (There are two: 1 and 2.)

c $P(multiple\ of\ 3) = \dfrac{2}{6} = \dfrac{1}{3}$

> Count the outcomes that are multiples of 3. (There are two: 3 and 6.)

Example 8

This two-way table shows the number of males and females in a group who are right- and left-handed.

A person from the group is chosen at random. Work out the probability that the person chosen is

a female

b left-handed

c a right-handed male.

	Male	Female	Total
Right-handed	17	20	37
Left-handed	7	6	13
Total	24	26	50

a $P(female) = \dfrac{26}{50} = \dfrac{13}{25}$

> To find the total number of females look where the 'Female' column meets the 'Total' row. (26)
> To find the total number of people look where the 'Total' column and 'Total' row meet. (50)

b $P(left\text{-}handed) = \dfrac{13}{50}$

> To fnd the total number of left-handed people look where the 'Left-handed' row meets the 'Total' column'. (13)

c $P(right\text{-}handed\ male) = \dfrac{17}{50}$

> To find the total number of right-handed males look where the 'Right-handed' row meets the 'Male' column. (17)

Results Plus
Build better answers!

Question: A farmer has two farms.
On one farm he has battery hens, on the other farm he has the same number of free-range hens.
One Saturday the sizes of the eggs collected from the two farms were as follows.

	Large	Medium	Small	Total number of eggs
Free-range hens	125	210	105	
Battery hens	75	210	125	
Totals				

(a) Complete the two-way table.   (2 marks)

An egg from those collected on the Saturday is chosen at random.

(b) Write down the probability that the egg chosen is
    (i) large,
    (ii) from a free-range hen and medium.   (2 marks)

(c) Compare and contrast the numbers of the different sizes of eggs laid by the free-range hens and the battery hens on these farms.   (2 marks)

(Total 6 marks)

■ **Basic   1–2 mark answers**
Complete the two-way table accurately in part a). Common mistakes are to miss out the overall total number of eggs in the bottom right-hand cell, or to add all the totals for the rows and columns to get the incorrect answer of 1700.

● **Good   3–4 mark answers**
Complete the two-way answer table in part a) and use your values to calculate the probabilities, leaving your answers as fractions. Make sure the denominator of each fraction is the total number of eggs from the bottom right-hand cell. You don't need to simplify your fractions.

▲ **Excellent   5–6 mark answers**
Complete the table and calculate the probabilities in parts a) and b). Part c) is worth two marks, so write two comments. Compare the size of egg and type of hen, for example 'free range hens lay more large eggs than battery hens'. Also compare the total number of eggs laid by each type of hen.

Exercise 7B

1 A fair six-sided dice, numbered 1 to 6, is rolled. Work out

a the probability of getting a four

b the probability of getting a number < 5

c the probability that the number is a multiple of 2, but not 2 itself.

2 The fair spinner shown is spun. Work out the probability of the arrow pointing to

a yellow (Y)

b green (G)

c red (R)

d blue (B)

e red or blue (R or B).

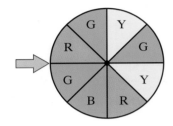

3 A card is selected at random from a pack of 52 playing cards.

a Work out the probability that the card is

   i) a heart

   ii) the King of Hearts

   iii) a black card

   iv) a Jack.

b Explain why the card had to be picked at random.

> **Examiner's Tip**
> A pack of cards consists of four suits. Two red (hearts and diamonds) and two black (clubs and spades). Each suit contains 13 cards: Ace, 2, 3, 4, 5, 6, 7, 8, 9, 10, Jack, Queen and King.

4 Six names are written on separate pieces of paper and put into a hat. The names are Ann Smith, Shulah Brown, Yves Black, Mai Jones, Jack Brown, and Jack Firth. One name is drawn at random from the hat. Work out the probability that the name drawn has

a the first name Jack

b the first name Yves

c the surname Brown.

5 This two-way table shows the numbers of males and females in a group of 50 wearing or not wearing glasses.

	Male	Female	Total
Wearing glasses	16	18	34
Not wearing glasses	9	7	16
Total	25	25	50

Work out the probability that a person chosen from this group at random is

a female

b not wearing glasses

c a male wearing glasses.

ResultsPlus

Watch out!

● When asked to calculate probabilities from a two-way table, some students gave them as ratios, such as 36:100, rather than correctly as the fraction $\frac{36}{100}$, decimal (0.36) or percentage (36%).

6   Shoppers in a supermarket are asked to taste two jams marked A and B, and say which they prefer. The two-way table shows the preferences of male and female shoppers.

	Male	Female	Total
Jam A	10	13	23
Jam B	2	20	
Total	12		45

   a Copy and complete the table.
   b Work out the probability that a shopper chosen at random is
      i) a female who prefers jam B
     ii) a male who prefers jam A
    iii) a person who prefers jam A.

7   Children going on a boat trip can choose to travel either on the Raven or on the Eagle. Their choices are shown in the two-way table.

	Boys	Girls	Total
Raven	11		51
Eagle	35	19	
Total			

   a Copy and complete the table.
   b Work out the probability that a child chosen at random is
      i) a girl
     ii) a girl who chooses Raven
    iii) someone who chooses Eagle.

8   There are eight volunteers in a medical trial. In treatment group A, three volunteers are given a new drug treatment. In treatment group B, the rest of the volunteers are given a sugar pill.

   The volunteers are randomly allotted to the treatment groups. Jean is one of the volunteers.

   Work out the probability that

   a Jean will be given the new drug

   b Jean will not be given the new drug.

7.4 Experimental probability

In real-life situations the probabilities of different outcomes are not always equal or possible to work out. For example, you cannot work out the theoretical probability of a seed germinating and growing into a plant. You have to sow some seeds and see how many germinate.

In such cases you may need to carry out an experiment or survey to estimate the probability of an event happening.

Each experiment (or response to a survey) is called a **trial**. The number of the trials with successful outcomes is recorded.

> The estimated probability that an event might happen $= \dfrac{\text{number of successful outcomes}}{\text{total number of trials}}$

Example 9

Supporters of Tingle Football Club want to find the probability of the club winning a game.

Supporters Pat and Nami work out their own probabilities of the club winning a game.

Pat says 'The club won one of their first two games so the probability of them winning is 0.5'.

Nami says 'I do not agree. After 10 games they have won eight games and lost two so the probability of them winning is 0.8.'

The following is a table of results of the last 100 games.

Number of games played	2	10	25	50	75	100
Number of games won	1	8	15	36	51	70

Draw a graph for these data and use it to estimate the probability of the team winning a game.

Number of games played	2	10	25	50	75	100
Number of games won	1	8	15	36	51	70
Probability of winning	$\frac{1}{2} = 0.5$	$\frac{8}{10} = 0.8$	$\frac{15}{25} = 0.6$	$\frac{36}{50} = 0.72$	$\frac{51}{75} = 0.68$	$\frac{70}{100} = 0.7$

> Add a row to the table and find the probabilities.

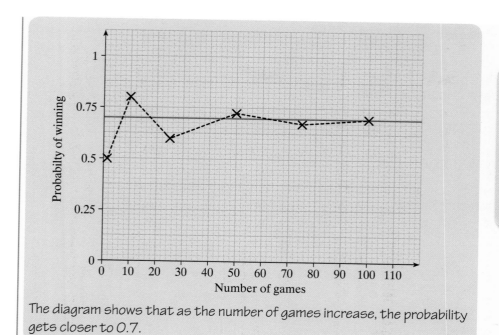

Draw a graph of the results and join the points with a dotted line.

The probability of winning gets closer to 0.7 as the number of games increase.

(A red line has been drawn at probability = 0.7 on the graph to illustrate this.)

The diagram shows that as the number of games increase, the probability gets closer to 0.7.

As the number of trials increase in experiments and surveys, the nearer an estimate for the probability should be to the true value.

Simulation

It may not be possible to carry out an experiment to estimate the probability of an event happening. This may be because it is too complex or it is undesirable to carry out the study. For example, it would not be sensible to study the spread of an infectious disease by practical experiment.

In such cases you can **simulate** the problem.

Simulation is the imitation of the conditions of a situation by doing a theoretical study.

Example 10

A central warehouse has two loading bays and five lorries. They are buying a sixth lorry.

Is another loading bay needed?

Explain why simulation would be used to answer this question.

It would be expensive to build the bay if it is not needed, but there might be costly delays if the lorry is bought and the bay not built. Several runs of a computer simulation with six lorries and two loading bays would tell the owners if hold-ups are likely to occur. This model could easily be altered and re-run with three bays.

Make comments in context on simulation being

- quick and cheap
- easily altered
- repeatable.

Higher Statistics

There are several ways of introducing randomness to a simulation.

- Use coins.
- Use a dice or several dice.

- Generate random numbers using a calculator or spreadsheet.

Example 11

A shop sells ice creams of four different flavours: vanilla, chocolate, raspberry and strawberry.

Over a long period of time, out of ten customers,

- three choose vanilla
- two choose chocolate
- one chooses raspberry
- four choose strawberry.

The shop has an average of 20 customers each day.

a Run a simulation to show a typical day.

b Repeat your simulation.

c Explain how repeated simulations could help the shopkeeper to estimate the minimum amount of each flavour he should have in stock at the beginning of each day, if he is to satisfy all the possible demands of his customers.

a Allocate numbers as follows.

- 1, 2 and 3 = vanilla (V).
- 4 and 5 = chocolate (C).
- 6 = raspberry (R).
- 7, 8, 9 and 10 = strawberry (S).

Here is the result of one simulation. Yours will be different.

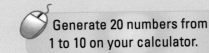

Generate 20 numbers from 1 to 10 on your calculator.

Random number	9	7	2	4	9	5	2	1	1	8	4	1	4	10	6	10	8	9	3	5
Flavour	S	S	V	C	S	C	V	V	V	S	C	V	C	S	R	S	S	S	V	C

This run gives eight strawberry, six vanilla, five chocolate and one raspberry.

b Here is the result of a second simulation. Yours will be different.

Random number	7	1	9	8	5	9	2	3	10	10	5	1	10	5	7	10	2	4	4	2
Flavour	S	V	S	S	C	S	V	V	S	S	C	V	S	C	S	S	V	C	C	V

This run gives nine strawberry, six vanilla, five chocolate and zero raspberry.

c From the two runs, the maximum sales of each variety are nine strawberry, six vanilla, five chocolate and one raspberry. If the simulation is repeated many times the shopkeeper will have a better idea of the maximum daily demand for each flavour.

Comparing expected frequencies and actual frequencies

Example 12

Nimer gives Josh a dice and says he will give him £10 if he scores more than 10 sixes in 100 rolls of the dice.

a If the dice is fair how many sixes would Josh expect to score in 100 rolls of the dice?

The actual results are in the table.

Score	1	2	3	4	5	6
Frequency	40	13	15	14	13	5

b How do these results compare with the expected ones?

a Expected number of sixes is $\frac{100}{6} = 16.7$.

> If the dice is fair then every number has an equal chance of being rolled.

b The number 1 occurs many more times than the 16.7 expected. This suggests that 1 has a higher probability of being rolled. The numbers 2, 3, 4 and 5 occur fewer times than expected but have a similar chance of being rolled. The number six occurs a much lower number of times than expected. The dice appears to be biased in favour of the number 1 and against the number 6.

> Comment on the probabilities and give an overall view.

Exercise 7C

1 In an experiment 50 poppy seeds are sown and 30 produce plants. Work out an estimate for the probability of a poppy seed producing a plant.

2 20 children each plant a bean seed in a jam jar. 17 of the bean seeds grow into bean plants.

Work out an estimate for the probability that a bean seed will grow into a plant.

3 Under what circumstances would you use a simulation rather than an experiment?

4 The four meals offered in a canteen are a salad, a roast, pasta and a vegetarian meal. Over a period of time it was found that out of every ten workers

- two chose salad
- five chose roast
- one chose vegetarian
- two chose pasta.

Every lunchtime the canteen serves 30 workers.

A simulation is run using random numbers on a calculator. The simulation is repeated.

Explain how repeated simulations like this could help the chef to decide how many of each type of meal he should produce.

5 A country park wants to provide sufficient wet weather activities. It is going to create four activity centres. A survey of visitors shows that out of every ten people

- five would choose centre A
- one would choose centre B
- two would choose centre C
- two would choose centre D.

The park normally has 50 visitors on a rainy day.

a Use a calculator or random number tables to generate 50 random numbers from 1 to 10.

b Suggest how you would use these numbers to simulate a rainy day with 50 visitors.

c Why would it be sensible for the park to run several simulations?

6 During one week Jonah recorded the timeliness of 50 trains. The table shows the number of trains that were early, on time, late or cancelled.

Timeliness	Early	On time	Late	Cancelled
Frequency	5	27	15	3

Jonah wishes to simulate the timeliness of the next 10 trains. He uses random numbers between 00 and 99.

The table below shows the numbers he gives each result.

Timeliness	Early	On time	Late	Cancelled
Frequency	00–09	10–63	64–93	94–99

a Explain why the numbers 64–93 were given to the number of times a train is late.

Jonah uses random numbers to get the timeliness of the next 50 trains. He puts the first 40 on a tally chart. He then uses the following random numbers to simulate the next 10 trains.

40   31   65   05   07   07   85   11   86   76

b Copy the tally chart and complete it by putting in the final 10 simulations.

Timeliness	Tally	Frequency
Early	||||	
On time	ⅢⅢ ⅢⅢ ⅢⅢ ⅢⅢ |	
Late	ⅢⅢ ⅢⅢ |||	
Cancelled	||	

The actual results of the next 50 trains are in the table below.

Timeliness	Early	On time	Late	Cancelled
Frequency	2	15	23	10

c How do the actual results compare with the simulated ones?

d What does this suggest is happening to the timeliness of the trains?

7 **Activity**

a Flip a coin 25 times and record the number of heads. Repeat the experiment three more times.

Experiment	1	2	3	4
Number of heads				

b Redraw the table using cumulative totals.

Number of coins	25	50	75	100
Number of heads				
Probability				

c Use your results to estimate the probability of the flipped coin showing heads.

Using probability to assess risk

What is the risk of your house burning down this year?

If you look at the number of houses like yours that are burned down each year and compare it with the total number of houses like yours, you can estimate the probability of your house burning down.

The estimated probability of your house burning down

$$p = \frac{\text{number of successful outcomes}}{\text{total number of trials}} = \frac{\text{number of houses burned}}{\text{total number of houses}}$$

The risk of your house burning down in any year is p. The insurance company should charge you £1000 $\times p$ for each £1000 of house value (i.e. house value $\times p$). (It will probably be a bit more than this to allow for running costs and profits.)

> cost of risk insurance = money at risk $\times$ risk assessment (p)

Example 13

Using past records, an insurance company assesses the yearly risk of a house in a certain area being flooded.

During the last 50 years, flooding in that area has occurred only once. The company uses this risk assessment to work out the cost of the flood risk part of its annual premiums.

a What is the risk assessment?

b The average pay-out by this insurance company when a house is flooded is £1200. How much would you expect to pay for insurance against it being flooded?

a   Risk assessment = $\frac{1}{50}$

  = 0.02

> Divide the number of times the area floods (number of successful outcomes) by the number of years (number of trials).

b   Cost of flood insurance = £1200 × 0.02
  = £24

> Multiply the money at risk by the risk assessment.

Exercise 7D

1   John is 18 and has just passed his driving test. An insurance company finds that young men of John's age and living in John's postal district had four accidents in the past year. There were 150 drivers like John living in the area during the last year. What is the risk assessment for John having an accident this year?

2   An insurance company that insures all types of boats has found that, out of 800 boats insured, 20 had a shipping accident of some sort in the past year. What risk assessment would you make of a particular boat having an accident?

3   The risk assessment of a particular type of company going bankrupt during a one-year period is $\frac{2}{27}$. If a small country has 54 companies of this type how many are likely to end up bankrupt this year?

4   A bank finds that in the past, 1.5% of its customers have been unable to pay their mortgage payments each year. The bank has an insurance policy to cover this happening. If a customer has mortgage repayments of £3200, what is a fair yearly charge for the insurance to cover the customer's mortgage repayments?

5   When going on holiday abroad you are told to take out insurance to cover the cost of possible medical treatment. The average cost of medical treatment abroad last year was £5000, and 2% of people going abroad made claims. Using these figures, how much would you expect to pay for this insurance? Assume you do not include the insurance company's expenses and profits.

7.5 Sample space

To help find the probability of one, two or more events occurring you can list all the possible outcomes.

For example, there are six possible outcomes if a six-sided dice is rolled. The **sample space** is S = (1, 2, 3, 4, 5, 6).

> A list of all possible outcomes is called a **sample space**.

> **Examiner's Tip**
> To use a sample space, the outcomes must be equally likely.

If there are two events, a table can represent the sample space. This is often referred to as a **sample space diagram**.

Example 14

A fair coin is flipped and a fair dice is rolled.

a Draw a table of all possible outcomes.

b Work out the probability of getting a head and a six.

c Work out the probability of getting a head and an even number.

a

		Dice					
		1	2	3	4	5	6
Coin	Head (*H*)	*H*, 1	*H*, 2	*H*, 3	*H*, 4	*H*, 5	*H*, 6
	Tail (*T*)	*T*, 1	*T*, 2	*T*, 3	*T*, 4	*T*, 5	*T*, 6

Put the outcomes of the dice along the top. Put the outcomes of the coin down the side. Fill in the outcomes in the middle.

b $P(H, 6) = \dfrac{1}{12}$

Count how many possible outcomes there are. (12)
Count how many times the outcome $(H, 6)$ occurs. (1)

c $P(H, even) = \dfrac{3}{12}$

$= \dfrac{1}{4}$

Count how many times the outcomes $(H, 2)$, $(H, 4)$ and $(H, 6)$ occur. (3)

Example 15

Two fair dice are rolled and the scores added together.

a Draw a sample space diagram showing all the possible outcomes.

b Work out the probability of a total of 10.

c Work out the probability of an even total.

a

		Dice 1					
		1	2	3	4	5	6
Dice 2	1	2	3	4	5	6	7
	2	3	4	5	6	7	8
	3	4	5	6	7	8	9
	4	5	6	7	8	9	10
	5	6	7	8	9	10	11
	6	7	8	9	10	11	12

Put the outcomes of one dice along the top. Put the outcomes of the other dice down the side. Fill in the outcomes in the middle by adding the two scores together, e.g. $4 + 1 = 5$.

b $P(10) = \dfrac{3}{36}$

$= \dfrac{1}{12}$

> Count how many possible outcomes there are. (36)
> Count how many times 10 appears. (3)

c $P(even) = \dfrac{18}{36}$

$= \dfrac{1}{2}$

> Count how many times 2, 4, 6, 8, 10 and 12 appear. (18)

Example 16

Three fair coins are flipped.

a Write down the sample space.

b Work out the probability of flipping exactly three heads.

c Work out the probability of flipping exactly two heads.

a $S = \begin{array}{|l|} \hline HHH \\ HHT \ HTH \ THH \\ HTT \ THT \ TTH \\ TTT \\ \hline \end{array}$

> Write down all the possible outcomes with three heads, two heads, one head and no heads.

b $P(3\ heads) = \dfrac{1}{8}$

> Count how many possible outcomes there are. (8)
> Count how many outcomes include exactly three *H*s. (1)

c $P(2\ heads) = \dfrac{3}{8}$

> Count how many outcomes include exactly two *H*s. (3)

Exercise 7E

1 Two fair coins are flipped.
   a Draw a table to find the sample space and fill in all possible outcomes.
   b Work out the probability of getting a head and a tail.
   c Work out the probability of getting two heads.

2 Two fair dice, each with sides numbered from 1 to 6, are rolled and the total of the scores noted.
   a Draw a table to find the sample space and fill in the total scores.
   b Write down the probability of a total score of 12.
   c Write down the probability of a total score ≤ 7.
   d Write down the probability of a total score > 10.
   e Write down the probability of a total score that is even.

ResultsPlus
Watch out!

When asked to write down all the ways of scoring 7 from two dice, some students missed out half of the possible pairings. Remember that (1, 6) is not the same as (6, 1).

3 A team from a tennis club consists of four men players A, B, C and D, and two women players, X and Y. A fair spinner and a fair coin are used to decide who will represent the club in mixed doubles.

- The spinner, lettered A, B, C and D, is spun to represent the men.
- The coin is flipped to represent the two women players (a head representing X and a tail representing Y).

a Draw a table to find the sample space.

b Write down the probability that the man selected is D.

c Write down the probability that X and B form a pair for mixed doubles.

d Write down the probability that Y and either B, C or D form a pair for mixed doubles.

4 A fair spinner, numbered from one to four, is spun and a fair six-sided dice is rolled. The total score is noted.

a Draw a table to find the sample space.

b What is the probability that the score is 8?

c What is the probability that the score < 2?

d What is the probability that the score ⩾ 6?

5 Pens are equally likely to be blue, red or green. Pen caps are equally likely to be blue, red or green. Caps are allocated to the pens randomly.

a Draw a sample space diagram to show all possible outcomes.

b Work out P(*a red pen has a red cap*).

c Work out P(*a pen has a matching cap*).

d Work out P(*a blue pen has a green cap*).

6 The total score on the three spinners shown is used for a board game.

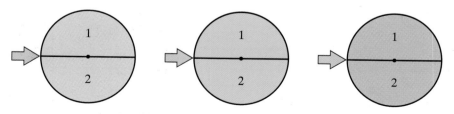

a Write down the sample space for the outcomes.

b Work out the probability of a total score of 6.

c Work out the probability of a total score of 2.

d Work out the probability of a total score of 4.

e Work out the probability of a total score < 6.

7 Ann, Brenda and Carol go shopping together in a busy supermarket.

They each join a different queue. All the queues are the same length.

  a Write down the sample space for the order in which they leave the supermarket.

  b Work out the probability that Ann leaves before Carol.

  c Work out the probability that Brenda is the last to leave.

7.6 Venn diagrams

> A **Venn diagram** is used to represent a sample space and may be used to calculate probabilities. Sets are represented as circles and common elements of the set are represented by the areas where the circles overlap.

Each region of a **Venn diagram** represents different sets of data.

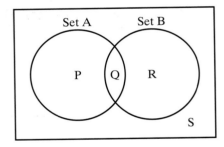

Area P represents the data that are in set A but not in set B.

Area R represents the data that are in set B but not in set A.

Area Q represents the data that are in set A and also in set B.

Area S represents the data that are not in set A and not in set B.

Example 17

In a medical trial there are 70 patients. 24 receive treatment B, 30 receive treatment A and 20 receive both treatment A and treatment B.

  a Draw a Venn diagram to represent these data.

  b What is the probability that a patient chosen at random is receiving neither treatment?

a

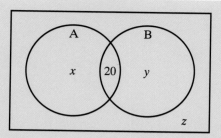

Draw and label a Venn diagram.

Fill in any known values.

The intersection is 20 because that many people receive treatment A and treatment B.

Use letters to label any areas where the value is unknown.

There are 30 in group A, so

$$x + 20 = 30$$

There are 24 in group B, so

$$y + 20 = 24$$

Therefore $x = 10$ and $y = 4$.

The total number must be 70, so

$$z = 70 - 10 - 20 - 4 = 36$$

The final Venn diagram is

Calculate x and y by rearranging the formulae.

$$x = 30 - 20$$
$$= 10$$

and $y = 24 - 20$
$$= 4$$

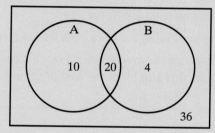

The number that does not lie in the circles is the number not taking either drug. The probability of this is 36 divided by the total number of patients.

b $P(\text{neither drug A nor drug B}) = \dfrac{36}{70} = \dfrac{18}{35}$

Probabilities can be used instead of numbers in a Venn diagram.

If probabilities are entered instead of numbers the diagram in Example 17 looks like this.

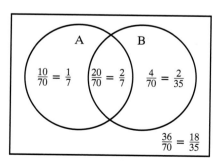

Examiner's Tip

To change to probabilities, every number in the diagram is divided by the total number. In this case, 70.

Example 18

The prime minister makes a speech.

Two newspapers differ in their editorial policies.

The probability that paper G will put the speech as their main headline is 0.2.

The probability that paper L will put the speech as their main headline is 0.8.

The probability that neither put the prime minister's speech as their main headline is 0.1.

a Draw a Venn diagram to represent these probabilities.

b Use it to find the probability that both use the prime minister's speech as their main headline.

> Draw and label the Venn diagram.
>
> Remember to fill in the intersection. A value has not been given, so a letter is used.
>
> Remember to label the area outside the circles.
>
> Enter any other known numbers (in this case the probability 0.1) and label the rest of the areas with letters.

a

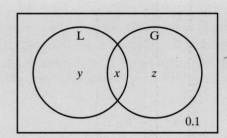

b $y = 0.2 - x$

$z = 0.8 - x$

$x + y + z + 0.1 = 1$

> The probability needed is x.
>
> Calculate y and z in terms of x.

> Add all the probabilities together. The total of them is 1.

Examiner's Tip
Remember the sum of the probabilities of all possible outcomes must equal 1.

$x + (0.2 - x) + (0.8 - x) + 0.1 = 1$

$1.1 - x = 1$

so $\qquad x = 0.1$

> Make substitutions to calculate x.

The probability of both newspapers using the prime minister's speech as their main headline is 0.1.

> Comment on the probability in the context of the question.

Example 19

There are 30 people in an office.

Twelve each have an A level in Art (A).

Eight each have an A level in Biology (B).

Eight each have an A level in Latin (L).

Three each have an A level in Art and in Biology.

Three each have an A level in Biology and in Latin.

Four each have an A level in Latin and in Art.

Two each have an A level in Art, in Biology and in Latin.

a   Draw a Venn diagram to represent these data.

b   One person is chosen at random. Calculate the probability that they have an A level in

   i) at least one of the three subjects

   ii) only one of the three subjects

   iii) Latin but not Biology.

a

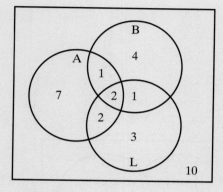

Three circles are needed.
Start by filling in those that are in three circles (i.e. the 2 in the middle of the diagram). Then work outwards to those that are in two circles (e.g. the intersection of L and B should have a total of 3, so the missing figure is 3 − 2 = 1).
Fill in those that are in one circle only (e.g. there are 12 − 2 − 2 − 1 = 7 in A only).

Find how many people have none of the A levels. This is 'total' − 'all the numbers in the circles'.

b i   There are 30 people altogether.
    Ten have A levels in none of the three subjects, so 20 must have A levels in at least one subject.

Use 'at least one' =
'total number of people' − 'the number of people with none'

$$P(A \text{ level in at least one of the three subjects}) = \frac{20}{30}$$
$$= \frac{2}{3}$$

To get the probability, divide the number of people in the office with at least one A level by the total number of people.

  ii   Seven people have an A level only in Art; four have an A level only in Biology; three have an A level only in Latin.

    The number of people who have only one A level
$$= 7 + 4 + 3$$
$$= 14$$

Add together the number of people in the office who have only an Art A level, only a Biology A level and only a Latin A level. Divide the result by the total number of people.

$$P(A \text{ level in only one subject}) = \frac{14}{30}$$
$$= \frac{7}{15}$$

  iii $$P(A \text{ level in Latin but not Biology}) = \frac{5}{30}$$
$$= \frac{1}{6}$$

Add together those in the Latin circle but not in the Biology circle (3 + 2) and divide the result by the total.

Exercise 7F

1 The Venn diagrams show subjects taken by a group of 50 children.

Work out the value of x in each case and say what x represents.

a
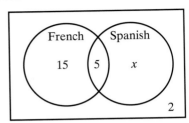

b
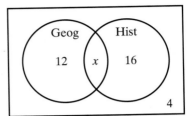

c
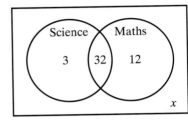

2 This Venn diagram represents the fur colour of 60 rabbits. B represents the rabbits with 'some black fur'. W represents the rabbits with 'some white fur'.

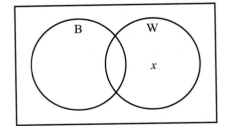

6 rabbits have only black fur, 20 rabbits have black and white fur, 16 rabbits have no black or white fur.

a Copy and complete the diagram.

b Work out the value of x.

c How many rabbits have some black fur?

d What is the probability that a rabbit picked at random has no black or white fur?

3 Copy each Venn diagram and fill in the missing probabilities.

a
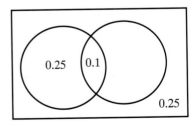

b
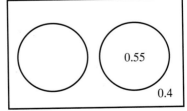

Examiner's Tip
Remember all of the probabilities added together must equal 1.

c

b
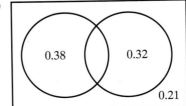

4   120 people were questioned about cold cures they had used.

60 had taken aspirin; 28 had taken coldex; 24 had taken both.

  a   Draw a Venn diagram to represent these data.

  b   Work out the probability that a person picked at random from the group takes neither aspirin nor coldex.

5   A town council was awarded a grant to build two sports centres. During the first year the centres were open it was found that 36% of the town's population had been to centre A, 22% had been to centre B and 10% had been to both.

  a   Draw a Venn diagram to represent these data.

  b   Work out the percentage of the town's population that had been to neither centre.

  c   Write down the percentage that had been to centre A only.

6   There are 30 children in a nursery.

15 children like orange squash, 10 like both orange squash and milk, and 8 like neither orange squash nor milk.

  a   Draw a Venn diagram to represent these data.

  b   How many children like milk?

  c   Work out the probability that a child chosen at random likes milk but dislikes orange squash.

7   An aviary contains 30 parrots.

The probability of a parrot having only green feathers is 0.3.

The probability of a parrot having both red and green feathers is 0.1.

The probability of a parrot having no green or red feathers is 0.45.

  a   Draw a Venn diagram to represent these probabilities.

  b   Calculate the probability that a parrot chosen at random will have some red feathers.

  c   Work out how many parrots will have green feathers and no red feathers?

8   A music school has 100 students. Fifty students play the piano, 20 play the violin, 60 play the oboe, 10 play all three instruments, 15 play both the piano and the oboe, 5 play both the piano and violin, and 2 play both the violin and the oboe.

  a   Copy the Venn diagram opposite and complete it by putting in appropriate numbers.

A student is chosen at random.

  b   What is the probability that the student does not play any of the three instruments?

  c   Work out the probability that the student plays only the piano.

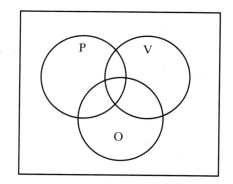

9 The Venn diagram opposite shows the probability of events A, B and C happening.

P(B) = 0.4 and P(C) = 0.35.

Work out x, y and z.

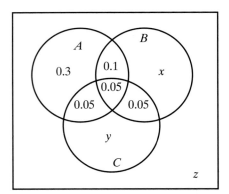

10 A market researcher asks 100 school children which of three different chocolate bars they liked.

Twenty five liked bar A, 30 liked bar B, 20 liked bar C, 6 liked both bars A and B, 7 liked both bars B and C, 5 liked both bars C and A, and 40 liked none of the bars.

a Draw a Venn diagram for these data.

b A child is selected at random. Work out:

i) P(*the child liked at least one of the bars*)

ii) P(*the child liked only one of the bars*)

iii) P(*the child liked none of the bars*).

ResultsPlus
Watch out!

■ Most students were unable to complete a Venn diagram with three intersecting circles.

7.7 Mutually exclusive events

When a dice is rolled the outcome is either even or odd. It is not possible to get an odd number and an even number at the same time. The event 'odd number' and the event 'even number' are **mutually exclusive**.

> Outcomes/events are **mutually exclusive** if they cannot happen at the same time.

Example 20

Which of the following are mutually exclusive events?

A   Getting a 2 and an odd number on a single roll of a dice.

B   The next car you see is red and three years old.

C   The sun will shine and the temperature will be below freezing.

Only the events in A are mutually exclusive.

It is possible to see a three-year-old red car.

It is possible for the sun to shine when it is freezing.

The addition law for mutually exclusive events

If two events A and B are mutually exclusive, then the probability that either one happens is written P(A or B).

> For two mutually exclusive events, *A* and *B*
> P(*A* or *B*) = P(*A*) + P(*B*)

This is called the **addition law for mutually exclusive events**.

Examiner's Tip
It is sometimes called the **Or rule**.

Example 21

A card is drawn from a pack of 52 cards. Work out the probability that the card drawn is either an ace or a king.

$P(ace) = \dfrac{4}{52}$

$= \dfrac{1}{13}$

> There are 52 cards and four are aces.

$P(king) = \dfrac{4}{52}$

$= \dfrac{1}{13}$

> There are 52 cards and four are kings.

$P(ace \text{ or } king) = P(ace) + P(king) = \dfrac{1}{13} + \dfrac{1}{13}$

$= \dfrac{2}{13}$

> Use the addition law to calculate P(ace or king).
> $P(A \text{ or } B) = P(A) + P(B)$

The addition law can be extended for three or more mutually exclusive events. Simply add the probabilities.

$P(A \text{ or } B \text{ or } C) = P(A) + P(B) + P(C)$

Example 22

In cricket, hitting the ball over the boundary scores four runs. Hitting it over the boundary without it bouncing gets six runs. If the ball is not hit over the boundary, runs are scored by running between the wickets.

It is the last ball of a cricket match. Jack's side needs four runs to win.

The probability of Jack scoring four runs by hitting the ball over the boundary is 0.08. The probability that he scores six runs by hitting the ball over the boundary without it bouncing is 0.03. The probability that Jack gets four runs by running between the wickets is 0.01.

What is the probability of Jack

a   scoring four or six runs by hitting the ball over the boundary?

b   scoring exactly four runs?

c   winning the match for his team?

a $P(hitting\ 4\ or\ 6) = 0.08 + 0.03$
$= 0.11$

> Add together P(*hitting a 4*) and P(*hitting a 6*).

b $P(scoring\ exactly\ 4) = 0.01 + 0.08$
$= 0.09$

> Add together P(*running 4*) and P(*hitting a 4*).

c $P(winning) = 0.08 + 0.03 + 0.01$
$= 0.12$

> Add together P(*running 4*), P(*hitting a 4*) and P(*hitting a 6*).

7.8 Exhaustive events

A set of events is exhaustive if the set contains all possible outcomes.

Example 23

A fair dice is rolled.

A is the event 'a score ⩽ 3'.

B is the event 'a score > 3'.

C is the event 'an even number'.

Decide whether the following pairs of events are exhaustive.

a A and C

b A and B

c B and C

A is the set {1, 2, 3}.

B is the set {4, 5, 6}.

C is the set {2, 4, 6}.

a *A* and *C* are not exhaustive as 5 is not included.

b *A* and *B* are exhaustive as all the numbers are included.

c *B* and *C* are not exhaustive as 1 and 3 are not included.

Start by writing down the numbers in each set.

7.9 The sum of the probabilities for a set of mutually exclusive, exhaustive events

For a set of mutually exclusive, exhaustive events, the sum of all the probabilities = 1.

This is written as $\sum p = 1$.

Remember: $\sum$ means 'sum of' and p stands for probability.

The possible outcomes for a fair dice are 1, 2, 3, 4, 5 and 6. Each outcome is equally likely and therefore the probability of each outcome is $\frac{1}{6}$.

P is used for the word probability, so the probability of a 1 is P(1).

$P(1) + P(2) + P(3) + P(4) + P(5) + P(6) = \frac{1}{6} + \frac{1}{6} + \frac{1}{6} + \frac{1}{6} + \frac{1}{6} + \frac{1}{6} = 1$

The probability of an event not happening

> The probability of an event A not happening is written as P(not A).
>
> Since an event either happens or does not happen P(A) + P(not A) = 1.
> So, for an event P(not A) = 1 − P(A).

Example 24

In a fairground game, you throw a dart at a playing card on a revolving wheel to win a prize. The prizes are £10 for an Ace, and a cuddly toy for a picture card.

When Petra plays the game there are three possible events.

A Petra's dart hits an Ace: P(A) = 0.02.

B Petra's dart hits a picture card: P(B) = 0.08.

a What is the probability of Petra winning a prize?

b What is the probability of Petra not winning a prize?

a Let C be the event 'Petra wins a prize'.

   P(C) = P(A or B)

   $\quad$ = 0.02 + 0.08

   $\quad$ = 0.1

b P(not C) = 1 − P(C)

   $\quad$ = 1 − 0.1

   $\quad$ = 0.9

> Petra wins a prize for event A or event B, so add together P(A) and P(B).

> Since Petra will either win or not win, calculate 1 − P(C).

Exercise 7G

1 Which of these are mutually exclusive events?

   A A drawing pin falling head down and a drawing pin falling point down.

   B A student studying mathematics and studying French.

   C Rolling a dice once and getting a six and a one.

2 A, B and C are mutually exclusive events.
   P(A) = 0.2, P(B) = 0.4 and P(C) = 0.3.
   Work out

   a P(A or B)

   b P(A or C)

   c P(C or B)

3 Which of the following events are exhaustive?

   A The event 'getting a head' and the event 'getting a tail' when flipping a coin.

   B The event 'getting two heads' and the event 'getting a head and a tail' when flipping a fair coin twice.

   C The event 'winning a game of darts' and the event 'losing a game of darts'.

4 This fair spinner is spun.
The event X is a score > 4.
The event Y is an even number.
The event Z is a score ≤ 4.
Which of these are exhaustive events?

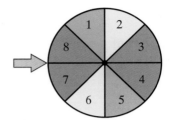

   A X and Y

   B X and Z

   C Y and Z

5 A random number table is used to give numbers between 1 and 8.

   a Work out the probability of getting a 1.

   b Work out the probability of getting a 2.

   c Work out the probability of getting any of the numbers 1 to 8.

6 A coin is biased in such a way that the probability of getting a head is 0.6. Work out the probability of not getting a head.

7 In a class of 30 children, 10 have a cooked breakfast only and 5 have cereal only. The rest have just toast.
A is the event 'have a cooked breakfast'.
B is the event 'have cereal for breakfast'.
C is the event 'have toast for breakfast'.

Work out

   a how many children have just toast

   b the probability that a child has a cooked breakfast

   c the probability that a child has just cereal

   d P(A or B)

   e P(A or C)

   f P(not A)

8 The probability of having black hair is 0.6.
The probability of having red hair is 0.2.
The probability of having blonde hair is 0.15.
These events are mutually exclusive.

   a Work out the probability of having either black or blonde hair.

   b Work out the probability of having black, red or blonde hair.

   c Work out the probability of not having red hair.

7.10 The multiplication law for independent events

> Two events are **independent** if the outcome of one event does not affect the outcome of the other event.

The probability of two independent events, A and B, happening is written P(A and B).

> For two independent events, *A* and *B*
>
> P(*A* and *B*) = P(*A*) × P(*B*)
>
> This is called the **multiplication law** for independent events.

Example 25

Helen likes music.

The probability that her grandmother buys her a CD for her birthday is 0.7.

The probability that her mother buys her a CD for her birthday is 0.5.

What is the probability that she gets a CD as a present from both her mother and her grandmother. (Assume that her mother and grandmother do not discuss what they are each going to buy Helen for her birthday.)

G is the event that 'she gets a CD from her grandmother' and *M* is the event that 'she gets a CD from her mother'.

> Give each event a letter.

P(*G*) = 0.7 and P(*M*) = 0.5

P(*G* and *M*) = P(*G*) × P(*M*)

$\qquad$ = 0.7 × 0.5

$\qquad$ = 0.35

> Use the multiplication rule because the two events are independent. (The question states the mother and grandmother do not talk about what they are going to buy.)

Examiner's Tip
The events must be independent to use this multiplication rule.

> The multiplication law can be extended for three or more independent events. Simply multiply the probabilities.
>
> P(*A* and *B* and *C*) = P(*A*) × P(*B*) × P(*C*)

Higher Statistics

Example 26

An office has bought three computers. The probability of a computer breaking down in the first year is 0.03. What is the probability that all three break down in the first year? (Assume that the computers do not fail because of a faulty part that is built into each one.)

The three events are independent.

A is the event the first computer breaks down in the first year.

B is the event the second computer breaks down in the first year.

C is the event the third computer breaks down in the first year.

> Give each event a letter.

$$P(A \text{ and } B \text{ and } C) = P(A) \times P(B) \times P(C)$$
$$= 0.03 \times 0.03 \times 0.03$$
$$= 0.000\ 027$$

> Using the multiplication law, multiply $P(A)$ by $P(B)$ by $P(C)$.

Exercise 7H

1   Which of these are independent events?
   A Rolling a six on a dice and picking an ace from a pack of cards.
   B Having red hair and being very tall.
   C Being male and being bald.

2   A, B and C are independent events.
   P(A) = 0.3, P(B) = 0.2 and P(C) = 0.4.
   Work out
   a P(A and B)
   b P(B and C)
   c P(A and C)

3   The probability of a student taking a packed lunch to school is 0.7. The probability of a student walking to school is 0.6.
   a Are these events independent?
   b What is the probability of a student walking to school and taking a packed lunch?

4   The probability of Yoko going sailing on a Tuesday is $\frac{1}{7}$. The probability that she will have pasta for dinner on a Tuesday is $\frac{4}{5}$. These are independent events. Calculate
   a the probability that she will go sailing and have pasta for dinner on a Tuesday
   b the probability that she will not sail on a Tuesday
   c the probability that she will not sail and not have pasta on a Tuesday.

5 The probability of a football player scoring a goal is $\frac{1}{12}$. The probability that he will be injured in a match is $\frac{1}{40}$. The probability of his team winning is $\frac{2}{3}$. These events are independent. Work out the probability that

  a he scores and is injured

  b he scores and his team wins.

6 An alarm system has a stand-by battery that keeps the system working when the main electrical supply fails. The probability of a supply failure in any given week is 0.04. The probability of a battery failing in any given week is 0.15. What is the probability of both failing in any given week?

7 The probability of a woman having toast for breakfast is 0.3. The probability of her newspaper being delivered on time is 0.75. The probability that she will go to work in her car is 0.2. Assuming these are independent, work out the probability that

  a she will have toast for breakfast and her newspaper will be delivered on time

  b she will not go to work by car

  c she will have toast for breakfast and not go to work by car.

7.11 The addition law for events that are not mutually exclusive

P(A) is the probability of A occurring. This is represented on the Venn diagram by the shaded area.

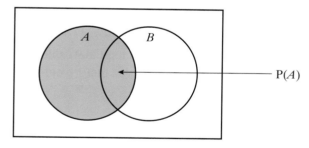

P(A and B) is the probability that both A and B occur.
This is the **intersection** of A and B, which is shaded on the Venn diagram. It is often written P(A ∩ B).

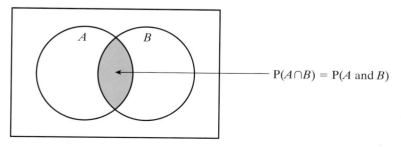

The events are not mutually exclusive, so P(A or B) is the probability that either A or B occur, or that both occur. This is the **union** of A and B, which is shaded on the Venn diagram. It is often written P(A ∪ B).

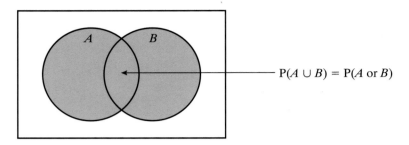

$$P(A \cup B) = P(A \text{ or } B)$$

> The addition law for events that are not mutually exclusive is
>
> $P(A \cup B) = P(A) + P(B) - P(A \cap B)$

Example 27

A card is chosen at random from a pack of 52 playing cards. R is the event 'the card is red'. Q is the event 'the card is a picture card'.
Find P(R ∪ Q).

Note: only the Jack, Queen and King are picture cards.

$P(R) = \dfrac{26}{52}$

$P(Q) = \dfrac{12}{52}$

Work out P(R), P(Q) and P(R ∩ Q).

$P(R \cap Q) = \dfrac{6}{52}$

$P(R \cup Q) = \dfrac{26}{52} + \dfrac{12}{52} - \dfrac{6}{52}$

$\qquad\quad = \dfrac{32}{52}$

$\qquad\quad = \dfrac{8}{13}$

Use the addition rule for events that are not mutually exclusive.

$P(R \cup Q) = P(R) + P(Q) - P(R \cap Q)$

(The events are not mutually exclusive because there are red picture cards.)

Exercise 7I

Answer these questions using the addition law for non-mutually exclusive events.

1   The probability that a person wears glasses is 0.4.

The probability that a person is right-handed is 0.8.

The probability of a person wearing glasses and being right-handed is 0.3.

A person is selected at random.

What is the probability they are right-handed or wear glasses?

2 A survey of 100 people is carried out to find out whether they took holidays in Britain or abroad in 2008.

Sixty-seven said they went abroad, 23 said they stayed in Britain, and 13 said they stayed in Britain and went abroad.

A is the event 'they went abroad' and *B* is the event 'they stayed in Britain'. Find P($A \cup B$).

3 In a litter of 12 collie puppies there are seven females, five tri-coloured puppies and two tri-coloured females. A puppy is selected at random.

a Work out

  i) P(*tri-coloured puppy*)

  ii) P(*female*)

  iii) P(*female and tri-coloured*)

b Calculate the probability that the puppy selected is a female or tri-coloured.

4 Two hundred people who live in Trumpingdon regularly visited the library.

Fifty-five visit on weekdays, 155 visit on Saturday, and 10 visit both on weekdays and a Saturday.

A is the event 'they visit on weekdays' and *B* is the event 'they visit on Saturday'. Find P($A \cup B$).

5 A survey shows that 90% of the households in Tovill own a TV, 58% own a laptop, and 50% of the households have both.

A household is chosen at random.

a Find the probability the household owns either a TV or a laptop or both.

b Work out the probability that the household has neither a TV nor a laptop.

7.12 Tree diagrams

A **tree diagram** can be used for combined events.

A **tree diagram** can be used to make calculations easier.

Each branch of the tree represents an outcome. The probability of the outcome is written on the branch.

Example 28

A bag contains five red balls (R) and four green balls (G). A ball is chosen at random, the colour noted and the ball is then replaced in the bag. A second ball is then chosen and the colour noted.

a Draw a tree diagram to represent this information.

b What is the probability of getting one ball of each colour?

Draw the branches and write the outcome at the end of each branch.

Write the probability on each branch.

Then write the outcome of each path.

Finally calculate the probability of each outcome by multiplying together the probabilities on the branches of the path taken.

a

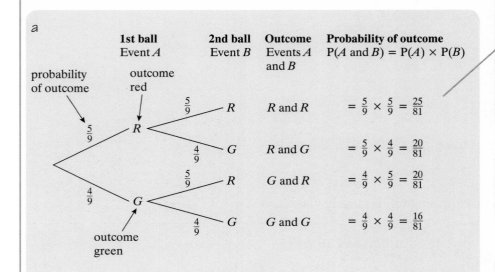

	1st ball Event A	2nd ball Event B	Outcome Events A and B	Probability of outcome P(A and B) = P(A) × P(B)
		R	R and R	$= \frac{5}{9} \times \frac{5}{9} = \frac{25}{81}$
	R	G	R and G	$= \frac{5}{9} \times \frac{4}{9} = \frac{20}{81}$
	G	R	G and R	$= \frac{4}{9} \times \frac{5}{9} = \frac{20}{81}$
		G	G and G	$= \frac{4}{9} \times \frac{4}{9} = \frac{16}{81}$

Examiner's Tip

Remember the sum of the probabilities of the outcomes is 1.

$\frac{25}{81} + \frac{20}{81} + \frac{20}{81} + \frac{16}{81} = 1$

b P(*getting one of each colour*) = P(*R and G*) + P(*G and R*)

$$= \frac{20}{81} + \frac{20}{81}$$

$$= \frac{40}{81}$$

Use the addition law.

P(A or B) = P(A) + P(B)

Add together the outcomes at the end of the branches for 'R and G' and 'G and R'.

Note that each path through the tree branches from left to right and produces a different outcome. There is a differentiation between R followed by G and G followed by R.

Example 29

A company is going to employ three new recruits and it interviews equal numbers of males and females. The recruits are equally likely to be male or female because of the company's equal opportunities policy.

Draw a tree diagram and use it to find the probability of all three recruits being female.

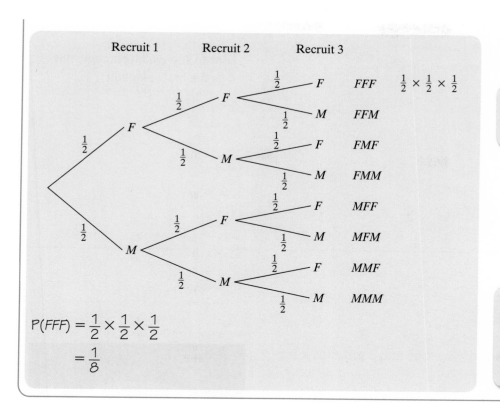

Draw the tree diagram and write on it the outcomes and the probability of each outcome.

Look for any outcomes that have three *F*s. There is only one.

Calculate the probability by multiplying the probabilities of each individual event.

$$P(FFF) = \frac{1}{2} \times \frac{1}{2} \times \frac{1}{2}$$

$$= \frac{1}{8}$$

Sometimes one branch of a tree may end earlier than others.

Example 30

There are 4 balls in a bag. One ball is Red, the rest are Black. A person is allowed to select a ball from the bag. If they select a Red ball the person wins the game. If they select a Black ball it is replaced and the person tries again. They are allowed 3 tries. If they have not selected the red ball after 3 tries they lose the game.

a   Draw a tree diagram to represent this information.

b   Work out the probability

   i) winning the game on the first go   ii) losing the game.

Work out the probability of selecting the Red ball, P(*R*), and the probability of selecting a Black ball, P(*B*).

Draw the branches and label them.

Stop drawing a set of branches when *R* is selected or three *B*s are found.

Write the probability on each branch, then write the outcome of each path.

Finally calculate the probability of each outcome.

a $P(R) = \dfrac{1}{4}$

$P(B) = 1 - \dfrac{1}{4} = \dfrac{3}{4}$

1st go	2nd go	3rd go	Outcome	Probability of outcome
$\frac{1}{4}$ R			R	$= \frac{1}{4}$
	$\frac{1}{4}$ R		BR	$= \frac{3}{4} \times \frac{1}{4} = \frac{3}{16}$
$\frac{3}{4}$ B		$\frac{1}{4}$ R	BBR	$= \frac{3}{4} \times \frac{3}{4} \times \frac{1}{4} = \frac{9}{64}$
	$\frac{3}{4}$ B			
		$\frac{3}{4}$ B	BBB	$= \frac{3}{4} \times \frac{3}{4} \times \frac{3}{4} = \frac{27}{64}$

b　i　P(wins first go) = P(R)

$$= \frac{1}{4}$$

Look to see which outcome has
R as the first selection.

ii　P(lose)　　= P(BBB)

$$= \frac{27}{64}$$

Exercise 7J

1　A boy picks a card from a pack of 52 cards and a girl flips a fair coin.

　　a　Work out the probability that the boy picks a red card.

　　b　Work out the probability that the girl gets a head.

　　c　Copy and complete the tree diagram.

Card colour	Result of coin flip	Outcome	Probability of outcome

```
                          H        B H
              B
      0.5            
                          T        B T

                          H
              R
                          T
```

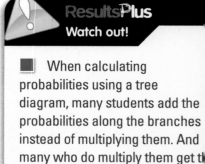

　　d　What is the probability of the boy getting a red card and the girl getting a tail?

2　A box contains three red beads and four green beads. A bead is drawn from the box and its colour is noted. The bead is returned to the box and after the box is shaken a second bead is drawn.

　　a　Draw a tree diagram to show all the different outcomes for the two bead colours.

　　b　Work out the probability of getting two red beads.

　　c　Work out the probability of getting one red and one green bead.

3 A researcher randomly selected a factory for study from a list of 20 factories.

Eight of the factories were in the north of the country and the rest in the south.

Six of the northern factories do heavy engineering, the rest do light engineering.

Six of the factories in the south do heavy engineering.

a Work out the probability that a factory in the north will do heavy engineering.

b Work out the probability that a factory in the south will do heavy engineering.

c Draw a tree diagram to represent the information given.

d Work out the probability that a factory chosen at random will be doing heavy engineering in the north of the country.

e Work out the probability that the factory chosen will be doing light engineering.

4 In a survey of 60 women and 40 men, 60% of each gender said that they smoked cigarettes.

A person chosen at random is sent a follow-up questionnaire. Draw a tree diagram to show the probability of the person chosen being a man/woman who smokes/does not smoke.

5 This tree diagram shows the events

It will rain tomorrow (R)

It will not rain tomorrow (NR)

John will be selected to play cricket for the first team (S)

John will not be selected for the first team (NS)

The team only plays if it does not rain.

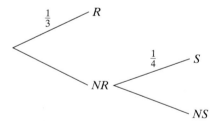

a Complete the tree diagram.

b Work out the probability of John playing cricket tomorrow.

c Work out the probability of the game being played, but John not being in the team.

6 Three fair plastic discs have a 1 on one side and a 6 on the other.

a Use a tree diagram to show all the possible results when the three discs are flipped.

b Work out the probability of getting exactly two sixes.

c Work out the probability of getting at least two sixes.

7 A small factory employs 5 men and 15 women. They make bags in two types of leather: A and B.

Out of every 20 bags, 14 are made of leather A.

Two different types of bag are made: a shoulder bag and a handbag. Both types are made in equal numbers.

The government minister for trade is to visit the factory. The manager cannot decide which bag the minister would like to take away as a reminder of her visit, so he picks a bag at random.

a Draw a tree diagram to show the probability of the bag being made by a man/woman from type A/B leather and being a shoulder bag/handbag.

b Work out the probability that the minister will get a handbag in type A leather made by a man.

c Work out the probability that she will either get a shoulder bag made by a man in type B leather, or a handbag made by a woman in type A leather.

ResultsPlus
Exam Question Report

Question: David applies for two jobs.
The probability that he will get an interview for the first job is 0.4.
The probability that he will get an interview for the second job is 0.3.

(a) Work out the probability that he will **not** get an interview for the first job. (2 marks)

(b) Work out the probability that he will **not** get an interview for the first job and that he will **not** get an interview for the second job. (2 marks)

Poor   53%

Most students scored zero marks because they worked out the probabilities incorrectly in both parts (a) and (b).

Good   1%

Most students either answered the whole question correctly or were very unsure. A common incorrect answer to part b) was:

$$0.6 + 0.7 = 1.3, \frac{1.3}{2} = 9.65.$$

Excellent   46%

Part a) correct. In part (b) few students drew a tree diagram. They correctly calculated the probability as $0.6 \times 0.7 = 0.42$.

7.13 Conditional probability

When the probability of an event happening depends on a previous event having happened it is called **conditional probability**.

> **Conditional probability** is the probability of *A* given that *B* has already happened.
>
> It is written as P(*A*|*B*).

Example 31

John takes his raincoat to school on some days but not on others.
He is more likely to take his raincoat if it is raining or if it looks like rain.

The probability of John taking his raincoat if it is raining or it looks like rain is 0.9.

The probability of John taking his raincoat if it looks as if it is not going to rain is 0.2.

Write these as conditional probabilities.

P(*John takes his raincoat **given** that it rains or looks like rain*) = 0.9

P(*John takes his raincoat **given** that it does not look like rain*) = 0.2

> Write the probability that John takes his raincoat given that it rains or looks like rain.

> Write the probability that John takes his raincoat given that it does not look like rain.

> **Examiner's Tip**
> Notice the word 'given' in each of these statements.
> This is often used to imply conditional probability.

Example 32

A delivery from a sub-contractor contains nine similar components. Four of the components are faulty. A component is chosen at random, and not replaced. A second component is then chosen and both components are checked for faults.

a Work out the probability that
  i) the second component is accepted given the first is accepted,
  ii) the second component is accepted given the first is faulty.
b Draw a tree diagram to represent this information. Include all the possible outcomes and their probabilities.
c What is the probability of
  i) two acceptable components being selected?
  ii) the two components both being either accepted or both being rejected?
  iii) at least one component being accepted?

a **i** If the first component is accepted there are eight components left of which four are acceptable.

P(*second acceptable given first acceptable*)

$$= \frac{4}{8}$$

$$= \frac{1}{2}$$

> Work out how many are left after the first is checked (8) and how many of those left are acceptable (4).

ii If the first component is faulty there are eight components left of which five are acceptable.

$$= P(\textit{second acceptable given first is faulty})$$

$$= \frac{5}{8}$$

> Work out how many are left after the first is checked (8) and how many of those left are acceptable (5).

b Use *F* for 'faulty' and *A* for 'acceptable'.

	1st component	2nd component	Probability of outcome

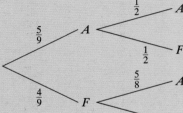

$$AA = \frac{5}{9} \times \frac{1}{2} = \frac{5}{18}$$

$$AF = \frac{5}{9} \times \frac{1}{2} = \frac{5}{18}$$

$$FA = \frac{4}{9} \times \frac{5}{8} = \frac{20}{72} = \frac{5}{18}$$

$$FF = \frac{4}{9} \times \frac{3}{8} = \frac{12}{72} = \frac{1}{6}$$

> Draw the branches and write the outcome at the end of each branch.
> Write the probability on each branch using the ones you calculated in part **a**.
> Write the outcome of each path.
> Calculate the probability of each outcome.

c **i** $P(AA) = \dfrac{5}{18}$

> Give P(*AA*).

ii $P(AA \text{ or } FF) = P(AA) + P(FF)$

$$= \frac{5}{18} + \frac{1}{6}$$

$$= \frac{8}{18}$$

$$= \frac{4}{9}$$

> Add together P(AA) and P(FF).

iii $P(\textit{at least one acceptable}) = 1 - P(\textit{neither acceptable})$

$$= 1 - \frac{1}{6}$$

$$= \frac{5}{6}$$

> Add together the probabilities P(*AA*), P(*FA*) and P(*AF*) or calculate 1 − P(*FF*) as shown here.

7.14 Multiplication law for events that are not independent

Two events are not independent if the outcome of one event affects the outcome of the other event.

If A and B are NOT independent then $P(A \cap B) = P(B|A) \times P(A)$

Example 33

The attendance at a village fête depends on the weather.

The probability of a high attendance at the fete is 0.9.

This is reduced to 0.4 if it is raining.

The probability of it raining is 0.2.

What is the probability of it raining **and** there being a high attendance?

R is the event 'it rains'.

H is the event 'high attendance'.

> Define the events using letters.

$P(H|R) = 0.4$
$P(R) = 0.2$

> State the probabilities.

$P(R \cap H) = P(H|R) \times P(R)$
$\qquad = 0.4 \times 0.2$
$\qquad = 0.08$

> Use the multiplication rule for events that are not independent. The events are not independent because the weather affects the probability of a high attendance.

Exercise 7K

1 A bag contains 5 blue balls and 7 red balls. The balls are taken from the bag one at a time and are not replaced.

  a Work out the probability that the first ball is blue.

  b Work out the probability that of the first three balls removed at least two are blue.

2 A bag contains 20 marbles. 5 of the marbles are red and the rest are white. Marbles are taken from the bag one at a time without replacement. Work out the probability that

  a the second marble is red if the first is not red

  b the first three marbles are red

  c the first two marbles are red and the third is white.

3   A bowl contains 27 coloured glass pebbles.

12 of the pebbles are blue, and the rest are red.

A customer selects 3 pebbles at random. The pebbles are not replaced.

a Draw a tree diagram to represent this information.

b Work out the probability that the pebbles are all the same colour.

c Work out the probability that exactly two pebbles are red?

4   In a group of 30 adults, 7 have blonde hair.

a Two adults are chosen at random. Work out the probability that they both have blonde hair.

b Three adults are chosen at random. Work out the probability that exactly two have blonde hair.

5   An investment group of ten financiers have a meeting to vote on their investment strategy for the following three months. Three vote to sell their investments. The rest of the financiers vote not to sell their investments. After the meeting two of them are chosen at random and interviewed on television one after the other.

a Draw a tree diagram to show the ways in which the financiers interviewed could have voted.

b What is the probability that

i) neither voted to sell

ii) one of them voted to sell?

6   A certain medical disease occurs in 2% of the population. A simple screening procedure is available. In 9 cases out of 10 where the patient has the disease the test gives a positive result.

If the patient does not have the disease the test gives a positive result in 5 out of 100 cases.

a Draw a tree diagram to represent this information.

Use the tree diagram to find the probability that a randomly selected person

b does not have the disease and gives a positive result

c gives a positive result

d has the disease.

7   Two computers in a batch of six are known to be faulty but it is not known which two they are. Two machines are picked at random.

a) Draw a tree diagram to show all possible results.

b) Find the probability that

i) they are both faulty

ii) at least one is faulty.

Chapter 7 review

1 A magician puts a set of ten numbered counters in a hat.

The counters are numbered 1 to 10.

A woman takes a counter at random from the hat.

The events A, B and C are

  A She takes the number 10.

  B She takes an odd number.

  C She takes a number greater than 6.

  a Copy this probability scale and mark the events A, B and C on it.

  b For this set of counters, suggest a different event that has the same probability as event B.   edexcel ⠿ *past paper question*

2 A packet contains 12 transistors, 3 of which are known to be faulty. One transistor is chosen at random and tested.

  a What is the probability that the transistor is faulty?

  b What is the probability that the transistor is not faulty?

3 A farmer wants to find out if a vaccine can stop his sheep getting foot rot.

He uses a sample of 100 sheep that do not have foot rot.

He vaccinates 60 of these sheep.

The two-way table below shows the results after a period of time.

The effect of vaccine on foot rot

	Number with foot rot	Number without foot rot	Total
Vaccinated	10	50	60
Not vaccinated	20	20	40
Total	30	70	100

He chooses one of the 100 sheep at random.

a Write down the probability that the sheep

  i) does not have foot rot

  ii) had been vaccinated and has foot rot.

b Did the vaccine help to stop foot rot? Write down the reason for your answer.

c The farmer did not give the vaccine to all of the 100 sheep.

Explain why.   edexcel ⠿ *past paper question*

4   This two-way table shows the result of a study by a dentist of 100 patients. Each patient was asked whether or not they smoked.

 *challenge*

The dentist then looked to see how many of their teeth had been taken out.

	Smokers	Non-smokers	Total
One or more teeth taken out	7	13	
No teeth taken out	15		80
Total	22		100

a   Copy and complete the table.

b   One of the patients is chosen at random. What is the probability that
   i) he/she is a non-smoker
   ii) he/she is a smoker who has had no teeth taken out?

c   What effect does smoking appear to have on the number of teeth taken out?

5   A health trust does a survey to see how many of its hospital beds have access to a radio and/or a television.

It finds that 21% have access to televisions only, 12% have access to radios only, and 2% have access to neither.

a   Copy and complete the Venn diagram.

 *challenge*

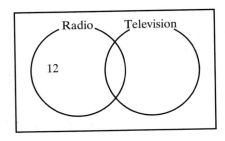

b   Find the probability that a bed chosen at random will have access to a radio or a television but not both.

c   Find the probability that a bed chosen at random will not have access to a television.

d   Given that a patient is in a bed with access to a radio, what is the probability that the bed also has access to a television?   **edexcel** ⣿ *past paper question*

 *challenge* *challenge* *challenge*

6   There are 23 different types of a particular plant (*Ranunculus*) growing in the wild in Britain.

14 of the types have yellow (Y) flowers, 14 have 30 or more stamens (S), and 13 have both yellow flowers and 30 or more stamens.

a Draw a Venn diagram to show this information.

b A plant is chosen at random. What is the probability of the plant having

    i) yellow flowers but less than 30 stamens

    ii) 30 or more stamens but not yellow flowers?

7 **a** What does it mean if two events are said to be 'independent'?

When a certain make of car breaks down a garage records the breakdown as being due to an electrical fault, due to a mechanical fault or due to other causes. (You can assume that electrical and mechanical faults do not occur together.) The probability of a breakdown in any year due to an electrical fault is 0.1 and to a mechanical fault is 0.05.

b **i)** What is the probability in any year that a car of this make will break down due to an electrical or mechanical fault?

    ii) What is the probability of the car not breaking down due to one or other of these faults?

The garage records the rest of the breakdowns, such as running out of petrol, as being due to other causes. The probability of a breakdown due to other causes in any one year is 0.01.

c What is the probability of the car not breaking down in any given year?

8 An aeroplane has two engines. The probability of an engine failing during any flight is 0.01. The plane is able to fly on one engine and the engines fail independently of each other.

a Draw a tree diagram to show the probabilities of the different combinations of engine failures.

b What is the probability of a successful flight?

A second aeroplane has an engine in each wing and a third one in the tail. This aeroplane flies if two engines are working. Assume that the engines have the same reliability as the engines on the two-engine planes.

c Draw a tree diagram to show the probabilities of the different combinations of engine failure.

d What is the probability of the plane making a successful flight?

e Would you prefer to travel in the two-engine or three-engine plane? Give a reason for your answer.

9 A box contains 50 resistors of which 5 are faulty. A resistor is taken from the box, and not replaced. A second resistor is then taken from the box.

  a What is the probability that the second resistor is faulty given that the first was faulty? ☆☆☆☆★ *challenge*

  b What is the probability that the second resistor is faulty given that the first was not faulty? ☆☆☆☆★ *challenge*

  c Draw a tree diagram to show the probability of the different combinations of faulty/non-faulty resistors. ☆☆☆☆☆ *challenge*

  d What is the probability of one or less resistors being faulty? ☆☆☆☆☆ *challenge*

Chapter 7 summary

Events and outcomes

1 A **trial** is the act of testing/doing something.

2 **Outcomes** are the possible results of a trial.

3 An **event** is a set of one or more successful outcomes.

The meaning of probability

4 **Probability** is a numerical measure of the chance of an event happening.
  - A probability of 0 means it is impossible for the event to happen.
  - A probability of 1 means the event is certain to happen.

5 Probabilities can be written as fractions, decimals or percentages.

6 When something is chosen 'at **random**', it means it is selected without a conscious choice.

7 When outcomes have the same chance of happening they are called **equally likely outcomes**.

The probability of an event

8 If all possible outcomes are equally likely, the **probability of an event**

$$= \frac{\text{number of successful outcomes}}{\text{total number of possible outcomes}}$$

Experimental probability

9 The estimated probability that an event might happen

$$= \frac{\text{number of successful outcomes}}{\text{total number of trials}}$$

10 As the number of trials increase in experiments and surveys, the nearer an estimate for the probability should be to the true value.

11 **Simulation** is the imitation of the conditions of a situation by a theoretical study.

12 cost of risk insurance = money at risk × risk assessment (p)

Sample space

13 A list of all possible outcomes is called a **sample space**.

Venn diagrams

14 A **Venn diagram** represents a sample space and may be used to calculate probabilities. Sets are represented as circles and common elements of the set are represented by the areas where the circles overlap.

Mutually exclusive events

15 Outcomes/events are **mutually exclusive** if they cannot happen at the same time.

16 For two mutually exclusive events, A and B
$P(A \text{ or } B) = P(A) + P(B)$.

17 The addition law can be extended for three or more mutually exclusive events. Simply add the probabilities:
$P(A \text{ or } B \text{ or } C) = P(A) + P(B) + P(C)$.

Exhaustive events

18 A set of events is **exhaustive** if the set contains all possible outcomes.

The sum of the probabilities for a set of mutually exclusive, exhaustive events

19 For a set of mutually exclusive exhaustive events, the sum of all the probabilities $= 1$. This is written as $\sum p = 1$.

20 The probability of an event A not happening is written as P(not A).

21 Since an event either happens or does not happen
P(A) + P(not A) = 1.
So for an event P(not A) = 1 − P(A).

The multiplication law for independent events

22 Two events are **independent** if the outcome of one event does not affect the outcome of the other event.

23 For two independent events, A and B
P(A and B) = P(A) × P(B)

24 The multiplication law can be extended for three or more independent events. Simply multiply the probabilities.
P(A and B and C) = P(A) × P(B) × P(C)

The addition law for events that are not mutually exclusive

25 The addition law for events that are not mutually exclusive is
$P(A \cup B) = P(A) + P(B) - P(A \cap B)$

Tree diagrams

26 A **tree diagram** can be used to make calculations easier.
Each branch of the tree represents an outcome. The probability of the outcome is written on the branch.

27 Sometimes one branch of a tree may end earlier than others.

Conditional probability

28 Conditional probability is the probability of A given that B has already happened.
It is written as P(A | B).

Multiplication law for events that are not independent

29 Two events are not independent if the outcome of one event affects the outcome of the other event.
If A and B are NOT independent then
$P(A \cap B) = P(B \mid A) \times P(A)$.

Test yourself

1 Draw a probability scale from 0 to 1. Mark on it the probability of being born on a Saturday or a Sunday. Use the letter X.

2 A fair dice is rolled. What is the probability of getting an odd number?

3 This two-way table shows the number of boys and girls studying art and music.

	Boys	Girls	Total
Art	10		22
Music		6	
Total	16		

   a Copy and complete the table.

   b A student is chosen at random. What is the probability that

      i) the student is a boy

      ii) given that the student is a boy, he studies music?

4 Jonas has 14 books. 5 of them are novels. He picks a book at random. What is the probability that the book he chooses is not a novel?

5 A fair coin is flipped and a fair four-sided spinner is spun.

   a List the sample space of all the possible outcomes.

   b What is the probability of getting a head and an even number?

6 There are 58 workers in a factory. 27 make screws and 44 make bolts. Some workers make both.

   a Draw a Venn diagram to represent these data.

   b A worker is picked at random. What is the probability that he makes both?

7 A fair octagonal dice is rolled.

A is the event 'an even number'.

B is the event 'an odd number'.

C is the event 'a multiple of 4'.

Write down any pair of events that are exhaustive.

8 A is the event 'red hair'. B is the event 'male'.

$P(A) = 0.1$

$P(B) = 0.5$

Work out the probability of A and B.

9   P(*studying science but not history*) = 0.9.

P(*studying history but not science*) = 0.5.

P(*studying science and history*) = 0.7.

Work out P(A ∪ B).

10 Jamal throws a ball twice. The probability that he hits a coconut is 0.5.

   a Draw a tree diagram to show his two throws.

   b Use your tree diagram to work out the probability that Jamal gets one hit.

11 A is the event 'it rains'. B is the event 'going to the park'.

P(A) = 0.3

P(B|A) = 0.1

Find P(A ∩ B).

Chapter 8:
Probability distributions and quality assurance

After completing this chapter you should be able to

- use simple cases of the binomial distribution
- use simple cases of the discrete uniform distribution
- know the shape and properties of the normal distribution
- use the normal distribution to model populations

- plot quality control charts for means and medians or ranges
- understand target values, and action and warning limits
- understand what actions need to be taken when values lie between/outside the action and warning limits.

Clothes companies need to know how many clothes of each size they should make. To do this they must look at how the sizes of people are distributed. By the end of this chapter you will understand that clothes sizes can be approximated by a normal distribution.

8.1 Probability distributions

Rolling a fair six-sided dice has six possible, equally likely outcomes.

The sample space S = (1, 2, 3, 4, 5, 6).

If x is a particular successful outcome, then

$$p(x) = \frac{\text{number of successful outcomes}}{\text{total number of outcomes}} = \frac{1}{6}$$

where $p(x)$ means the probability of getting a particular value, x.

All outcomes in this case are equally likely.

Here are all the outcomes (x) and their probabilities.

x:	1	2	3	4	5	6
$p(x)$:	$\frac{1}{6}$	$\frac{1}{6}$	$\frac{1}{6}$	$\frac{1}{6}$	$\frac{1}{6}$	$\frac{1}{6}$

This is called a **probability distribution**.

> **Examiner's Tip**
> $\sum p = 1$
> The sum of the probabilities is 1.

> A **probability distribution** is a list of all possible outcomes together with their probabilities.

Example 1

A spinner is designed so that the probability of a 1 occurring is $\frac{1}{2}$. The numbers 2, 3, 4 and 5 each occur with frequency k.

a Write down the probability distribution of X in terms of k, where X is the number showing when the spinner is spun.

b i) Work out the value of k.

ii) Write down the probability distribution of X.

> **Examiner's Tip**
> X describes the set of all outcomes.
> x describes a particular outcome of X.

a x: 1   2   3   4   5

  $p(x)$: $\frac{1}{2}$ k k k k

b i) $\sum p = 1$

  $p(2) = p(3) = p(4) = p(5) = k$

so $\frac{1}{2} + k + k + k + k = 1$

       $\frac{1}{2} + 4k = 1$

       $4k = 1 - \frac{1}{2}$

       $= \frac{1}{2}$

so $k = \frac{\frac{1}{2}}{4}$

       $= \frac{1}{8}$

ii) x: 1   2   3   4   5

   $p(x)$: $\frac{1}{2}$ $\frac{1}{8}$ $\frac{1}{8}$ $\frac{1}{8}$ $\frac{1}{8}$

> *X* is the set of values 1, 2, 3, 4, 5.
> Write out the possible values (x).
> Add in the probabilities.

> Add the probabilities and put them equal to 1

> Solve the equation to find *k*.

> Write out the probability distribution using the probabilities rather than letters.

Exercise 8A

1 Four cars are being filled with petrol at pumps 1 to 4 in a petrol station. X is the pump number of the first car to finish being filled. Here is a probability distribution of X.

x:	1	2	3	4
p(x):	k	k	k	k

Work out the value of k.

2 Clothing is sold in five different size ranges labelled a, b, c, d and e. X represents the size. The probability distribution of X is shown.

x:	a	b	c	d	e
p(x):	0.1	k	0.3	0.1	0.2

Work out the value of k.

3 There are five faults that can cause a machine to break down.

Faults 1 and 5 are likely to happen with probabilities of 0.2 and 0.3 respectively.

Faults 2 and 4 are twice as likely to happen as fault 3.

X is the set of the faults.

The probability of fault 3 happening is k.

a Write down the probability distribution of X in terms of k.

b Find the value of k.

4 A circular disc is divided into seven sectors.

Sector 5 has the same angle as sector 1.

Sectors 2 and 6 have twice the area of sector 1.

Sectors 3 and 7 have three times the area of sector 1.

Sector 4 has four times the area of sector 1.

The circular disc is spun, and Y is the set of areas that could come to rest opposite a pointer.

The probability of $y = 1$ is s.

a Write down the probability distribution of Y in terms of s.

b Work out the value of s.

c Find the probability that $y < 3$.

8.2 Discrete uniform distributions

Example 2 shows a probability distribution known as a **discrete uniform distribution**. It is called discrete because the numbers 1, 2, 3, 4, 5 and 6 are discrete, and uniform because the probabilities are all the same.

Example 2

A fair six-sided dice is rolled. There are six possible outcomes. The sample space is X = (1, 2, 3, 4, 5, 6). Write down the probability distribution of X.

Examiner's Tip
Each outcome is equally likely.

The dice is fair so all outcomes are equally likely.

If x is a particular outcome, then

$p(x) = \frac{1}{6}$ for all x.

The probability distribution is

x:	1	2	3	4	5	6
p(x):	$\frac{1}{6}$	$\frac{1}{6}$	$\frac{1}{6}$	$\frac{1}{6}$	$\frac{1}{6}$	$\frac{1}{6}$

Find the probability of each outcome using

$$p(x) = \frac{\text{number of successful outcomes}}{\text{total number of possible outcomes}}$$

Write out the probability distribution.

A **discrete uniform distribution** has n distinct outcomes. Each outcome is equally likely.

The probability of any given outcome $= \frac{1}{n}$.

Example 3

A taxi is equally likely to pick up 1, 2, 3 or 4 passengers.

a Draw a probability distribution for the number of passengers, A.

b What is P(A < 3)?

Examiner's Tip
P(A < 3) means the probability that you will get one of the subset of A with a value less than 3 (i.e. a 1 or a 2).

a The distribution is uniform so all probabilities are the same, i.e. $p(a) = \frac{1}{4}$ for all values of a.

a:	1	2	3	4
p(a):	$\frac{1}{4}$	$\frac{1}{4}$	$\frac{1}{4}$	$\frac{1}{4}$

b P(a < 3) = p(1) + p(2)

$= \frac{1}{4} + \frac{1}{4}$

$= \frac{1}{2}$

Find the probability p(a) using $\frac{1}{n}$.

Remember, a is a particular value of A.

Exercise 8B

1 Which of the following distributions of X are likely to be modelled by a discrete uniform distribution? Explain your answer.

A X = the height of a student selected at random from a class of 7-year-olds.

B X = the day of the week on which an adult selected at random was born.

C X = the last digit of a telephone number selected at random from a telephone directory.

2 Given that X is the number showing when a fair dice is rolled, name the distribution of X.

3 A fair octagonal dice has the numbers 1 to 8 on its faces and X is the number that shows when the dice is rolled.

a Draw a table to show the probability distribution of X.

b Write down the probability that X = 4.

c Work out the probability that X < 3.

4 A dartboard has 20 equal sized sectors numbered 1 to 20. A dart is thrown to land in the number 20 sector. If the dart misses it is thrown again. A discrete uniform distribution is suggested as a model to describe the sector in which the dart lands. Comment on this suggestion.

5 In a lottery draw each of the numbers 1 to 50 is equally likely to chosen. X is the number drawn.

a Write down the name of the distribution of X.

b Work out the probability that X = 1.

6 A box of sweets contains equal numbers of orange, red, green, purple and brown sweets. X is the colour of a sweet chosen at random from the box.

Work out the probability that X = purple.

7 An economist is simulating the daily movement, X points, of a stock exchange indicator. The economist rolls a fair six-sided dice and if an odd number is uppermost the indicator is moved down that number of points. If an even number is uppermost he moves it up that number of points.

a Write down the name of the distribution of X.

b What is the probability that X < −1 point?

8.3 Binomial distributions

There are six possible outcomes when rolling a fair six-sided dice.

If you want a six then rolling a six is a success (s) and any other number counts as a failure (f).

If you roll one dice there are two possible results:

s (success)   f (failure)

If you roll two dice, one after the other, the events are independent and there are four possible results:

ss   sf   fs   ff

> **Examiner's Tip**
> Two events are independent if the outcome of one event does not affect the outcome of the other.

For three dice there are eight possibilities:

sss   ssf   sfs   sff   fss   fsf   ffs   fff

If the order of success and failure is unimportant then terms like *ffs*, *sff* and *fsf* are the same, (one six, and two other numbers). In which case these results can be written more concisely:

One dice			s		f	
Two dice		ss		2sf		ff
Three dice	sss	3ssf		3sff		fff

> **Examiner's Tip**
> Remember *ssf, sfs, fss* is the same as 3*ssf*.

For four and five dice the results would be:

Four dice		ssss	4sssf	6ssff	4sfff	ffff	
Five dice	sssss	5ssssf	10sssff	10ssfff	5sffff	fffff	

Now label the probability of success, (s), as p and that of failure, (f), as q. Since a six is rolled or a six is not rolled, $p + q = 1$ and the probabilities can be written as:

One dice				p		q		
Two dice			p^2		$2pq$		q^2	
Three dice		p^3		$3p^2q$		$3pq^2$		q^3
Four dice	p^4		$4p^3q$		$6p^2q^2$		$4pq^3$	q^4
Five dice	p^5	$5p^4q$		$10p^3q^2$		$10p^2q^3$	$5pq^4$	q^5

The entries shown are:

For one dice: the terms in the expansion of $(p + q)^1 = p + q$.

For two dice: the terms in the expansion of $(p + q)^2 = p^2 + 2pq + q^2$.

For three dice: the terms in the expansion of $(p + q)^3 = p^3 + 3p^2q + 3pq^2 + q^3$.

Etc.

> **Examiner's Tip**
> In an exam you will be given the binomial expansions for three or more events, but will be expected to know the expansion for two.

The probabilities for the events when n dice are rolled will be the terms of the expansion of $(p + q)^n$.

The power to which p is raised is the number of successful outcomes and the power to which q is raised is the number of failures. For example, $10p^3q^2$ gives the probability of three successful outcomes and two failures.

Probability distributions with two possible outcomes like this are known as **binomial distributions**. The distribution is defined with two other pieces of information: the number of possible trials n and the probability of success p. The binomial distribution can then be written for short as B(n, p). (q is not given since $q = 1 - p$.)

Examiner's Tip
A binomial distribution is often written as B(n, p).

The distribution B(5, 0.6) is binomial (hence the B). It has five possible trials with a probability of success of 0.6.

> A **binomial distribution** has a fixed number of independent trials n. Each trial has only two outcomes (success and failure). The probability of success is p. The probability of failure is q.
>
> The probabilities for the events of n binomial trials will be terms of the expansion of $(p + q)^n$.

Example 4

The probability that a seed, from a particular supplier, produces flowers when it is planted is 75%. Four seeds are planted.

Calculate the probability that

a exactly three of the seeds produce flowers

b less than two of the seeds produce flowers.

You may use $(p + q)^4 = p^4 + 4p^3q + 6p^2q^2 + 4pq^3 + q^4$.

Examiner's Tip
This is a binomial situation since there are only two outcomes: flowers or no flowers.

a $n = 4$

Write down the number of trials, n.

$p = 0.75$

$q = 1 - 0.75$
$\quad = 0.25$

Write down the probability of success, p, and calculate the probability of failure, $q = 1 - p$.

Probability of three flowers $= 4p^3q$

$\qquad\qquad = 4 \times 0.75^3 \times 0.25$

$\qquad\qquad = 0.422$

Write down the term required.

The number of successes required is three, so the term with p^3 in it is needed.

Substitute in the values of p and q.

b Probability of less than two flowers

$\quad = \text{P(0 flowers)} + \text{P(1 flower)}$

$= q^4 + 4pq^3$

$= 0.25^4 + (4 \times 0.75 \times 0.25^3)$

$= 0.0508$

Write down the terms required.

The number of successes required are 0 and 1, so the term with no ps in, and the term with one p in it are needed.

Exercise 8C

1 A distribution X is described as B(12, 0.325).

  a Write down the name of this distribution.

  b What do the numbers 12 and 0.325 represent?

2 A drug cures three people out of every five suffering from a disease.

  a Write down the probability of a person given the drug being cured.

  b Calculate the probability that if four people are given the drug exactly three will be cured.

    You may use $(p + q)^4 = p^4 + 4p^3q + 6p^2q^2 + 4pq^3 + q^4$.

> Let p = probability of cure. Then p^4 is the probability that all four are cured, $4p^3q$ is the probability that three are cured and one is not cured, etc.

3 85% of students who sit a statistics examination pass it. A group of three students sit the statistics examination. Calculate the probability that

  a all three pass

  b only one passes.

  You may use $(p + q)^3 = p^3 + 3p^2q + 3pq^2 + q^3$.

4 Five fair coins are flipped and the total number of heads shown is counted. Calculate the probability of

  a exactly one head showing

  b at least one head showing

  c the number of heads showing being greater than four.

  You may use $(p + q)^5 = p^5 + 5p^4q + 10p^3q^2 + 10p^2q^3 + 5pq^4 + q^5$.

5 Two people in ten will catch a cold this winter.

  a Write down the probability that a person will catch a cold this winter.

  b In a group of three people, calculate the probability that at most one catches a cold this winter.

  You may use $(p + q)^3 = p^3 + 3p^2q + 3pq^2 + q^3$.

6 On a particular road, the police stop cars in groups of three taken at random and check the tyres. One car in ten has faulty tyres. Work out the probability that

  a the first three stopped all have faulty tyres

  b none of the first three cars stopped have faulty tyres.

  You may use $(p + q)^3 = p^3 + 3p^2q + 3pq^2 + q^3$.

ResultsPlus
Watch out!

  ■ When asked to calculate the probability of exactly two out of three darts hitting a target and given the expansion for $(p + q)^3$, many students incorrectly used all the terms in their calculation, instead of the term for two successes and one failure: $3p^2q$. If your answer for a probability is greater than 1, you must have made a mistake!

7 The probability of a sheep producing twin lambs is 0.84. Two sheep are selected at random from a flock. Calculate the probability that

  a neither has twins

  b just one has twins.

8 The probability of a water pump being faulty is 0.05. A sample of five, randomly selected pumps are checked. Calculate the probability that

   a more than three are faulty

   b less than three are faulty.

You may use $(p + q)^5 = p^5 + 5p^4q + 10p^3q^2 + 10p^2q^3 + 5pq^4 + q^5$.

ResultsPlus
Watch out!

  ■ When asked to suggest a distribution to model the number of faulty electrical components in a sample, many students said 'mode' or 'frequency', suggesting that they didn't understand what 'distribution' means.

9 The probability that a certain make and model of car breaks down within two years of being bought is 0.1. A hire company buys five of these cars. Calculate the probability that

   a all five cars break down within the first two years

   b less than two of the cars break down within the first two years.

You may use $(p + q)^5 = p^5 + 5p^4q + 10p^3q^2 + 10p^2q^3 + 5pq^4 + q^5$.

10 The probability that a train travelling from Glasgow to Penzance is 30 minutes or more late is 0.15. Three travellers catch trains at different times and on different dates. Calculate the probability that

   a all three arrive less than 30 minutes late

   b only one arrives less than 30 minutes late.

You may use $(p + q)^3 = p^3 + 3p^2q + 3pq^2 + q^3$.

ResultsPlus
Exam Question Report

Question: Gordon is going to throw five stones in turn at the target. Gordon can hit the target with a probability of 0.8 with any one of these five stones.

(d) Name the probability distribution that models the number of times he will hit the target in the five throws.

(e) Work out the probability that he will hit the target with only one of the five stones.

[You may use $(p + q)^5 = p^5 + 5p^4q + 10p^3q^2 + 10p^2q^3 + 5pq^4 + q^5$.]

(f) Work out the most likely number of times he will hit the target.

[You may use $(p + q)^5 = p^5 + 5p^4q + 10p^3q^2 + 10p^2q^3 + 5pq^4 + q^5$.]

Poor   62%

Few candidates could identify the binomial distribution in part (d), even though the binomial expansion is given in the last two parts of the question. A common wrong answer was 'normal distribution'.

Good   11%

Most candidates couldn't calculate any probabilities in parts (e) and (f), but scored 1 mark for writing $q = 0.2$ in part (e).

Excellent   27%

Some students used the given binomial expansion in parts e) and f). In part f) a common mistake was to use the expectation formula $n \times p$ to calculate the most likely number of hits.

8.4 Normal distributions

Suppose you measure the weights of boys, the weight is a continuous variable. By grouping their weights you can draw a frequency density histogram.

A frequency density histogram for the weights of 100 boys might look like this.

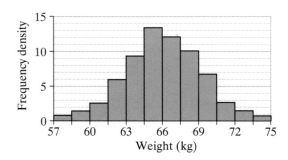

If the number of boys observed was increased to 200 and the class intervals halved the histogram would look something like this.

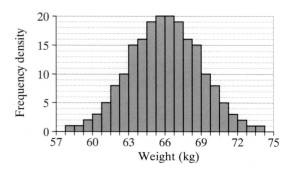

By doubling the number of boys observed and making the class intervals half the size, the outline of the histogram has become smoother. If this process is continued, then the outline of the histogram will eventually be a smooth curve like this.

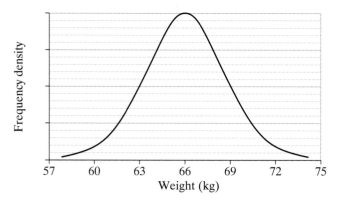

The curve is shaped like a bell and is called a bell-shaped curve. Weight is a continuous variable, so this smooth curve is needed to model boys' weights.

This is known as a **continuous** probability distribution.

This distribution is very important in statistics. Many variables have a bell-shaped distribution. A histogram showing the lengths of 100 oak leaves or the weights of Cox's apples would be roughly bell-shaped.

Observations like these are the results of natural processes, and natural processes lead to populations that have this bell-shaped curve.

Such variables are said to be normally distributed. The sketch opposite shows a **normal distribution** with mean μ.

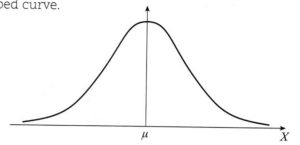

Two properties of a **normal distribution** are
- the distribution is symmetrical about the mean, μ
- the mode, median and mean are all equal (because the distribution is symmetrical).

8.5 Standard deviation and variance of a normal distribution

A normal distribution has a mean μ and a standard deviation σ.

Different values of μ and σ give different normal distributions.

Further properties of a normal distribution are
- 95% of the observations lie within $\pm$ two standard deviations of the mean.
- Virtually all (99.8%) lie within $\pm$ three standard deviations of the mean.

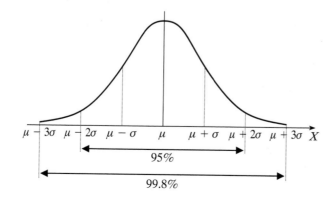

In each case half the area lays either side of the mean because of the symmetry.

47.5% lies between μ and $\mu + 2\sigma$ and 47.5% between μ and $\mu - 2\sigma$.

49.9% lies between μ and $\mu + 3\sigma$ and 49.9% between μ and $\mu - 3\sigma$.

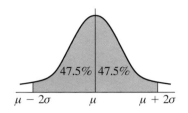 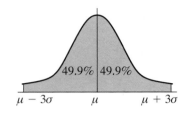

Examiner's Tip

It is often a good idea to sketch a normal distribution to show the area you are interested in.

The **variance** of a normal distribution is a measure of how spread out the data are.

Variance = (standard deviation)^2

Example 5

On the same axes, sketch the normal distributions, A and B, shown in the table.

	A	B
Mean	15	20
Standard deviation	3	5

A: $15 \pm (3 \times 3) = 6$ to 24.

B: $20 \pm (3 \times 5) = 5$ to 35.

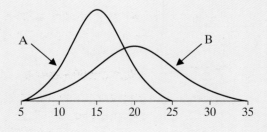

When sketching normal distributions you cannot sketch the whole curve since it goes from $-\infty$ to $+\infty$. Sketch three standard deviations either side of the mean.

Work out three standard deviations either side of the mean.

Sketch bell-shaped curves centred on the mean and ending at three standard deviations from the mean. Since the area under each curve has to be 1, draw the curve that has the larger range with a smaller maximum height.

Generally the number of standard deviations of a value from the mean can be worked out using

$$\text{number of standard deviations from mean} = \frac{\text{value} - \text{mean}}{\text{standard deviation}}$$

Example 6

The mark, X out of 100 in an examination is approximated by a normal distribution. The frequency distribution of X has a mean of 60 marks and a standard deviation of 6 marks. There are 1000 examinees.

a   **i)** Work out the mark that is two standard deviations above the mean.

    ii) Work out the two scores that you would expect almost all the scores to lie between.

b   How many examinees' marks would you expect to lie between

    i) 48 and 72 marks

    ii) 60 and 72 marks?

a i) $60 + (2 \times 6) = 60 + 12$

                  $= 72$

> Calculate $\mu + 2\sigma$.

  ii) $60 \pm (3 \times 6) = 42$ to 78.

> Calculate $\mu \pm 3\sigma$.

b i)   The mean score is 60, so 48 is $48 - 60$

     $= -12$ away from the mean.

     The standard deviation (SD) is 6, so 48 is $\dfrac{-12}{6}$

     $= -2$ from the mean.

     This can be written as $\dfrac{48 - 60}{6} = -2$.

     Work out how many SD 72 is from the mean.

     $\dfrac{72 - 60}{6} = 2$

> Work out how many standard deviations away from the mean 48 and 72 are.
>
> Use: $\dfrac{\text{score} - \text{mean}}{\text{standard deviation}}$

     48 is two standard deviations below the mean (indicated by the negative) and 72 is two standard deviations above the mean.

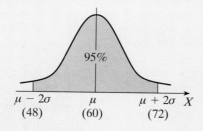

95%

$\mu - 2\sigma$ μ $\mu + 2\sigma$ X
(48)   (60)   (72)

> **Examiner's Tip**
> Sketch a diagram showing the area you want.

     There will be 95% of 1000 = 950 marks between 48 and 72.

  ii) Since the distribution is symmetrical, half the marks will be above 60 and half below 60.

     There will be $\dfrac{1}{2} \times 950 = 475$ between 60 and 72

> Work out 95% of 1000.
>
> (Remember 95% of marks lie within two standard deviations of the mean.)
>
> Work out 50% of 950.
>
> Half of the shaded area is between 60 and 72.

Example 7

Samples of size 10 are to be taken from a production line producing jars of coffee. The target weight of the coffee in a jar is 200 g. The means of the samples are normally distributed with a standard deviation of 2 g.

Calculate the limits between which you would expect

a 95% of the sample means to lie

b 99.8% of the sample means to lie.

a $\mu - 2\sigma = 200 - (2 \times 2)$
     $= 196\,g$
  $\mu + 2\sigma = 200 + (2 \times 2)$
     $= 204\,g$
  95% of the means will lie between 196 g and 204 g.

> Calculate $x = \mu - 2\sigma$ and $x = \mu + 2\sigma$.

b $\mu - 3\sigma = 200 - (3 \times 2)$
     $= 194\,g$
  $\mu + 3\sigma = 200 + (3 \times 2)$
     $= 206\,g$
  99.8% of the means will lie between 194 g and 206 g.

> Calculate $x = \mu - 3\sigma$ and $x = \mu + 3\sigma$.

Example 8

A long-life light bulb has a mean life of 12 000 hours and a standard deviation of 300 hours.

a Work out the probability that a light bulb chosen at random will

  i) last between 11 400 hours and 12 600 hours

  ii) last less than 11 400 hours.

b 5000 light bulbs are tested. Estimate how many of them would last longer than 12 600 hours.

a i) $\dfrac{12\,600 - 12\,000}{300} = 2$

   $\dfrac{11\,400 - 12\,000}{300} = -2$

Probability of lasting between 11 400 hours and 12 600 hours = 95%.

> Work out how many standard deviations away from the mean 12 600 and 11 400 are.
>
> Use number of SD
> $= \dfrac{\text{value} - \text{mean}}{\text{standard deviation}}$

> Work out how many standard deviations away from the mean 11 400 is.

ii) $\dfrac{11\,400 - 12\,000}{300} = -2$

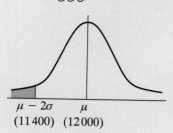

$\mu - 2\sigma$ μ
(11 400)   (12 000)

Probability of failing before $\mu - 2\sigma = \dfrac{100\% - 95\%}{2}$

$= 2.5\%\ (\text{or } 0.025)$

b   Probability of a bulb lasting more than 12 600 hours

$= \dfrac{100\% - 95\%}{2}$

$= 2.5\%$

2.5% of bulbs are expected to last more than 12 600 hours, so the number of light bulbs lasting more than 12 600 hours from a batch of 5000

$= 5000 \times 2.5\%$

$= 5000 \times \dfrac{2.5}{100}$

$= 125$

> The total area is 100%.
>
> The area between $\mu \pm 2\sigma$ is 95%.
>
> The curve is symmetrical, so the area for $< \mu - 2\sigma =$ the area for $> \mu + 2\sigma = \dfrac{5\%}{2} = 2.5\%$.

Exercise 8D

1   Which of the following might be modelled by a normal distribution? Explain your answer.

    A   The number of accidents each month on a stretch of road.

    B   The heights of adult females.

    C   The time it takes for a light bulb to burn out.

    D   The distance people travel to work.

2   **a**   What is the relationship between the mean, mode and median of a normal distribution?

    b   List the other properties of a normal distribution.

3   The adult lengths of a certain type of caterpillar are normal distributed with a mean length of 3.3 cm and a standard deviation of 0.8 cm.

    Calculate the two lengths between which nearly all the adu caterpillars lie.

4 The normal distributions A and B represent the weights of sacks of pre-packed potatoes from two different food producing companies. Sketch on the same axes the normal distributions A and B as described in the table below.

	A	B
Mean	20 kg	28 kg
Standard deviation	4 kg	6 kg

5 The random variable X represents the number of sweets in a packet. X has a distribution with a mean of 18 and a variance of 4X can be modelled by a normal distribution.

  a Write down the standard deviation of X.

  b Write down the value of x that is

    i) two standard deviations below the mean

    ii) three standard deviations above the mean.

> **Examiner's Tip**
> Remember:
> Standard deviation^2 = variance

6 The mean speed of vehicles on a road can be modelled by a normal distribution with a mean of 52.5 km/h and a standard deviation of 9 km/h.

Write down the speed of a vehicle that was travelling at

  a two standard deviations above the mean speed

  b two standard deviations below the mean speed

  c three standard deviations below the mean speed

  d three standard deviations above the mean speed.

7 The normal distribution shown below has a mean of 52 cm and a standard deviation of 6 cm. It represents the height of seedling beech hedging plants being sent out from a nursery to customers.

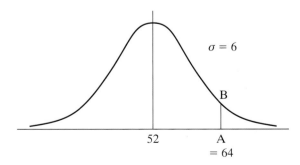

  a Calculate the percentage of the plants that will be less than 64 cm high (i.e. lie to the left of line AB).

  b Calculate the percentage of the plants that will be greater than 64 cm high (i.e. lie to the right of line AB).

  c Calculate the percentage of the plants that will have heights between 52 cm and 64 cm.

8   The length X of bamboo canes sold in a garden centre can be modelled by the normal distribution shown in the diagrams below. Work out the probability of a cane chosen at random falling in the shaded area of each diagram.

a

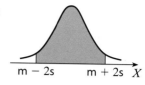

b

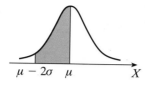

c

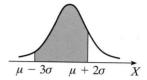

d

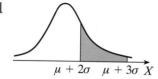

9   The heights of adult men are normally distributed with a mean of m and standard deviation of σ. Calculate the probability that a man chosen at random will have a height lying between $\mu + 2\sigma$ and $\mu + 3\sigma$.

10   The mean time it takes factory workers to get to a factory is 35 minutes. The time taken can be modelled by a normal distribution with a standard deviation of 6.5 minutes.

Calculate the percentage of workers that take

a   **i)** between 22 and 48 minutes to get to work

    ii) longer than 48 minutes to get to work.

b   There are 600 factory workers. Calculate how many will take between 22 and 48 minutes to get to work.

11   The weights of a group of 1000 school children were recorded. Their distribution can be modelled by a normal distribution with mean 42 kg and standard deviation 6 kg.

a   Calculate the percentage of children that you would expect to find with weights in the range

    i) 30 kg to 54 kg

    ii) 24 kg to 60 kg.

b   Calculate how many children you would expect to find in each of the size ranges in part **a**.

c   A child is selected at random. Work out the probability that the child's weight lies between 24 kg and 54 kg?

12 It has been found over many years that the temperatures, in °C, for June can be modelled by a normal distribution with a mean of 19 and a standard deviation of 3.5.

Estimate how many days in June will have a temperature:

a less than 26°C

b more than 26°C

c between 12°C and 26°C.

Give your answers to the nearest whole number.

13 Televisions have a mean life of 4000 hours and standard deviation 500 hours. Assume that their life can be modelled by a normal distribution. Estimate

a **i)** the probability of a television lasting less than 3000 hours

   ii) the probability that a television will last for between 3000 and 5000 hours.

A batch contains 10 000 televisions.

b Calculate after how many hours you would expect only $2\frac{1}{2}$% of the televisions to still be working.

14 The heights of a large number of students can be modelled by a normal distribution with mean 175 cm. 95% of students have heights between 160 and 190 cm. Work out the standard deviation of the students' heights.

15 Tennis balls are tested by dropping them from a given height and measuring their rebound height. Balls that rebound less than 128 cm are rejected. Assume that the rebound height can be modelled by a normal distribution with a mean of 134 cm and a standard deviation of 3 cm. Work out how many balls in a batch of 1000 will be rejected.

8.6 Quality assurance

A packet of crisps must have the weight of the contents marked on it. It might, for example, be marked 50 g. The manufacturers try to keep the weight as near as possible to this target value.

When products are made on a production line there will be some variation in the weight/size (quality) of the product.

The machinery will be set to produce a mean weight/size. This is called the target size. There is, however, bound to be some difference in the weights of crisps in a packet due to the manufacturing process.

These changes are unavoidable. The weight of crisps in any packet may not be exactly 50 g, however it is desirable that the mean weight of the packets produced should be 50 g. The difference between the minimum and maximum weights (range) should also stay constant.

Changes in the weight/size can also be caused by wear of tools etc. These changes will cause either the mean weight/size to change or the range of the weights/sizes to change.

If a change of either the mean or range occurs the process is stopped. Quality assurance gives warning of these changes.

To check the quality of the product, samples are taken at regular time intervals, and time series charts called **control charts** are constructed for both the sample mean and the sample range.

> A **control chart** is a time series chart that is used for process control.

Quality control charts for means

The diagram shows a control chart used for the mean weight of a packet of crisps.

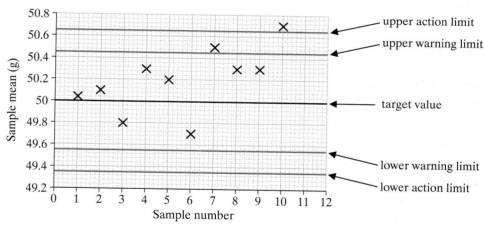

A line on the chart indicates the target value of 50 g for the mean weight.

Warning limits are set so that 95% of the means of the samples should lie between them. Mean sample weight is normally distributed, so 95% of values lay within two standard deviations of the mean. Therefore, the warning limits are set at $\mu \pm 2\sigma$ (here σ is the standard deviation of the sample mean).

This means that only 5% of the means should fall outside the warning limit. 5% is the same as saying one out of every 20 means will fall outside the warning limit.

If a sample mean falls between the warning limit and the action limit it is usual to take another sample just to check that nothing has gone wrong and that this is the 1 in 20 chance. If it falls between the warning limits the process is under control.

> Note: σ will be different for different sample sizes.

Warning limits are usually set at $\mu \pm 2\sigma$.

If a sample mean is between the warning limits the process is in control and the product is acceptable.

Action limits are set so that all the means should lie within them. As mean sample weight is normally distributed, 99.8% of values lie within three standard deviations from the mean. Therefore the action limits are set at $\mu \pm 3\sigma$. Only 2 in 1000 means fall outside the action limits. If a mean falls outside the action limits the process is assumed to have gone wrong and it is stopped so the machine can be reset.

Action limits are usually set at $\mu \pm 3\sigma$.

If a sample mean is between the warning and action limits another sample is taken immediately to see if there might be a problem.

If a sample mean is outside the action limits the process is stopped and the machinery is reset.

Example 9

For the control chart on p. 311, describe what actions would have been taken.

Sample 7 lies between the warning and action limits and would cause another sample to be taken immediately.

Sample 10 lies outside the action limits and would have caused the process to stop. The machine would have been reset.

ResultsPlus
Better build answers!

Question: A machine fills packets with sugar.
Every hour Peter takes a sample of 10 packets.

(a) Write down the name given to this method of sampling.

Peter calculates the mean weight for each sample of 10 packets. These mean weights are normally distributed with a mean of 505 g and a standard deviation of 1.6 g.

(b) Write down the percentage of the samples that have a mean weight between ± 2 standard deviations of 505 g.

The allowable limits for the mean weights of the samples are 505 g $\pm$ 3 standard deviations.
Peter takes a sample, it has a mean weight of 500 g.

(c) What should Peter do now? Show your working. (5 marks)

■ **Basic   1–2 mark answers**
In part a) recognise that this is systematic sampling. In part b) remember that in a normal distribution 95% of the means lie within ± 2 standard deviations.

● **Good   3–4 mark answers**
Complete parts a) and b) correctly. In part c) calculate the critical value for the allowable lower limit, using 505 g − 3 standard deviations.

▲ **Excellent   5 mark answers**
Complete parts a) and b) correctly and in part c) calculate the critical value for the allowable lower limit. Compare the sample mean with this limit. Suggest what Peter should do next: mean ± 3 standard deviations are action limits, as 99.8% of means should fall in this range. If not, the process should be stopped immediately and the machinery checked.

Quality control charts for medians

In some circumstances the median is plotted instead of the mean. The median is easier to find and can be plotted quicker. The control chart for medians looks the same as the control chart for means.

Quality control charts for ranges

As well as checking that the mean size does not vary it is also necessary to check that the range (maximum size − minimum size) does not vary.

Ideally the range would be zero (i.e. each part produced being equal in size). However, the range of sizes of the items produced depends on the accuracy of the machine. As the range depends on the machine being used, no target value for the range is set. If a sample range falls outside the action limits it is a sign that the machine is no longer producing parts to its normal accuracy and needs investigating.

> The setting of limits for range are more complex than those for means. You will only be asked to plot the points and state what action should be taken.

Quality control charts for ranges have action and warning limits, the same as the charts for means.

A control chart for sample ranges of the weight of crisp packets is shown below.

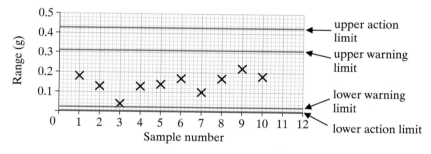

Sample ranges within the warning limits are acceptable. The range in this instance is under control because all points lie within the warning limits

Sample ranges between the warning and action limits would cause another sample to be taken.

If any range is outside the warning limits the process is stopped.

Sometimes the lower warning and action limits on a range chart are omitted.

Example 10

A machine produces pins whose target length is 5.04 cm. Samples of pins are taken and their lengths measured. The mean length of the samples is 5.04 cm and the standard deviation is 0.02 cm.

The mean length of the samples is normally distributed.

a Between what lengths would you expect these percentages of the mean sample lengths to lie?

   i) 95%   **ii)** 99.8%

The following samples were taken from a machine.

Sample	1	2	3	4	5	6	7	8	9	10
	4.94	5.17	5.02	5.16	5.03	5.09	4.93	5.11	4.97	5.00
Size	5.06	5.01	5.03	5.03	5.13	5.10	5.15	5.05	5.19	5.16
	5.12	5.03	4.98	5.14	4.99	4.99	5.10	4.90	5.05	5.02

b Work out the mean and range of each sample.

c Draw a control chart for the mean weight of samples. Use your answers to part **a** as the action and warning limits.

The range has to be less than 0.3 cm with a warning limit at 0.25 cm.

d Draw a control chart for the ranges.

e Look at both charts and comment on any action that would have been taken.

---

a i) $5.04 \pm (2 \times 0.02) = 5$ and 5.08

ii) $5.04 \pm (3 \times 0.02) = 4.98$ and 5.1

Calculate the action and warning limits using $\mu \pm 2\sigma$ and $\mu \pm 3\sigma$.

b The means and ranges of the samples are

Sample	1	2	3	4	5	6	7	8	9	10
Mean	5.04	5.07	5.01	5.11	5.05	5.06	5.06	5.02	5.07	5.06
Range	0.18	0.16	0.05	0.13	0.14	0.11	0.22	0.21	0.22	0.16

Calculate the means and ranges.
For example, for sample 1

$$\text{Mean} = \frac{4.94 + 5.06 + 5.12}{3}$$
$$= 5.04$$

$$\text{Range} = 5.12 - 4.94$$
$$= 0.18$$

c **Control chart for means**

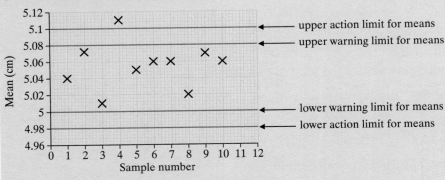

Draw the control chart and put in the calculated limits. Plot the mean points.

d **Control chart for range**

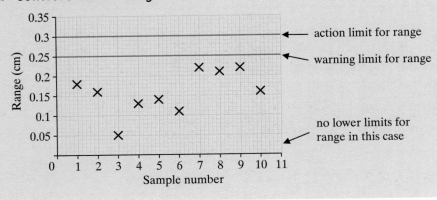

Draw the control chart for range and mark the limits given in the question. Plot the range points.

e   The sample mean for sample 4 was too large so the machine was reset. It came back into control after the machine was reset. The range was under control.

> Look to see if any points fell outside action limits or between warning and action limits.

Exercise 8E

1   What is a control chart used for?

2   Explain why it is necessary to have control charts for both mean and range.

3   A machine is used for cutting curtain rails to a length of 150 cm.

Samples of size 4 are taken at regular time intervals. The means of the samples have to be within the limits 149.2 cm and 150.8 cm, and their ranges have to be less than 3.28 cm. The first eight samples are shown.

Sample	1	2	3	4	5	6	7	8
Sizes	150.3	150.2	150.1	149.2	149.8	149.3	150.2	149.3
	149.9	149.7	149.7	149.7	150.7	149.7	150.0	150.7
	150.7	149.2	149.2	150.6	150.2	150.1	149.7	150.6
	150.7	150.9	150.2	150.9	149.3	149.9	149.1	149.4

   a   Work out the mean and range of each sample.
   b   Plot control charts for the mean length of the samples and the range.

4   A machine is being used to fill packets with crisps. The target weight of the packets is 50 g. Samples of size 3 are taken at regular time intervals.
The sample means action limits have to be 49 g and 51 g.
The sample warning limits have to be 48.33 and 51.67.

The action limit for the range is 2 g. The warning limit for the range is 1.6 g.

The first eight samples are shown.

Sample	1	2	3	4	5	6	7	8
Weight (g)	49.02	50.10	49.05	50.02	50.04	50.00	49.90	50.02
	50.02	50.03	49.88	50.00	50.00	49.88	49.70	49.89
	49.70	49.81	49.90	49.77	49.96	48.38	50.10	49.40

   a   Work out the mean and range of each sample.
   b   Plot control charts for the mean weight of the samples and the range.

5 Tennis balls are to be produced with a mean diameter of 65 mm. Samples are to be taken from the production line and the means of the samples are to have a standard deviation of 0.3 cm. Work out suitable action and warning limits so that 99.8% of the means lie within the action limits, and 95% of the means lie within the warning limits.

6 The manufacturer of electrical shafts wishes to control the diameter of the shafts. The target value of the shaft diameters is 38 mm. It has been found that the mean diameter of the samples is 38 mm and the standard deviation of the mean diameters of the samples is 0.3 mm. The mean diameters of the samples are normally distributed.

   a Between what limits would you expect these percentages of the sample means to lie?

   i) 99.8%

   ii) 95%

At the end of each hour a sample of four shafts is taken and the mean diameter of the sample is found. The mean diameters of the samples are plotted on the chart opposite.

   b Using your answers to part **a** as the values for the warning and action limits comment on any action that would have been taken during the 10 hours.

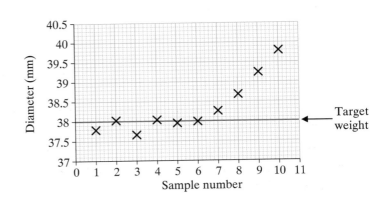

Chapter 8 review

1 Rashid has various numbers of 1p, 2p, 5p and 10p coins in his pocket. If he takes out a coin at random to put into a collection box, then the probability of the coin being a 1p, 2p, 5p or 10p is given by this probability distribution

Coin x:	1p	2p	5p	10p
p(x):	$\frac{1}{k}$	$\frac{5}{k}$	$\frac{3}{k}$	$\frac{1}{10}$

 ⭐⭐⭐⭐⭐ *challenge*

   a Work out the value of k.

   b He has 20 coins in his pocket. How many 5p coins are there?

   Emily has equal numbers of each coin in her pocket.

   c What name would be given to the distribution of the coins in her pocket?

2 An office is equipped with three new photocopiers. The company that makes the photocopiers states that the probability of any photocopier breaking down in the first three months is 0.01.

 ⭐⭐⭐⭐⭐ *challenge*

   a Work out the probability that none of the photocopiers break down in the first three months.

   b Work out the probability that two photocopiers break down in the first three months.

You may use $(p + q)^3 = p^3 + 3p^2q + 3pq^2 + q^3$.

3 A train company claims that 95% of their trains arrive on time. To test this claim four trains are selected at random. Assume that the train company's claim is true. Use a binomial model to predict the probability that

   a all four arrive on time

   b none of the four trains arrive on time.

   c more than two of the four trains arrive on time.

You may use $(p + q)^4 = p^4 + 4p^3q + 6p^2q^2 + 4pq^3 + q^4$.

4 John is going on a five-day holiday to Costa Packet. The travel brochure says that on average five out of every seven days are sunny there. Each day's weather is independent of the weather on all preceding days.

   a   **i)** Name the probability distribution that would model the number of sunny days in Costa Packet.

      There are two values, n and p, that you need to use in this probability distribution.

      ii) Write down the value of n and the value of p.

   b Calculate the probability that John has two, or less, sunny days on his holiday.

You may use $(p + q)^5 = p^5 + 5p^4q + 10p^3q^2 + 10p^2q^3 + 5pq^4 + q^5$.

   c What is the most likely number of sunny days that John will get on his holiday? Show your working.

               edexcel ::: *past paper question*

5 Chicken portions produced for a fast food restaurant have weights that are normally distributed with a mean of 160 g and a standard deviation of 10 g.

   a What percentage of the portions weigh between

     i) 140 g and 180 g

     ii) 130 g and 190 g?

Portions are packed in boxes of 100 portions.

   b How many portions in a box would you expect to weigh between 140 g and 190 g?

6 The weights of bags of potatoes for sale in a supermarket are normally distributed with a mean weight of 2 kg and a variance of 144 g. Find the probability that a bag chosen at random will weigh

   a between 1976 g and 2000 g

   b less than 2024 g.

7   The manager of a clothes factory is keen to ensure that skirt waistbands are made to the correct size. He proposes to use control charts for mean and range. The target size of the waistbands is 26 cm, measured from one side of the skirt waistband to the other when the skirt is laid flat.

☆☆☆☆☆ *challenge*

The mean size of the samples is 26 cm and the standard deviation of the mean size of the samples is 0.195 cm.

The mean sizes of the samples are normally distributed.

a   between what limits would you expect 95% of the sample means to lie?

The manager decides to take samples of four skirts at a time. The first eight samples are shown in the table below.

				Sample				
	1	**2**	**3**	**4**	**5**	**6**	**7**	**8**
Size (cm)	25.6	26.8	25.0	25.6	24.9	26.2	25.9	25.2
	25.8	25.3	26.0	26.2	26.3	25.7	25.5	25.8
	26.4	26.3	25.4	25.3	26.2	24.7	26.5	25.3
	25.4	26.0	26.4	26.1	26.2	26.2	25.3	26.5

b   Draw a control chart for the mean and plot the sample means. Use your answers to part **a** as your warning limits and use action limits of 26.62 cm and 25.38 cm.

c   Draw a range control chart and plot the sample ranges. Use warning limits of 0.24 cm and 1.59 cm, and action limits of 0.08 cm and 2.12 cm.

d   Comment on the state of the process based on the two charts. Describe any actions taken.

Chapter 8 summary

Probability distributions

1   A **probability distribution** is a list of all possible outcomes together with their probabilities.

Discrete uniform distributions

2   A **discrete uniform distribution** has n distinct outcomes.
Each outcome is equally likely. The probability of any given outcome $= \frac{1}{n}$.

Binomial distributions

3   A **binomial distribution** has a fixed number of independent trials n. Each trial has only two outcomes (success and failure). The probability of success is p. The probability of failure is q.

4   The probabilities for the events of n binomial trials will be terms of the expansion of $(p + q)^n$.

Normal distributions

5   Two properties of a **normal distribution** are

- The distribution is symmetrical about the mean, μ
- The mode, median and mean are all equal (because the distribution is symmetrical).

6   Further properties of a normal distribution are

- 95% of the observations lie within ± two standard deviations of the mean
- Virtually all (99.8%) lie within ± three standard deviations of the mean.

7   The **variance** of a normal distribution is a measure of how spread out the data are.

     Variance = (standard deviation)^2

Quality assurance

8   A **control chart** is a time series chart that is used for process control.

9   **Warning limits** are usually set at $\mu \pm 2\sigma$.

10 If a sample mean is between the warning limits the process is in control and the product is acceptable.

11 **Action limits** are usually set at $\mu \pm 3\sigma$.

12 If a sample mean is between the warning and action limits another sample is taken immediately to see if there might be a problem.

13 If a sample mean is outside the action limits the process is stopped and the machinery is reset.

Test yourself

1 In a TV game a contestant is asked to choose randomly one of five boxes labelled A, B, C, D and E. X is the chosen box.

  a Write down the probability distribution of X.

  b What is the name of the distribution X.

  The chosen box contains an amount of money. The probability of the amount being £5, £10, £20 or £50 is given by the following probability distribution.

Amount, x	£5	£10	£20	£50
p(x)	k	$2k$	$2k$	k

  c Work out the value of k.

  d What is the probability of the box containing at least £20?

2 John plays a game on the computer five times. He either wins the game or loses. He wins on average four out of every nine games. Each game is independent of previous games.

  a Name the probability distribution that would model the number of games John wins.

  b Write down the value of n and p.

  c Calculate the probability that John wins exactly two games.

  d Calculate the probability that John wins more than half of the games.

  You may use $(p + q)^5 = p^5 + 5p^4q + 10p^3q^2 + 10p^2q^3 + 5pq^4 + q^5$.

3 The mean weight of bags of apples are normally distributed with mean weight 2 kg and a standard deviation of 150 g. Find the probability that a bag chosen at random will weigh

  a between 1700 g and 2300 g

  b between 1700 g and 2450 g

  c more than 2300 g.

4 On a production line in a factory, tomatoes are put in tins. The target weight of the contents of each tin is 400 g.

  The line manager wishes to check the machine is working properly. Samples of tins are taken at hourly intervals and the weights of the contents are found. He then proposes to use control charts.

  a If one of the plotted points falls between a warning and an action limit what action should the manger take?

b If one of the plotted points falls outside the action limits what
action should the manger take?

The means of the samples are normally distributed with a
standard deviation of 3 g.

c Between what limits would you expect these percentages of
tins to lie?

   i) 95%

   ii) 99.8%

d What action and warning limits would you use on a control
chart for the mean sample weight of tomatoes in the tins?

Welcome to examzone

Revising for your exams can be a daunting prospect. In this section of the book we'll take you through the best way of revising for your exams, step-by-step, to ensure you get the best results that you can achieve.

Zone In!

Have you ever become so absorbed in a task that suddenly it feels entirely natural and easy to perform? This is a feeling familiar to many athletes and performers. They work hard to recreate it in competition in order to do their very best. It's a feeling of being 'in the zone', and if you can achieve that same feeling in an exam, the chances are you'll perform brilliantly.

The good news is that you can get 'in the zone' by taking some simple steps in advance of the exam. Here are our top tips:

UNDERSTAND IT
Understand the exam process and what revision you need to do. This will give you confidence but also help you to put things into perspective. These pages are a good place to find some starting pointers for performing well in exams.

FRIENDS AND FAMILY
Make sure that they know when you want to revise and even share your revision plan with them. Learn to control them so you don't get distracted and so you can have more quality time with them when you aren't revising, because you aren't worrying about what you should be doing.

DEAL WITH DISTRACTIONS
Think about the issues in your life that may interfere with revision. Write them all down. Then think about how you can deal with each so they don't affect your revision.

COMPARTMENTALISE
You might not be able to deal with all issues. For example, you may be worried about an ill friend, or just be afraid of the exam. In this case, you can employ a useful technique. Put all of these things into an imagined box in your mind at the start of your revision (or in the exam) and mentally lock it, then open it again at the end of your revision session.

DIET AND EXERCISE
Make sure you eat well and exercise! If your body is not in the right state, how can your mind be? A substantial breakfast will set you up for the day, and a light evening meal will keep your energy levels high.

BUILD CONFIDENCE
Use your revision time not just to revise content but to build your confidence for tackling the exam. Try tackling a short sequence of easy tasks in record time.

More on ActiveTeach CD-ROM

The key to success in exams and revision often lies in the right planning. Knowing what you need to do and when you need to do it is your best path to a stress-free experience. Here are some top tips in creating a great personal revision plan.

Tip 1: Get yourself organised
First get all your revision materials together. Check they are complete and organised in the right order.

Tip 2: Identify your strengths and weaknesses
To do this you could:
- For each topic, take the relevant multiple-choice question test in 'Know Zone'
- Work through the end-of-chapter summaries and the 'Test yourself' exercises in this Student Book.

Tip 3: Set your goals
Highlight any weaknesses you have identified on your revision planner – it is your goal to get to grips with these!

Tip 4: Divide up your time and plan ahead
Plot on the revision planner the exact date that you will start your revision and also the date of your exam. Note any dates that you will be unable to revise. Divide this time into revision chunks.

Tip 5: Know what you are doing
You can use the topic checklists in the Know Zone section to kick-start your revision.

Tip 6: Be honest
Be honest with yourself about how much time you spend actually revising and how much time you have spent organising your files, looking at Facebook and doodling in the margin of your file.

Tip 7: Concentrate on what you are doing
Do your revision when you concentrate best, for example early in the morning. Find yourself a workplace without distractions.

Tip 8: Revise actively
Keep a pen in your hand to make notes on revision cards and do lots of practice questions. It doesn't really matter which way you approach your revision, just so long as you are taking an active part in the process.

Tip 9: Check your own progress
Make sure that you are making progress by testing yourself from time to time. Tick the boxes in the Know Zone checklists to record your progress from 'Needs more work' to 'Confident'.

Tip 10: Follow the plan
Finally, follow your revision plan. Good Luck!

29

Find out your exam dates. Go to the Edexcel website to find all final exam dates, and check with your teacher.

1

Be realistic about how much time you can devote to your revision, but also make sure you put in enough time. Give yourself regular breaks or different activities to give your life some variety. Revision need not be a prison sentence!

7

8

Draw up a calendar or a list of all the dates from when you can start your revision through to your exams. Or use the revision planner on the Activeteach CD-ROM.

14

Make sure you allow time for assessing progress against your initial self-assessment. Measuring progress will allow you to see and celebrate your improvement, and these little victories will build your confidence.

Chunk your revision in each subject down into smaller sections. This will make it more manageable and less daunting.

27

28
EXAM DAY!

29

You can use these topic checklists to help you identify your strengths and weaknesses.

Each checklist tells you where in the Student Book you can revise different topics.

You can also look at the summaries at the end of each chapter in this Student Book, which list all the key points you should know.

The Test yourself quizzes in the Know Zone on the ActiveTeach CD-ROM can be used to help you to identify strengths and weaknesses, or as tests when you have done some revision.

Topic 1: Data types and accuracy

Topic	Revise this in Section...	Confident	Fair	Needs work!
hypothesis	1.1			
qualitative data	1.2			
quantitative data	1.2			
discrete data	1.2			
continuous data	1.2			
ranked data	1.3			
bivariate data	1.4			
rounding and accuracy	1.5			

Topic 2: Data collection

Topic	Revise this in Section...	Confident	Fair	Needs work!
populations	1.6			
census data	1.6			
sampling	1.6, 1.7			
primary and secondary data	1.9			
pilot surveys	1.9			
questionnaire design	1.9			
interviews	1.9			
statistical experiments	1.9			
capture–recapture method	1.9			
systematic, quota and cluster sampling	1.8			

Higher

More on ActiveTeach CD-ROM

Topic 3: Presenting and analysing data 1: tables, charts and diagrams

Topic	Revise this in Section...	Confident	Fair	Needs work!
tallying	2.1			
frequency tables	2.1 (discrete), 3.1 (continuous)			
cumulative frequency tables	2.2 (discrete), 3.3 (continuous)			
grouping discrete data	2.3			
two-way tables and databases	2.5, 2.6			
pictograms	2.8			
pie charts	2.12 (discrete), 3.2 (continuous)			
stem and leaf diagrams	2.11 (discrete), 3.2 (continuous)			
choropleth maps	3.9			
bar charts and vertical line graphs	2.9, 2.10, 2.15			
frequency tables with open and unequal classes	2.4 (discrete), 3.1 (continuous)			
comparative pie charts	2.16			
misleading pie charts	3.10			

Higher

Topic 4: Presenting and analysing data 2: graphs

Topic	Revise this in Section...	Confident	Fair	Needs work!
histograms	3.4			
frequency polygons	3.4			
cumulative frequency diagrams	3.3			
population pyramids	3.8			
skewness	3.5			
cumulative frequency step polygons	2.14, 4.8			
histograms with unequal class width; frequency density	3.6, 3.7			

Higher

Topic 5: Averages

Topic	Revise this in Section...	Confident	Fair	Needs work!
mode, median and mean for raw data	4.2			
mode, median and mean for discrete data in a frequency table	4.3			
mode, median and mean for grouped data	4.4			
deciding which average to use	4.6			
transforming data	4.5			
weighted mean	4.7			

Higher

Topic 6: Measures of spread

Topic	Revise this in Section...	Confident	Fair	Needs work!
range	4.8			
quartiles and inter-quartile range	4.8			
percentiles	4.8			
box plots	4.9			
skew	4.9			
outliers	4.9			
comparing data sets	4.11			
deciles	4.8			
outliers	4.9			
variance	4.10			
standard deviation	4.10			
standardised score	4.11			

Higher (variance, standard deviation, standardised score)

Topic 7: Scatter diagrams and correlation

Topic	Revise this in Section...	Confident	Fair	Needs work!
scatter diagrams	5.1			
association and correlation	5.2			
causal relationships	5.3			
line of best fit	5.4			
interpolation and extrapolation	5.5			
equation of the line of best fit	5.6			
non-linear models	5.7			
Spearman's rank correlation coefficient	5.8, 5.9			

Higher (non-linear models, Spearman's rank correlation coefficient)

Topic 8: Time series and index numbers

Topic	Revise this in Section...	Confident	Fair	Needs work!
line graphs	6.1			
time series	6.2			
trends and trend lines	6.3			
seasonal variation	6.4			
moving averages	6.5			
index numbers	4.12			
seasonal variation at a point and mean seasonal variation	6.6			
predictions from a trend line	6.7			
equation of a trend line	6.8			
chain base index numbers	4.12			
weighted index numbers	4.12			

Higher (seasonal variation at a point and mean seasonal variation, predictions from a trend line, equation of a trend line, chain base index numbers, weighted index numbers)

Topic 9: Probability

Topic	Revise this in Section...	Confident	Fair	Needs work!
trials, outcomes and events	7.1			
sample space	7.5			
expressing probability numerically	7.2			
theoretical probability	7.3			
experimental probability	7.4			
mutually exclusive events	7.7			
exhaustive events	7.8			
sum of probabilities	7.9			
independent events	7.10			
risk insurance	7.4			
tree diagrams	7.12			
simulation	7.4			
Venn diagrams	7.6			
addition law for more than two events	7.7			
multiplication law for more than two events	7.10			
addition law for events that are not mutually exclusive	7.11			
conditional probability	7.13			
multiplication law for events that are not independent	7.14			

Higher

Topic 10: Probability distributions

Topic	Revise this in Section...	Confident	Fair	Needs work!
probability distributions	8.1			
uniform distributions	8.2			
binomial distributions	8.3			
normal distributions	8.4, 8.5			
quality assurance	8.6			

Higher

Don't Panic Zone

Once you have completed your revision in your plan, you'll be coming closer and closer to The Big Day.

Many students find this the most stressful time and tend to go into panic-mode, either working long hours without really giving their brain a chance to absorb information, or giving up and staring blankly at the wall.

Last minute learning tips:

1. Know where your exams are, when they start and how long they are.

2. If you find that reading over revision notes just before an exam relaxes you, feel free to do so, but be aware that in most cases it could make you more nervous; any new information is not normally absorbed at this stage.

3. Try to memorise formulae as close to the exam time as possible, and when the exam begins, make quick notes somewhere on the paper that will help your recall.

4. Prepare items needed for the exam the evening before. Make sure you have the correct equipment needed for your exam: black ink or ball-point pen, ruler, protractor, compasses, pencil, pencil sharpener, eraser, calculator.

5. You can find last minute crib sheets and quick quizzes in the Don't Panic Zone on the ActiveTeach CD-ROM.

Don't panic

Exam Zone

What to expect in the exam paper

You will have one GCSE Statistics written exam paper – and you will take either the Foundation or the Higher tier. If you aren't sure which tier you are entered for, check with your teacher.

EXAM TIMINGS

The Foundation exam is 1.5 hours long.

The Higher exam is 2 hours long.

Questions

Both exam papers have a mixture of short and long questions. You will need to attempt them all.

The questions will be easier at the beginning and then get more challenging.

Some of the questions will involve real-life data.

The practice exam papers in the Exam Zone have been written by examiners, so will give you a good idea of the type of question you may be asked. So will the questions marked **edexcel** *past paper question* in the chapters in this book, as these have been taken from past exam papers.

Another Foundation and another Higher paper can be found in the Exam Zone on the ActiveTeach CD-ROM.

FORMULAE SHEETS

Each exam paper has a formulae sheet.

The formulae sheet for the Higher paper is included in the Higher practice exam paper in the Exam Zone (page 340). The formulae sheet for the Foundation paper will include only the first two formulae from this sheet.

Familiarise yourself with the formulae sheet you will have in your exam – so you know which formulae it gives you. You don't need to learn these, but you should make sure you know how to use them.

 More on ActiveTeach CD-ROM

Meet the exam paper

This diagram shows the front cover of the exam paper. These instructions, information and advice will always appear on the front of the paper. It is worth reading it carefully now. Check you understand it. Now is a good opportunity to ask your teacher about anything you are not sure of here.

Print your surname here, and your other names afterwards. This is an additional safeguard to ensure that the exam board awards the marks to the right candidate.

Ensure that you understand exactly how long the examination will last, and plan your time accordingly.

Foundation: 1 hour 30 minutes

Higher: 2 hours

This tells you the equipment you need to take to the exam.

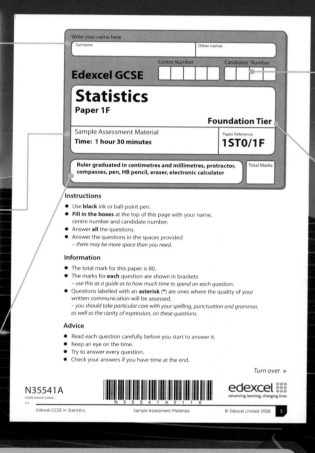

Here you fill in your personal exam number. Take care when writing it down because the number is important to the exam board when writing your score.

This tells you whether it is the Foundation or the Higher paper. This is a Foundation paper.

Understanding the language of the exam paper

When a question says...	It means...
Calculate...	Some working out is needed – make sure you show it!
Work out... OR Find...	A written or mental calculation is needed.
Write down...	Working out is not usually required. This is a useful hint that you can write the answer directly using information from the question or from one of your previous answers.
Estimate...	Work out an answer from a table or graph. Estimating means that you can't work out an exact answer but can use the data to get a good idea of the answer.
Diagram NOT accurately drawn...	Often used for pie charts. So don't measure angles from the diagram as they will be wrong. If you are asked to find angles, you need to work them out from information given.
Give reasons for your answer OR Explain why...	You need to write an explanation. Show any working out.
Plot...	Draw the points **accurately** on the graph.
Draw...	Use the data to draw an **accurate** graph or chart.
Use your (the) graph...	Read values from your graph and use them to answer the question.
Predict...	Use your line of best fit, trend line, data, etc. to find the answer. Don't just guess.
Compare...	Make your comparison explicit. You need to explain how the items you're comparing relate to each other – just describing them will not get you any marks.

Project work for controlled assessment

The controlled assessment project is worth 25% of the total GCSE Statistics assessment. This section will give you help and guidance about how to produce a good project.

The project has three stages.

1 **Planning** – This is completed in the classroom with your teacher.

2 **Collecting, processing and representing data** – This is completed outside the classroom.

3 **Interpreting and evaluating the data** – This is completed in the classroom with your teacher.

Each of the three stages is broken down into separate tasks as shown in this flowchart.

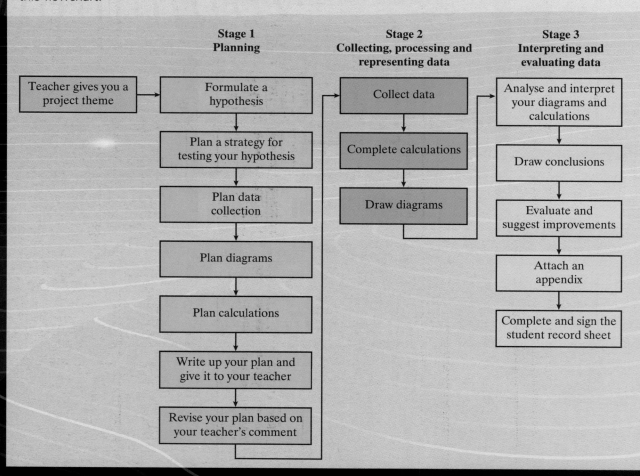

Stage 1 Planning	Stage 2 Collecting, processing and representing data	Stage 3 Interpreting and evaluating data
Teacher gives you a project theme → Formulate a hypothesis	Collect data	Analyse and interpret your diagrams and calculations
Plan a strategy for testing your hypothesis	Complete calculations	Draw conclusions
Plan data collection	Draw diagrams	Evaluate and suggest improvements
Plan diagrams		Attach an appendix
Plan calculations		Complete and sign the student record sheet
Write up your plan and give it to your teacher		
Revise your plan based on your teacher's comment		

Stage 1 Planning

Ask yourself the questions below for each planning task to ensure you have covered everything that is required. This stage must be completed in the classroom.

Formulate a hypothesis

1 What are you going to investigate?

- Some ideas will be offered by your teacher but to achieve the highest marks, it is important to be original and think of your own ideas.

2 What is your hypothesis going to be?

- You must turn your ideas into a question or hypothesis to investigate. For example, when investigating the times when goals are scored in football matches, the hypothesis may say

  'A football team is most likely to score a goal just before half time.'

  or

  'A football team is most likely to concede a goal just after it has scored.'

Remember: Your project can be based on just one question but it would be better if it also answered one or two related hypotheses.

Plan a strategy for testing your hypotheses

How will you test your hypotheses?

- Your hypotheses should be chosen so that you can use a variety of statistical techniques.

Remember: The hypotheses you choose and the strategies for testing them will help you to decide what data you need to collect. It is important to think about what answers you are expecting and why you expect those answers.

Plan data collection

1 Where will your data come from?

- They could be primary data and come from surveys, observations or experiments.
- They could be secondary data and come from sources such as websites, newspapers or government statistics.

2 How reliable is your source?

- Reliable sources include official bodies such as the Office for National Statistics, government departments or local/county councils.

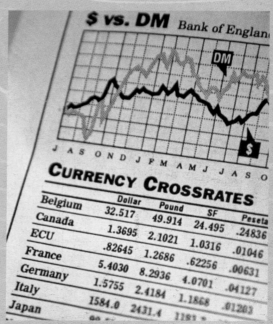

3 How big a sample will you need?

- Discuss a suitable sample size with your teacher. The sample size you use will depend on the questions you are trying to answer but try to collect at least 30 pieces of data for each variable you are investigating.

4 How are you going to collect the data?

- Choose a suitable sampling method. You may work with others to share the data collection but you must write down exactly what you did and what the others were asked to do.

5 How are you going to record the data you collect?

- Data may be collected in a table such as this.

Gender	Year group	Time (s)
M	10	1.76
F	7	1.34

- Questionnaires, tally charts, grouped frequency tables and data loggers could also be used.

6 What are you going to do about outliers or anomalies (i.e. items of data that appear to be out of place or incorrect)?

- You may wish to remove outliers before analysing your data.

Remember: For some themes, secondary data or primary data may be suitable while for others primary data may be the only option.

Plan diagrams

Which charts or graphs will give the most information to answer your hypotheses? Give reasons for your choice.

- For example, use a scatter diagram to look for an association (relationship) between two variables. Use moving averages to look for trends over time. Use a frequency density histogram to look at the shape of a distribution.

Remember: You should always choose the most appropriate diagram for each hypothesis. There is no need to draw several diagrams to represent the same data. The diagram you choose will depend upon whether the data are categorical, continuous or discrete.

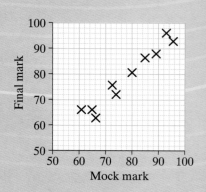

Plan calculations

Which calculations will best summarise your data and answer your hypotheses? You must consider outliers.

Remember: You should not do calculations that do not answer your initial questions. You will get no extra marks and will be wasting your time.

Write up your plan and give it to your teacher

Your plan should:

- state the hypotheses you are going to investigate
- describe your plans for data collection
- state the diagrams and calculations you intend to use.

Remember: You must write down your reasons for the choice of diagrams and calculations.

Revise your plan based on your teacher's comments

What suggestions has my teacher made for improving my plan? What do I need to change or improve?

Remember: You will need to give the plan back to your teacher for final assessment.

Stage 2 Collecting, processing and representing data

Complete the list of activities for each task to ensure your data are collected, processed and represented correctly. This stage is completed away from the classroom.

Collect data

1 Collect the data as you planned in Stage 1.

- Use the data collection sheets or questionnaires you designe
- Remember that data may be collected individually or in group

2 Write down how you collected the data and any problem you had.

Complete the calculations and draw diagrams

Process and represent data

- Carry out the calculations and draw the diagrams you planned.
- You may use ICT as there is no need to show evidence of calculations or hand-drawn diagrams (these skills are assessed in the written paper examination).

- It is essential that diagrams are given titles, that sensible scales are used and that they are labelled correctly. A key must be provided, if appropriate.

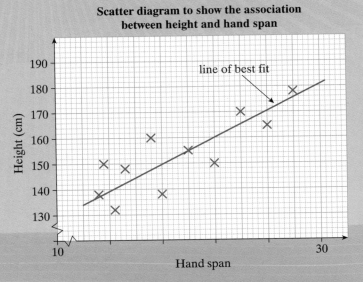

Scatter diagram to show the association between height and hand span

During this stage you may adapt your plan but you must discuss this with your teacher first.

If a diagram shows your hypothesis to be obviously incorrect, stop and form a new hypothesis.

Separate diagrams may be needed if you have more than one hypothesis. You may find it easier to draw diagrams to represent the data needed for one hypothesis and then analyse and interpret the data before drawing diagrams for the next hypothesis.

Always use a sensible degree of accuracy for numerical results. Three significant figures is usually suitable.

Stage 3 Interpreting and evaluating data

This final stage of your project must be completed in the classroom.

You will have completed most, if not all, of your diagrams and calculations already. Now you need to put your project together to make it into a complete report. This list shows the headings for the parts of your project, the order in which they should appear and what should be included in each section.

1 **Plan** Your plan is the first item in your report.

2 **Data collection** This should include your data collection tables. You must also provide any extra information about the data that were not in your plan. You should write about any problems you had and how you resolved them.

3 **Analysis** Present your diagrams and calculations in a form that helps you to interpret them.

4 **Interpretation** You should interpret your various tables, charts, graphs and calculations, explaining what they tell you about the data you collected.

5 **Conclusion** This section should draw together your findings. Your conclusions should directly relate to your original hypotheses that are detailed in your plan. You must explain any relationships you found and make comparisons of and between data. You must state whether your findings either support or reject your hypotheses.

6 Evaluation Discuss whether you have got the results you expected. Suggest improvements and discuss limitations. For example, discuss what new hypothesis you could use and why, or suggest how you could collect your data in a different way if you needed to improve your sampling. Would more data have improved the work?

7 Appendix Your raw data should be included in an appendix at the end of your report.

Once your report is finished you must complete your student record sheet and sign the authentication form. These must be included at the front of your report.

Finally, number the pages of your report, fix them together and hand it in to your teacher.

Exam Zone

Printable versions of these papers are available on the ActiveTeach CD-ROM.

Examination practice paper

FOUNDATION

1 The pictogram shows information about the number of accidental fires that took place on one day in 2005.

Time	Number of fires
00 00 – 05 59	
06 00 – 11 59	👑 👑 👑 👑 👑
12 00 – 17 59	👑 👑 👑 👑 👑 👑 👑 👑 👑
18 00 – 23 59	👑 👑 👑 👑 👑 👑 👑 👑 👑

Key: 👑 represents 4 fires

Source: Department for Communities and Local Government

The pictogram is not complete.

Twelve accidental fires started between 00 00 and 05 59.

a Copy and complete the pictogram. **(1)**

b Write down the number of accidental fires between 18 00 and 23 59. **(1)**

c Between what times of the day did most fires start? **(1)**

d Suggest a reason why few fires start between 00 00 and 05 59. **(1)**

(Total 4 marks)

2 The bar chart shows the number of British films shown on three main TV channels in one particular month.

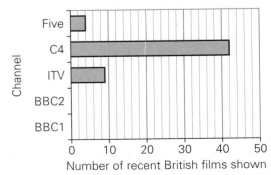

Source: DGA metrics. UK Film Council RSW

The number of British films shown on BBC1 was 30. On BBC2 it was 20.

a Copy and complete the bar chart. **(2)**

b Write down the number of British films shown on channel C4. **(1)**

c Write down the name of the channel that showed the smallest number of British films. **(1)**

(Total 4 marks)

3 The managers of a doctors' surgery want to find out whether patients are satisfied with the waiting room.

They decide to send a questionnaire to all their patients.

a Write down **one** disadvantage of using a census in this case. **(1)**

b Write down **one** advantage of using a census in this case. **(1)**

One of the questions they decide to put on the questionnaire is "What colour should the waiting room be painted?"

c Discuss whether this is a good or a bad question. Could it be improved in any way? **(5)**

(Total 7 marks)

4 A number of men and a number of women were asked if they prefer visiting a theatre or going to a dance.

The table gives information about their choices.

	Prefer a theatre	Prefer a dance	Total
Men	65	70	
Women	20		50
Total			185

The table is not complete.

a Copy and complete the two-way table. **(2)**

A person is chosen at random.

b Write down the probability that this person will be

i a man who prefers to visit a theatre

ii a woman. **(2)**

(Total 4 marks)

5 The table shows information about single vehicle accidents, on all types of road, during 2005. The percentages of injuries and the number of people injured, organised by accident type, are given.

Object hit / type of accident	Fatal (%)	Serious (%)	Slight (%)	All (number)
None	1.5	18.4	80.1	41 110
Road sign or traffic sign	3.2	17.4	79.4	1561
Lamp post	3.6	18.4	78.0	1871
Telegraph or electricity pole	2.6	16.7	80.7	804
Tree	7.1	24.0	68.9	3445
Bus stop or shelter	5.6	15.4	79.0	162
Crash barrier	2.8	15.1	82.2	2409
Submerged	15.2	24.2	60.6	33
Entered ditch	1.8	18.1	80.0	1846
Other permanent objects	2.5	18.2	79.3	6553
Total (number)	1293	11 058	47 445	59 796

Source: Department for Transport

a Write down the percentage of people who had a collision with a crash barrier and were fatally injured. **(1)**

b Write down the type of accident in which there was the greatest percentage of serious accidents. **(1)**

John says: 'More people were involved in fatal accidents where no objects were hit than those in part **a**'.

c Explain why John is right. **(2)**

(Total 4 marks)

6 The numbers of strokes taken by 11 randomly selected golfers in the first round of the US PGA national tournament were

68 68 67 70 70 71
70 71 69 66 70

a Work out the mean number of strokes taken by these 11 golfers. **(2)**

b Work out the median of the data. **(2)**

c Write down the mode of the data. **(1)**

d Work out the range of the data. **(2)**

The expected number of strokes that should be taken to complete one round of this course (par) is 70.

e Discuss how the answers in (a), (b) and (c) relate to the expected number of strokes. **(2)**

(Total 9 marks)

7 A dice is rolled 50 times.

Every 10 rolls the probability of getting a six is calculated using totals to that time.

The graph shows the number of rolls and the probability of getting a six.

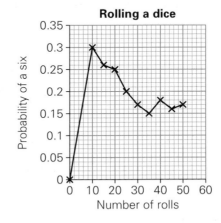

a Explain why the last result is the most accurate. **(2)**

The theoretical probability of getting a six using an unbiased dice is 0.167 (to 3 d.p.).

b Is this dice biased or unbiased? Give a reason for your answer. **(2)**

(Total 4 marks)

8 The table gives some information about the number of male and female prisoners.

It also shows the percentage that committed certain crimes.

	Robbery	Burglary	Theft and handling	Number of offenders
Males	38%	43%	19%	15 716
Females	29%	24%	47%	863

Source: National Offender Management Service

a Use the information given to copy and complete the composite bar charts below.

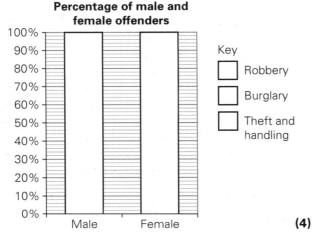

Percentage of male and female offenders

Key
☐ Robbery
☐ Burglary
☐ Theft and handling

(4)

b Write down the crime which was least often committed by females. **(1)**

c Did more males commit 'Theft and handling' crimes than females? Give a reason for your answer. **(2)**

(Total 7 marks)

9 A bag contains five red discs and three yellow discs.

A disc is randomly chosen from the bag and then replaced.

A second disc is then randomly chosen.

The tree diagram shows the possible outcomes and their probabilities.

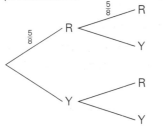

Key
R = Red
Y = Yellow

a Copy and complete the diagram. **(2)**

b Find the probability that both a red disc and a yellow disc are chosen. **(3)**

(Total 5 marks)

10 A town is divided into 25 regions of equal size.

The number of pizza wrappers dropped as litter in each region is counted.

The results are shown in the following diagram.

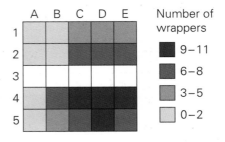

	A	B	C	D	E
1	0	0	3	4	5
2	1	2	6	6	7
3	1	2	8	9	8
4	2	7	10	11	10
5	1	5	8	9	8

| 10 | Means 10 wrappers in this region |

a Use the information in the diagram to copy and complete the choropleth map below. (Twenty regions have already been shaded.)

Number of wrappers
■ 9–11
▨ 6–8
▨ 3–5
☐ 0–2

(2)

b Write down the sector in which you think the pizza shop is situated. Give a reason for your answer. **(2)**

(Total 4 marks)

11 A call centre supervisor thinks that, of the telephone operators in the centre, men answer calls quicker than women but women give better answers.

She decides to investigate this.

a Suggest **two** suitable hypotheses she could use. **(2)**

She decides to look at a sample of workers.

b Describe the population from which she will get her sample. **(1)**

c Discuss the advantages and disadvantages of using a sample. **(5)**

(Total 8 marks)

12 A company works out its sales figures every four months. The results over a three-year period are shown in the table.

a Calculate the last two three-point moving averages. **(2)**

Year	Period	Sales (in £10 000)	Moving average
1	Jan–Apr	32	
	May–Aug	48	39
	Sep–Dec	37	42
2	Jan–Apr	41	43
	May–Aug	51	45
	Sep–Dec	43	47
3	Jan–Apr	47	
	May–Aug	54	
	Sep–Dec	46	

The sales and some of the moving averages have been plotted as a time series.

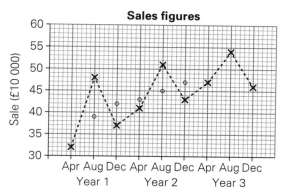

Sales figures

b Copy and complete the time series by plotting the last two moving averages. **(2)**

c Draw a trend line on the time series chart. **(1)**

d Describe the trend. **(1)**

e Write down what you can deduce from the graph about the seasonal nature of the product being made by this company. **(2)**

(Total 8 marks)

13 A student thinks that aircraft are as wide as they are long.

The table shows information about the aircraft that are used at London City Airport.

Aircraft	Wing span (m)	Length (m)
ATR 42	25.0	23.0
BAe 146	26.5	31.0
BAe 4100 Jetstream	18.5	19.0
DHC Dash 7	28.5	24.5
Bombardier Q series	26.0	22.0
Dornier Fairchild 228	17.0	16.5
Dornier Fairchild Do328	21.0	21.0
Embraer 135	20.0	26.5
Fokker 50	29.0	25.0
Fokker 70	28.0	31.0
Saab 340	21.5	20.0
Saab 2000		
Short 360-300		

Some of these data are shown on the scatter diagram.

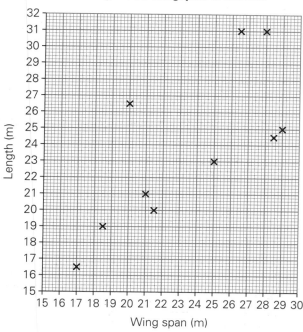

Length and wingspan of aircraft

The Saab 2000 has a 25 m wing span and is 27.5 m long.

The Short 360-300 has a 23 m wing span and is 21.5 m long.

a Copy the scatter diagram and plot these two points on it. **(2)**

b Draw what you think would be a line of best fit if the student is correct. **(2)**

c How well do these data fit the student's belief? **(2)**

A new aircraft is being built. The aircraft will be 27 m long.

d Suggest the wing span it is likely to have judging from these data. **(2)**

A Boeing 747 is 66.6 m long.

e Give two reasons why an estimate of its width using a line of best fit might be incorrect. **(2)**

The wing span of a Boeing 747 is in fact 64.4 m.

f Write down whether or not this fits the student's belief. Give a reason for your answer. **(2)**

(Total 12 marks)

Examination practice paper

HIGHER

1 As part of an investigation into children's weights and diets, children were asked to write down their weight and the number of packets of crisps they had eaten in the past week.

a Put crosses in one or more boxes as appropriate.

  i Weight is

Discrete   Quantitative   Continuous   Qualitative

☐   ☐   ☐   ☐

  ii The number of packets of crisps is

Discrete   Quantitative   Continuous   Qualitative

☐   ☐   ☐   ☐

                                              (4)

As part of the investigation each child was asked their height to the nearest centimetre.

Fiona gave her height as 144 centimetres.

b Write down the interval within which Fiona's height lies.   **(2)**

                                (Total 6 marks)

2 The pie chart gives information about the ways in which people travel to work.

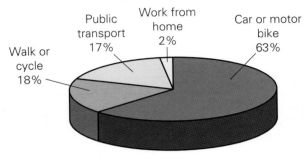

How people travel to work

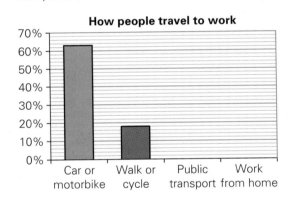

Source : Department for Environment, Food and Rural Affairs

a Write down one feature of the pie chart that can be misleading.   **(1)**

The ways in which people travel to work can also be shown as a bar chart.

b Use the data from the pie chart to copy and complete this bar chart.

How people travel to work

| 70% | 60% | 50% | 40% | 30% | 20% | 10% | 0% |

Car or motorbike   Walk or cycle   Public transport   Work from home   **(2)**

c Write down the most popular way used by people to travel to work.   **(1)**

                                            (Total 4 marks)

3 The table shows information about fuel consumption for different forms of transport for the years 2000 to 2006.

Fuel consumption million tonnes: by form of transport and fuel type: UK: 2000–2006

	2000	2001	2002	2003	2004	2005	2006
Motor spirit							
Cars and taxis	20.12	19.77	19.74	18.93	18.57	17.88	17.32
Light goods	0.89	0.76	0.67	0.58	0.52	0.44	0.43
Motor cycles	0.13	0.13	0.14	0.15	0.14	0.14	0.14
Diesel							
Cars and taxis	2.92	3.08	3.42	3.67	4.05	4.38	4.54
Light goods	3.45	3.72	3.94	4.27	4.53	4.87	5.05
Heavy goods	8.14	8.16	8.45	8.60	8.82	9.04	9.37
Buses and coaches	1.11	1.10	1.11	1.16	1.11	1.14	1.18
Propane	0.02	0.05	0.09	0.10	0.11	0.12	0.13
All road transport as a percentage of the total	36.79	36.79	37.55	37.47	37.84	38.01	38.15

Source: Department for Transport

a Write down how much diesel 'Heavy goods' used during 2004. **(1)**

b Describe the trend in the consumption of motor spirit by 'Cars and taxis' and 'Light goods' between 2000 and 2006. **(1)**

c Describe the trend in the consumption of diesel by 'Cars and taxis' and 'Light goods' between 2000 and 2006. **(1)**

d Write down the conclusion that you can draw from the answers to (b) and (c). **(1)**   **(Total 4 marks)**

---

4 The table shows the total rainfall for 11 consecutive quarters between 2005 and 2007 at Eastbourne. It also shows some of the four-point moving averages.

Year	Quarter	Rainfall (mm)	Four-point moving average
2005	1	119	
	2	106	
			153
	3	153	
			162
	4	234	
			163
2006	1	155	
			166
	2	110	
			175
	3	165	
			195
	4	270	
2007	1	235	
	2	150	
	3	153	

Source: Eastbourne Borough Council

a Calculate the two missing four-point moving averages. **(3)**

The data from the table are plotted as a time series graph.

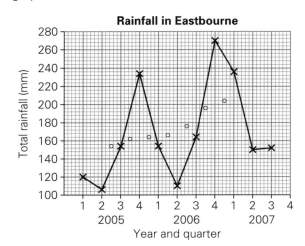

b Copy the time series graph and plot the last two moving averages on it. **(1)**

c In which quarter would you choose to visit Eastbourne if you wish to have the least rainfall? **(1)**

d Draw in a trend line for the moving averages. **(1)**

e Describe the trend in the rainfall between 2005 and 2007. **(1)**

(Total 7 marks)

5 Twelve breeds of dog were ranked for playfulness and curiosity. The results are shown in the table below.

Breed of dog	Playfulness	Curiosity	d	d^2
German shepherd	1	4		
Rottweiler	2	3		
Border collie	3	8		
Labrador	4	1		
Doberman	5	7		
Jack Russell	6	2		
Boxer	7	5		
Golden retriever	8	10		
Dalmatian	9	6		
Poodle	10	12		
Springer spaniel	11	9		
Rhodesian ridgeback	12	11		

Source: Stockholm University

a Copy the scatter diagram below and complete it by plotting the last three points. **(1)**

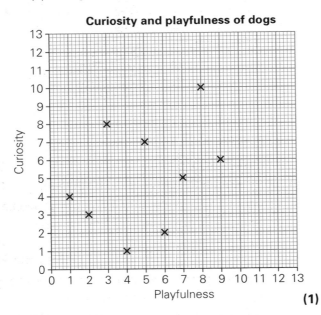

Curiosity and playfulness of dogs

(1)

b Calculate Spearman's rank correlation coefficient between playfulness and curiosity in dogs. **(2)**

c Describe the type of correlation shown in part (b). **(1)**

d What does this correlation mean in the context of the question? **(1)**

(Total 6 marks)

6 In a study on the weights of red deer the following summary statistics were obtained.

	Minimum weight (kg)	Lower quartile (kg)	Median weight (kg)	Upper quartile (kg)	Maximum weight (kg)
Male (stag)	92	123	134	144	180
Female (hind)	56	73	80	89	112

A box plot has been drawn on the grid to show the distribution of the weights of female red deer.

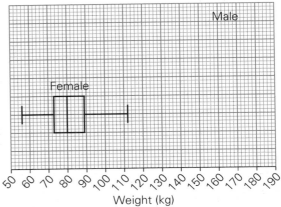

Weight of red deer

a Copy the grid and then draw a box plot to show the distribution of weights of male red deer. **(3)**

b Work out the inter-quartile range for female red deer. **(1)**

c Compare the distribution of the weights of female red deer with the distribution of the weights of male red deer. **(2)**

(Total 6 marks)

7 In a survey the number of road accidents was recorded per day at a busy road junction. The results are shown in the table.

Accidents per day	Number of days
0	26
1	90
2	90
3	57
4	19
5	5
6	3

a Work out the mean number of accidents per day. **(2)**

b Work out the standard deviation of the number of accidents per day. **(3)**

An earlier study at the same road junction produced the following results.

Mean: 3.3

Standard deviation: 1.15

c Compare the results of the two surveys. **(2)**

(Total 7 marks)

8 The national results of a survey into multiple channel TV showed
- 10% of all households do not receive multiple channel TV
- 36% receive it by satellite TV
- 18% receive it by cable TV
- 36% receive it by digital terrestrial TV.

Trevor is going to survey the 20 houses in his road to see how they receive multiple channel TV.

a Copy and complete the table below to show the expected result of Trevor's survey using the national percentages.

Method of reception	Satellite	Cable	Digital terrestrial	None of these
Frequency				

(2)

He decides to simulate the result.

The table below shows the numbers he gives to each method of receiving multiple channel TV.

Method of reception	Satellite	Cable	Digital terrestrial	None of these
Numbers given	00–35	36–53	54–89	90–99

b Explain why he gives the numbers 00–35 to Satellite TV. **(2)**

Trevor uses the following extract from a random number table to obtain 20 random numbers. He starts at the top left.

335217   045178   627341   532715

823859   482082   342173

c Copy and complete the table below to show the result of Trevor's simulation.

Method of reception	Satellite	Cable	Digital terrestrial	None of these
Frequency				

(3)

d Describe how Trevor's simulated result compares with the expected results. **(1)**

e Write down whether or not you would expect a second simulation to give the same frequencies as Trevor's first simulation. **(1)**

f Write down whether or not you would expect the actual figures for the first 20 houses to be the same as those predicted from the national percentages. **(1)**

(Total 11 marks)

9 A survey of the ages of 100 people who wear contact lenses was carried out. The results are shown in the table below.

Age group	$15 \leqslant a < 20$	$20 \leqslant a < 25$	$25 \leqslant a < 35$
Frequency	10	11	30
Frequency density			

Age group	$35 \leqslant a < 45$	$45 \leqslant a < 55$	$55 \leqslant a < 65$
Frequency	24	15	10
Frequency density	2.4	1.5	1

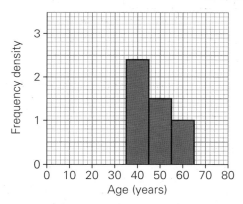

a Use the table to copy and complete the histogram above. **(3)**

b Comment on the skew of the histogram. **(1)**

c Calculate an estimate for the number of people between the ages of 30 and 42 who wear contact lenses. **(3)**

(Total 7 marks)

10 A factory bags up potatoes for selling in supermarkets. The bags are produced on a production line.

The mean weight of potatoes in samples of five bags is normally distributed with a mean weight of 1.06 kg and a standard deviation of 0.02 kg.

a Work out the weight limits between which you would expect 95% of the sample means to lie. **(2)**

b Work out the weight limits between which you would expect almost all of the sample means to lie. **(2)**

A sample of five bags of potatoes is taken every half hour and the mean weight of the contents is found.

The mean weights of the first eight samples are plotted on the chart below.

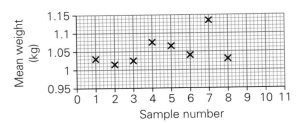

c Copy the graph and add suitable warning and action limits. **(2)**

The 9th and 10th samples had means of 1.025 kg and 1.08 kg respectively.

d Plot the 9th and 10th samples on the chart. **(2)**

e Describe the actions that would have been taken after the 2nd and 7th samples. **(2)**

(Total 10 marks)

11 A university wants to carry out a survey of the weekly rents paid for accommodation by the students. They decide to give a questionnaire to a sample of the students.

a Describe the population from which they will select their sample. **(1)**

b Should they use a closed or an open question on the questionnaire? Give a reason for your answer. **(2)**

c Write down a suitable question for using on the questionnaire. **(2)**

They decide to do a pilot survey.

d Discuss the advantages of doing a pilot survey. **(5)**

(Total 10 marks)

12 Ahmed wants to estimate the number of fish in a lake.

He catches 30 fish, marks them and puts them back in the lake. Later he catches 40 fish and finds that five of them are marked.

a What does this method of sampling assume about the population of fish in the lake. **(1)**

b Calculate an estimate for the number of fish in the lake. **(2)**

(Total 3 marks)

13 Mai is going to shoot four arrows at target. She claims that she hits the bull's-eye eight times out of 10.

a Name the probability distribution that models the number of times she hits the bull's-eye with her four arrows. **(1)**

b Write down what this model assumes about the probability of hitting the bull's-eye with each of the four arrows. **(1)**

Assuming that Mai's claim is true:

c i Work out the probability that she hits the bull's-eye with only one of the arrows.

You may use
$(p + q)^4 = p^4 + 4p^3q + 6p^2q^2 + 4pq^3 + q^4$. **(2)**

ii By calculating the probabilities of hitting the bull's-eye two, three and four times, find the most likely number of times she hits the bull's-eye. **(3)**

Mai fires four arrows at the target. She hits the bull's-eye only once.

d Write down whether Mai's claim is likely to be true. Give a reason for your answer. **(2)**

(Total 9 marks)

14 A survey was done of 100 children to see whether or not they were given pocket money and whether they had to do odd jobs to earn it.

The results of the survey were summarised as follows.

	Given pocket money	Had to earn pocket money	Did not get pocket money	Total
Girls	35	15	5	
Boys	35	8	2	
Total				

a Copy and complete the two-way table. **(2)**

A child is chosen at random.

b Write down the probability that the chosen child is

i given pocket money

ii a girl who had to earn her pocket money. **(2)**

c Discuss the way in which boys were treated compared with how girls were treated. **(2)**

The table below gives information about the average pocket money for children in England.

Year	2003	2004	2005	2006	2007
Mean amount of pocket money	£3.60	£5.79	£7.82	£8.36	£8.01

Source: Halifax pocket money survey

d Using 2003 as the base year, work out the index numbers for the mean amount of pocket money in 2006 and 2007. **(2)**

e Write down the percentage by which the mean amount of pocket money increased between 2003 and 2006. **(1)**

f Describe how the amount of pocket money changed over the years 2003 to 2007 inclusive. **(2)**

(Total 11 marks)

Formulae Sheet

Higher Tier

(Remember, you only need to familiarise yourself with the first two formulae if you're taking the Foundation paper.)

Mean of Frequency distribution
$$= \frac{\sum fx}{\sum f}$$

Mean of grouped frequency distribution
$$= \frac{\sum fx}{\sum f}, \text{ where } x \text{ is the mid-interval value.}$$

Variance
$$= \frac{\sum (x - \bar{x})^2}{n}$$

Standard deviation (set of numbers)
$$\sqrt{\left[\frac{\sum x^2}{n} - \left(\frac{\sum x}{n} \right)^2 \right]} \text{ or } \sqrt{\left[\frac{\sum (x - \bar{x})^2}{n} \right]} \text{ where } \bar{x} \text{ is the mean set of values.}$$

Standard deviation (discrete frequency distribution)
$$\sqrt{\left[\frac{\sum fx^2}{\sum f} - \left(\frac{\sum fx}{\sum f} \right)^2 \right]} \text{ or } \sqrt{\left[\frac{\sum f(x - \bar{x})^2}{\sum f} \right]}$$

Spearman's Rank Correlation Coefficient
$$1 - \frac{6 \sum d^2}{n(n^2 - 1)}$$

Zone Out

This section provides answers to the most common questions students have about what happens after they complete their exams. For more information, visit the Zone Out section on the ActiveTeach CD-ROM.

Got your results? What next?

Exam grades should be a fair reflection of your work, knowledge and performance in a subject. Sometimes, however, mistakes are made. If you think your result is unfair, you can ask (through your school or college) for a re-mark.

Can I have a re-mark of my examination paper?

Yes, this is possible, but remember that only your school or college can apply for a re-mark, not you or your parents/carers.

First of all, you should consider carefully whether or not to ask your school or college to make a request for a re-mark. You should remember that very few re-marks result in a change to a grade - not because Edexcel is embarrassed that a change of marks has been made, but simply because a re-mark request has shown that the original marking was accurate. Check the closing date for re-marking requests with your Examinations Officer.

Discuss your results with your subject teachers. Look at your results slip and think about the units that you have not done so well in. Was that expected or not? Would it be better to accept a rather disappointing result for one unit and get on with your next units, which you might do well in? Sometimes, students don't do as well on the next units because their mind is focusing on getting a higher uniform mark for an earlier unit instead. Remember, it is much more important to get a good mark on a unit worth 20% than on a unit worth 15%.

If you and your teachers decide to make an application for a re-mark the following options are available:

- **A clerical re-check**. Your paper(s) will be re-checked (a check has already taken place before your result was published) to ensure that all parts of your paper have been marked, that the totalling of marks is correct, that the marks have been correctly recorded on our computer system and, if appropriate, special consideration has been applied.

- **A full re-mark of any externally marked paper**. This will also include the clerical re-check described above. Here you must remember that your marks could go up or down, as well as stay the same as originally published. This type of re-mark can take quite a long time to complete (up to 35 days, but see priority re-marks below). Your paper(s) will be re-marked by a senior examiner - not by the same examiner who marked it originally.

- **Re-moderation of controlled assessments**. Controlled assessments can be re-moderated. The way this is done is rather different from the written paper procedures and your Examinations Officer will have all the necessary information. Normally, the need for controlled assessment re-moderation is decided on by the subject teacher and not by an individual student. Your controlled assessment cannot be re-moderated without all the students who were entered originally going through the moderation process again. This is a centre decision and may not take place in every instance.

How do I get a re-mark of my papers?

Your school or college can request a remark for you, but these must be received by Edexcel before the re-mark deadline. You should check what this deadline is with your school or college when you pick up your results so you can make your request in good time if you need to.

I've missed the deadline for re-marks. Is there anything I can do?

Unfortunately not. The deadlines are strict and imposed by the Awarding Bodies' regulator and so Edexcel cannot accept further requests once it has passed.

When I asked for a re-mark of my paper, my subject grade went down. What can I do?

There is no guarantee that your grades will go up if your papers are re-marked. They can also go down or stay the same. After a re-mark, the only way to improve your grade is to take the examination again. Your school or college Examinations Officer can tell you when you can do that.

I am not sure that my controlled assessment has been given the correct mark by my teacher. What can I do?

All controlled assessment marks are moderated by Edexcel. Generally, after this check, a centre's marks are approved and are accepted unchanged. Sometimes, a centre's marks have to be reduced. Occasionally, they are raised. You can check with your centre what has happened to its marks after this moderation exercise.

Moderation is necessary to make sure that all candidates have their controlled assessment assessed according to the same nationally agreed standard.

If your coursework mark has not been adjusted by Edexcel and it is still not the mark you believed your work had been given, then you must go back to your centre. Every centre that enters students for external examinations must have an appeals procedure in place. Students who do not believe that a correct mark was reported for their controlled assessment must first of all make use of this appeal process. Only when this appeal process has been completed within the centre can Edexcel become involved in any dispute.

What is involved in having an examination paper re-marked?

We send your examination paper to a different examiner and ask them to re-mark it using the same mark scheme. Not all examiners are involved in re-marking examination papers. Senior examiners are responsible for this.

If your grade changes, your Examinations Officer will receive, on your behalf, a new results slip. The original grade will either be confirmed, go up or go down. If the grade has fallen you cannot refuse it and ask for the original higher grade to stand. You must be aware that this is the chance you take when asking for a re-mark. Because of this, some students apply for a photocopy of their paper before applying for a re-mark.

As it takes time to get the photocopied paper back to your Examinations Officer and for you and your teacher to decide what to do, you ought to ask for a photocopy of your examination paper as soon as possible after the results are published. This will give you, your teacher and the Examinations Officer enough time to apply for the re-mark if this is the chosen option.

There is a charge for the re-marking of examination papers so you need to check whether you or your school/college will pay for this service.

If I have had a re-mark and the mark has not changed or there has been a minor change to the mark, is it possible to ask for another re-mark if my teacher and I are still not satisfied?

If you are still not satisfied after the re-mark, you can ask your centre to appeal on your behalf. Appeals do not normally involve the remarking of work.

Work will only be remarked at the appeals stage if there is evidence of procedures not being followed during the original re-mark.

Answers

Chapter 1

Exercise 1A

1 EITHER: More males than females buy the products.
OR: More females than males buy the products.
OR: Equal numbers of males and females buy the products.

2 EITHER: Drug A has a better cure rate than Drug B.
OR: Drug B has a better cure rate than Drug A.
OR: The cure rates for Drug A and Drug B are the same.

3 A Continuous B Discrete C Continuous

4 A Qualitative B Quantitative C Quantitative

5 A Discrete, Quantitative B Qualitative

6 A and B

7 Any **two** from: colour; type; make; engine size; model (others possible).

8 a Weight b Salary or wage c Value

9 0–<10
10–<20 14, 15, 15, 16, 17, 17, 18, 18, 19
20–<30 21, 22, 22, 23, 25, 26, 28, 29
30–<40 30, 31, 31, 31, 32, 33, 33, 34, 34, 34, 36, 37, 38
40–<50 40, 40, 42, 42, 42, 42, 44, 47
50–<60 56, 58, 58
60–<70 60, 60, 61, 63, 65
70–<80 72, 72, 77
80–<90 82

10 Appears discrete because age is usually a whole number. Actually continuous because age is not exactly 16 years but could be 16 years, 3 months, 5 days etc.

11 a 54 km b 5 kg c 61 litres

12 | | Upper bound | Lower bound |
|---|---|---|
| a | £202.50 | £197.5 |
| b | £10 250 | £9750 |
| c | 4.5 m | 3.5 m |
| d | 5.5 m | 4.5 m |
| e | $125\frac{1}{2}$ miles | $124\frac{1}{2}$ miles |
| f | $175\frac{1}{2}$ cm | $174\frac{1}{2}$ cm |
| g | 64 y 0 m 0 d | 63 y 0 m 0 d |
| h | 202.5 miles | 197.5 miles |

Exercise 1B

1 a All the students.
 b A student.
 c Advantage: all the students views will be used. Disadvantage: time consuming or lots of data to handle or expensive.

2 a All the house prices in the city.
 b Too difficult to get data or too time consuming or too expensive.
 c It does not take into account the other estate agent's housing or she might not have a good cross section of the city's houses on her books.

3 a A list (numbered or alphabetical) of all the students.
 b A student.

4 a **Two** from: cheaper, quicker, less data to handle.
 b **Two** from: unbiased, accurate, uses all the population.

5 a All the people in Britain.
 b His village may not be representative of the people in Britain. For example, they might nearly all be retired or rich or families, etc.

6 A Simple random sampling B Stratified sampling

7 $\frac{60}{120} \times 30 = 15$ from first age group.
$\frac{40}{120} \times 30 = 10$ from second age group.
$\frac{20}{120} \times 30 = 5$ from third age group.

8 a Six strata – each year group/male and each year group/female is a stratum.
 b 25 each sex, since there are 1500 males and 1500 females.
Year group 1: 9 males 7 females
Year group 2: 10 males 11 females
Year group 3: 6 males 7 females

9 a 33, 17, 04, 41, 27, 15, 38, 48, 20, 34
 b 33, 45, 26, 17, 45, 39, 19, 04, 14, 38

10 Any **two** from:
Some may never have a chance of being picked. Any student numbered more than $25 \times 6 = 150$ can never be picked.
OR Not everyone has an equal chance of being picked; Student 150 will only be picked if he rolls 25 sixes; student 5 could be selected when he rolls a five or a one then a four, etc.
OR On the first throw only the first 6 can be selected.

11 A Systematic B Quota C Cluster

Exercise 1C

1 A Primary B Secondary C Secondary D Primary

2 a Secondary b James' data are more recent.

3 a Ill-defined responses.
 b How old are you? 0 to 5 ☐ 6 to 10 ☐ etc.

4 a The boxes do not allow for all responses.
 b How often do you watch a cricket match?
Less than once a week ☐
Once a week ☐
More than once a week ☐

5 A Closed B Open C Closed

6 It suggests that you should agree.

7 Any **three** from:
Needs to be as short as possible.
Needs to be closed.
Needs to be free from bias.
Needs to be straightforward to answer.
Needs to be relevant to the survey you are doing.

8 Advantages: Any **two** from:
Can make question clear.
Can ensure enough answers.
Can ensure correct people asked.
Disadvantages: Any **two** from:
Can be time consuming.
Can be difficult if people refuse to answer.
Interviewer may introduce bias.

9   Ensures questions are clear.
Ensures you get the answers you require.
Ensures no errors in questions.

10 A personal interview is more likely to produce answers, but it will take longer and be more expensive.

11 $\frac{2}{40} = \frac{20}{x}$, so $x = 400$ birds.

12 **a** 400 fish.
b Any **two** from:
The marked fish did not die.
The marked fish had mixed well with the other fish.
The experiment was done outside the breeding season.
Every fish has an equal probability of being caught.

13 This is a randomly selected group that is not subjected to the factors you will test.

Chapter 1 review

1 **a** Qualitative   **b** Discrete or quantitative

2 A Continuous   B Discrete   C Continuous

3 A Quantitative   B Quantitative   C Qualitative

4 A Secondary   B Secondary   C Primary

5 It is trying to persuade people to agree.

6 A Any **one** of: expensive; time consuming; a lot of data to handle.
B Any **one** of: not completely representative; may be biased.

7 **a** Register or list of all children in the school.
b It is biased or it is persuading people to agree.

8 Any **one** of:
There are more men than women visiting the A&E Department;
There are more women than men visiting the A&E Department.

9 **a** B   **b** C
c Advantage: quick and easy to get.
Disadvantage: may be biased; do not know how it was collected.

10 **a** Record the number of hours of sunshine himself one June.
b Could get it from a website, the Met Office, town records, tourist information centre.
c Quantitative

11 **a** Adult patients on the doctor's patient list.
b Stratified sampling.
c People may not tell the truth and people may have changed their smoking habits after the database was set up.

12 **a** Any **one** of:
Twineasy is stronger than Plasuper.
Plasuper is not as strong as Twineasy.
b All the ropes of these two types that the manufacturer has.
c If he did a census all the ropes would be destroyed.
d A rope.

13 **a** No – not everyone would have an equal chance of being selected or all Class 4 would be chosen.
b 14

Chapter 1 Test yourself

1 **a** continuous and quantitative
b qualitative
c quantitative and discrete
d continuous and quantitative
e secondary
f primary and qualitative

2 **a** Advantages: Known accuracy and know how it was obtained.
Disadvantages: Expensive and time-consuming.
b Advantages: Easy and cheap.
Disadvantages: May have errors, may be out of date, unknown collection method.

3 **a** Everything or everybody who could be involved in the investigation.
b Gets information on every member of a population.

4 Any **two** of:
Cheaper, takes less time and has less data than a census.

5 Simple random sample

6 Open. It does not restrict answers. Respondents can say as much or as little as they wish.

7 Tables, calculator, computer

8 Identifies problems in questions and checks they get the responses required.

9 A control group is randomly selected and is used to help test the effect of various factors in an experiment. It is not subjected to any of the factors under test.

10 **a** Stratified   **b** Systematic

Chapter 2

Exercise 2A

1

Colour	Tally	Frequency
black	卌 II	7
blue	卌 卌	10
red	卌 卌 II	12
white	卌 卌 卌 III	18
yellow	卌	3
Total		50

2

Number of goals	Tally	Frequency
0	卌 I	6
1	卌 II	7
2	卌 IIII	9
3	IIII	4
4	III	3
5	I	1
Total		30

3 a

Number of pins	Tally	Frequency
48	I	1
49	III	3
50	IIII IIII	9
51	III	3
52	III	3
53	I	1
Total		20

b Sixteen of the boxes contain 50 or more, but there are four boxes with less than 50, so their claim is false.

4

Number of letters	Tally	Frequency
1	II	2
2	IIII IIII	9
3	IIII	5
4	IIII	4
5	III	3
6	IIII	4
7	II	2
8	IIII	4
9	I	1
10		0
Total		34

5 a

Mark	Tally	Frequency
20–29	III	3
30–39	IIII II	7
40–49	IIII IIII III	13
50–59	IIII IIII IIII I	16
60–69	IIII IIII III	13
70–79	IIII	5
80–89	III	3
Total		60

b 50 students

6 a

Papers sold	Tally	Frequency
40–44	IIII I	6
45–49	IIII	5
50–54	IIII II	7
55–59	IIII III	8
60–64	IIII	4
65–69	I	1
Total		31

b 4 days

7 a

Score	Tally	Frequency
1–10		0
11–20	II	2
21–30	III	3
31–40	IIII	5
41–50	IIII I	6
51–60	IIII II	7
61–70	I	1
71–80	II	2
81–90	I	1
91–100	III	3
101–110	IIII I	6
111–120	IIII	4
Total		40

b

Score	Frequency
1–20	2
21–40	8
41–60	13
61–80	3
81–100	4
101–120	10
Total	40

c

Score	Frequency
1–40	10
41–80	16
81–120	14
Total	40

d The table in part **b**. It shows the that marks are bunched for beginners and for experts.

e The class intervals are too wide in part **c**, so it looks like there is a uniform distribution. The intervals are too narrow in part **a** so all the frequencies are small.

8 Depends on the result of the experiment, but the frequency table should record the number of heads (0, 1, 2, 3) as a tally and as a frequency.

Exercise 2B

1 a

Amount	Tally	Frequency
£0–£1.00	III	3
£1.01–£1.50	IIII IIII	10
£1.51–£2.00	IIII IIII	9
£2.01–£3.00	IIII I	6
£3.01–	II	2
Total		30

2 a 15   **b** 50
c Various answers, for example

Score	Frequency
<25	10
25–29	15
30–34	17
35–39	4
>39	4

3 a Most of the data falls into one class. Almost half the classes have frequency of 1 or 0.
b When collecting data you do not know what the highest value will be.
c The person watched nine programmes per day, which is extreme.
d The most common numbers of programmes watched are 9–12 and 17–20, but 13–16 is not a common number of programmes watched.

4 a Sue has equal class intervals, but John's are varied.
b There are only three people below 30.
c John's table. Most ages are between 30 and 49, and John has smaller intervals in this range.
d He did not know how old the oldest person would be.
e Use varied class intervals. Leave the last class open.

5 a

	Adults	Children	Total
Right-handed	32	18	50
Left-handed	15	22	37
Total	47	40	87

b 18
c Yes, over half the children are left-handed, but only one third of the adults are.

6

	Lemonade	Orange juice	Total
Girls	3	9	12
Boys	10	6	16
Total	13	15	28

7 a

	'Butter-side down'	'Butter-side up'	Total
Dropped	26	11	37
Thrown	21	24	45
Total	47	35	82

b Butter-side down.

8 a

	Car	Bus	Cycle	Walk	Other	Total
English	15	4	0	1	0	20
Games	3	1	18	7	3	32
Geography	8	4	1	18	1	32
Maths	28	3	1	1	1	34
Science	16	5	7	6	4	38
Total	70	17	27	33	9	156

b 38   **c** 70   **d** 1

Exercise 2C

1 a i) 475 km   **ii)** 174 km   **iii)** 245 km
b 595 km (= 125 + 470)
c 1315 km (= 245 + 323 + 277 + 470)

2 a 07:30   **b** 08:45   **c** 11:19   **d** 20:40
e i) 12 min   **ii)** 16 min   **iii)** 11 min   **iv)** 50 min
f 13:54

3 a 1172   **b** reduced or got less
c 2469   **d** 5.6%

4 a i) £317   **ii)** £446   **iii)** £266
b 51–
c £62 (£300 − £238)
d 17–25-year-old male from Area C.
e 36–50-year-old female from Area B.

5 a Vienna   **b** Switzerland
c Italy, Greece and Austria   **d** Norway
e Ireland

6 a Portendales and The Town
b Hillstone
c Portendales and The Marion
d Hillstone, The Marion
e Portendales

7 a 22.81 thousand or 22 810
b Rounding errors
c i) Dropping/going down   **ii)** Level trend

8 Student's own answer.

Exercise 2D

1 a Theme park   **b** 240   **c** 408
d 156   **e** 1176 (= 408 + 240 + 156 + 372)

2 a Two circles for Cardiff.
b London   **c** 7

3
A pictogram to show the number if library books in the classroom

Parkfield
Easedale
Whinton
Graymans
Harris

Key: ☐ represents 5 books

4
A pictogram to show the hometowns of supermarket customers

Upshaw
Bunton
Chuckleswade
Newtown
Shenford

Key: represents 5 customers

5

A pictogram to show areas of public spending

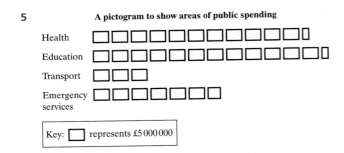

Health

Education

Transport

Emergency services

Key: ☐ represents £5 000 000

Exercise 2E

1 a 4 **b** 7 **c** 3
d Sizes 3 and 11, and sizes 6 and 9.
e 32 (= 0 + 1 + 3 + 4 + 5 + 7 + 6 + 5 + 0 + 1)
f

A pictogram to show shoe sizes

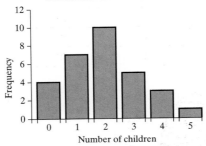

Size 3

Size 4

Size 5

Size 6

Size 7

Size 8

Size 9

Size 10

Size 11

Key 👟 = 2 students

2 a 7 **b** 9
c 31 (= 7 + 6 + 2 + 3 + 2 + 5 + 3 + 1 + 1 + 1)
d 0
e Yes. It is a very easy to read the frequecies

3

The number of children in families

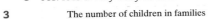

4 a

A Pictogram to show length of words

Number of letters

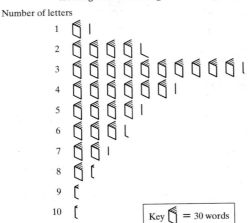

1
2
3
4
5
6
7
8
9
10

Key 📖 = 30 words

b A matter of opinion.
No. The figures need a more accurate method.
Yes. A good impression is given of the common lengths of words.
c A vertical line graph displays the figures more clearly as they fit better.

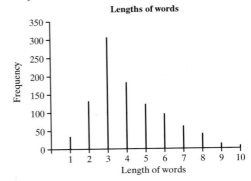

Lengths of words

5 Student's own answers.

Exercise 2F

1 a It is impossible to do 239 sit-ups in this time. The member either lied or made a mistake.

b **The number of sit-ups done in one minute**

0	
1	2 5 6
2	3 6 6 7 8 9 9
3	2 3 3 3 5 7 8 9
4	0 0 1 2 5 8
5	3 9
6	8
7	2 5

Key

3 | 2 = 32 sit-ups

c 33

2 a 30 **b** 32 **c** 12 **d** 2
e

Age	Frequency
0–4	3
5–9	5
10–14	8
15–19	6
20–24	4
25–29	3
30–34	1
Total	30

f **Ages of customers**

0	1 2 4 5 6 6 8 9
1	1 1 2 2 2 3 4 4 5 5 6 7 8 8
2	2 3 3 4 5 5 8
3	2

Key

3 | 2 = 32 years old

3

Milk yield

```
0 | 5 7 9 9
1 | 3 4 5 6 7 8 8 9 9
2 | 1 2 3 3 3 4 4 4 5 5 5 5 5 6 6 6 6 7 8 8 9
3 | 1 2 2 4 4 4 4 5 5 6 9 9
4 | 0 1 2
```

> Key
>
> 2 | 1 = 21 litres

4 a 26   **b** 52   **c** 23

5

The number choices of 60 of Carol's friends

```
10 | 0 2 2 3 3 4 5 6 6 7 9 9
11 | 2 4
12 | 2 3 4 7 8 9
13 | 2 4 4 5
14 | 0 2 3 4 4 5 5 6 7 7 8 8 9 9
15 | 1 2 2 3 4 5 5 6 6 7
16 | 2 3
17 | 0 1 1 2 2 3 5 7 8 9
```

> Key
>
> 10 | 2 = 102

Exercise 2G

1 a $\frac{20}{120} \times 360° = 60°$

b

Country	Frequency	Angle
UK	20	60°
France	15	45°
Germany	48	144°
Italy	5	15°
Japan	32	96°

c

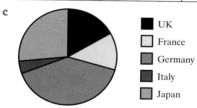

- UK
- France
- Germany
- Italy
- Japan

2 a

Area of spending	Amount (£million)	Angle
Health	59	118°
Education	67	134°
Transport	17	34°
Emergency services	37	74°

b

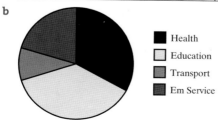

- Health
- Education
- Transport
- Em Service

3 a $\frac{4}{60} \times 360° = 24°$

b

Treatment	Angle in pie chart
Check up	144°
Filling	132°
Clean/Scale	60°
Cap	24°

c

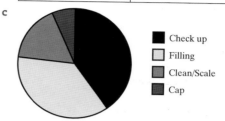

- Check up
- Filling
- Clean/Scale
- Cap

4 b

Ingredient	Weight (g)
Flour	70
Butter	74
Eggs	36
Sugar	60

5 a $15° \left(= \frac{360°}{24}\right)$

b $75° (= 360° - (120° + 30° + 135°))$

c 8 hours $\left(= \frac{120°}{15°}\right)$ **d** 9 hours $\left(= \frac{135°}{15°}\right)$

6 a

Colour	Frequency	Angle
Grey	22	110°
Green	16	80°
Patterned	10	50°
Red	6	30°
Other	18	90°

b 72   **c** Grey

d

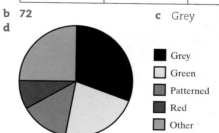

- Grey
- Green
- Patterned
- Red
- Other

7 Students' own answers.

Exercise 2H

1 a Blue   **b** Black

c 10 (= 2 + 3 + 5)   **d** 25 (= 7 + 3 + 2 + 5 + 8)

e Wednesday

2 Bar chart

Advantages: Easy to draw; values can be read from a scale; fairly accurate.

Disadvantages: Relationships not so clear as in pie chart.

Pie chart

Advantages: Shows proportions well.

Disadvantages: Not easy to draw; values cannot be read.

3 Gives a visual comparison; retains detail; uses a small space.

4 a Oranges   **b** Apples and grapes
c 23 kg   **e** 12 kg
e 24 kg

5 a Rounding errors. The published numbers were rounded to one decimal place.

b

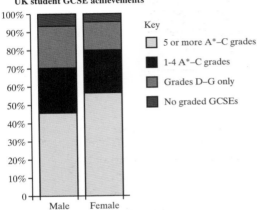

UK student GCSE achievements

Key
☐ 5 or more A*–C grades
■ 1-4 A*–C grades
▨ Grades D–G only
▩ No graded GCSEs

Reverse acceptable as long as both in same order

6

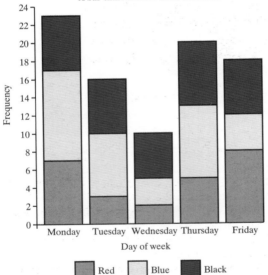

A bar chart to show sales of scarves

Red   Blue   Black

7

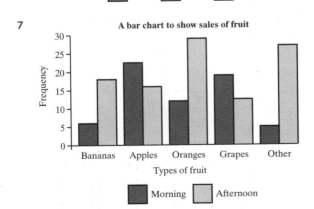

A bar chart to show sales of fruit

Morning   Afternoon

Exercise 2I

1 a French   **b** B or C
c French
A or A* → 53°   B or C → 160°
D or E → 107°
F or G → 40°

Spanish
A or A* → 59°   B or C → 213°
D or E → 66°   F or G → 22°

Ratio of radii is 1.29 :1

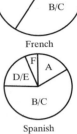

French

Spanish

2 a Urban   **b** Urban
c Southshire
Agriculture → 34°
Urban → 160°
Woodland → 139°
Water → 27°

Northshire
Agriculture → 38°
Urban → 204°
Woodland → 97°
Water → 21°

Ratio of radii is 1 :1.26

Southshire

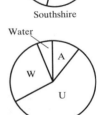

Northshire

3 a 320   **b** 63   **c** 27
d

Colour	Red	Blue	White	Yellow	Green	Other	Total
Frequency	62	45	30	20	13	10	180

e 115
f

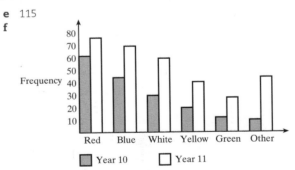

Frequency

Red   Blue   White   Yellow   Green   Other

☐ Year 10   ☐ Year 11

4 Germany
Beethoven → 38° Mozart → 223°
Handel → 62° Saint-Säens → 0°
Wagner → 22° Other → 15°

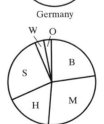

Germany

France
Beethoven → 82° Mozart → 103°
Handel → 61° Saint-Säens → 92°
Wagner → 12° Other → 10°

Ratio of radii is 1 :1.2

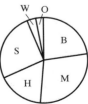

France

5 a A higher proportion of James' class walk, but there are more children in Alex's class overall and there are more walkers.

There are 25 children in James' class and 36 in Alex's. Using square roots, make the ratio of the radii of James' pie chart to Alex's 5:6.

b

James' pie chart   **Alex's pie chart**

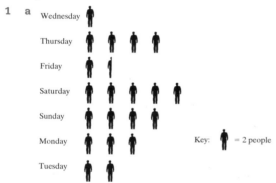

Ratio of radii is 1:1.2

Chapter 2 review

1 a

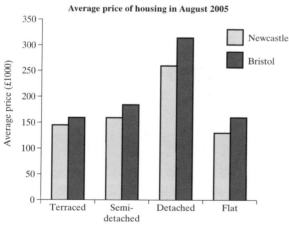

Key: ♟ = 2 people

b Saturday   **c** 8

2 a 48%   **b** 729.7 thousand or 729 700   **c** East

3 A, B and D

4 a

Average price of housing in August 2005

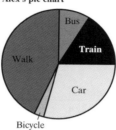

b EITHER: These are the most expensive types.
OR: They are more expensive in Bristol than Newcastle. (Or equivalent.)

OR: Average in Newcastle is £260 000 and in Bristol is £315 000. (Allow £310 000 to 320 000.)

c EITHER: Flat and terrace.
OR: Flat
OR: Terrace

5

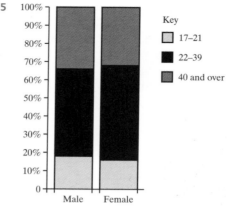

Key
17–21
22–39
40 and over

b 12 618

6 a 120°   **b** 30 acres

c Total acreage is 40 + 60 + 50 + 30 = 180

Angle for oats = $\frac{40}{180} \times 360° = 80°$

Angle for barley = $\frac{60}{180} \times 360° = 120°$

Angle for wheat = $\frac{50}{180} \times 360° = 100°$

Angle for others = $\frac{30}{180} \times 360° = 60°$

Average of crops on High Meadows farm

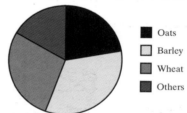

Oats
Barley
Wheat
Others

(Pie chart not to 13 cm scale.)

7 a 2   **b** 23   **c** 60

d 57 (= 11 + 23 + 14 + 6 + 3)

e **i)** 12 $\left(= \frac{90°}{360°} \times 48\right)$ **ii)** 18 $\left(= \frac{135°}{360°} \times 48\right)$

f Squib Street

g No. There are more five-bedroom houses on Round Street than four-bedroom houses.

h Crumple Street. It has the greatest fraction of two-bedroom houses.

i The bar chart is easy to read and you can see information at a glance.

OR: The pie chart shows the proportions of the types of houses more easily.

OR: The frequency table gives exact figures.

j **i)**

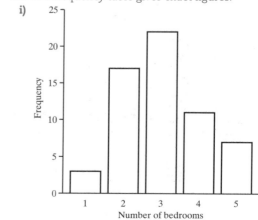

ii)

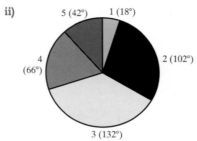

5 (42°) 1 (18°)

4 (66°) 2 (102°)

3 (132°)

iii)

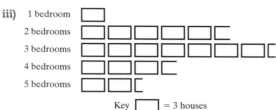

1 bedroom

2 bedrooms

3 bedrooms

4 bedrooms

5 bedrooms

Key ☐ = 3 houses

8 a

Age	Tally	Frequency
0–9	|||	3
10–19	||||	5
20–29	|||| |||| ||	12
30–39	|||| ||||		11
40–49	||||	6
50–59	|||	3
Total		**40**

b The groups would overlap.

c **The ages of customers in a supermarket**

```
0 | 3  7  8
1 | 1  2  6  7  7
2 | 1  2  2  4  5  5  6  7  8  8  9  9
3 | 1  3  3  3  5  6  6  6  7  8  8
4 | 1  1  2  4  6  8
5 | 5  6  6
```

Key

2 | 3 = 23 years old

d The exact ages of the customers.

9 a

	Year 7	Year 8	Year 9	Year 10	Year 11	Total
Boys	72	47	71	66	64	320
Girls	63	75	30	55	63	286
Total	135	122	101	121	127	606

b

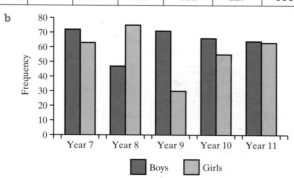

Boys Girls

Chapter 2 Test yourself

1

Absentees	Frequency
0–5	5
6–10	6
11–15	7
16–20	4
21–25	5
26–	3

2

	Vanilla	Chocolate	Strawberry	Total
Male	6	3	**11**	20
Female	**5**	8	5	18
Total	**11**	**11**	**16**	**38**

3 A vertical line graph uses lines to show the frequency of the data. A bar chart shows data frequency as bars.

4 a
```
0 | 9
1 | 6  7  9  9
2 | 2  3  5  6  7  7  7  8  8  9  9
3 | 0  1  2
4 | 1
```

b 27

c Able to show a lot of data in a small space; keeps details of the data; give a visual comparison.

5

Car colour	Frequency	Angle for pie chart
Silver	**14**	112°
Red	10	**80°**
Blue	6	**48°**
White	12	**96°**
Black	3	**24°**

6 A multiple bar chart has two or more bars for each class. A composite bar chart has each bar stacked showing its individual components.

7 8.485 cm

Chapter 3

Exercise 3A

1 a $1 \leqslant r < 2$

b

Rainfall, r (cm)	Tally	Frequency
$0 \leqslant r < 1$	||||		6
$1 \leqslant r < 2$	||||		6
$2 \leqslant r < 3$	||||		6
$3 \leqslant r < 4$	|||| ||	7
$4 \leqslant r < 5$	||	2
$5 \leqslant r < 6$	||||	4
Total		31

2 a 3.22 kg

b

Weight, w (kg)	Tally	Frequency
$0 \leqslant w < 0.5$	||	2
$0.5 \leqslant w < 1$	|||	3
$1 \leqslant w < 1.5$	⊬	5
$1.5 \leqslant w < 2$	⊬ ||||	9
$2 \leqslant w < 2.5$	⊬ |	6
$2.5 \leqslant w < 3$	⊬ |	6
$3 \leqslant w < 3.5$	|||	3
Total		**34**

3 a There is a gap between the group '1.0 to 1.4' and the group '1.5 to 1.9', so 1.46 does not fit into any group.
  b There is an overlap of groups. 1.5 fits into both the group '1.0–1.5' and the group '1.5–2.0'.
  c $0 \leqslant t < 0.5$ $0.5 \leqslant t < 1.0$ $1.0 \leqslant t < 1.5$
     $1.5 \leqslant t < 2.0$ $2.0 \leqslant t < 2.5$ $2.5 \leqslant t < 3.0$

4 a $47.5 \leqslant w < 48.5$
  b All numbers that round to 45 kg do not fit in the same class.
  c $34.5 \leqslant w < 39.5$, $39.5 \leqslant w < 44.5$ etc.
  d

Weight, w (kg)	Tally	Frequency
$34.5 \leqslant w < 39.5$	||	2
$39.5 \leqslant w < 44.5$	|||	3
$44.5 \leqslant w < 49.5$	⊬ |	6
$49.5 \leqslant w < 54.5$	⊬ ||||	9
$54.5 \leqslant w < 59.5$	⊬ |	6
$59.5 \leqslant w < 64.5$	|||	3
$64.5 \leqslant w < 69.5$	|	1
Total		**30**

5 a Joff's table has gaps.
  b His time could have been anything in the range $19.5 \leqslant t < 20.5$, so it may belong in the class interval $10 \leqslant t < 20$ or $20 \leqslant t < 30$.
  c He could use an open interval, $t \geqslant 50.5$

Exercise 3B

1 a

Unordered tree trunk data

0	2
1	
2	1 5 2 3 6 3 5 7 8
3	4 0 4 3 7 3 1 5 7 7 9 9 7 3
4	5 9 7 2 6 3 3 6 8 9
5	6 7 9 0 8 0 0 3 3 6 8 9
6	0 4 2 3

Key

5 | 6 = 5.6 cm

b

Ordered tree trunk data

0	2
1	
2	1 2 3 3 5 5 6 7 8
3	0 1 3 3 3 4 4 5 7 7 7 7 9 9
4	2 3 3 5 6 6 7 8 9 9
5	0 0 0 3 3 6 6 7 8 8 9 9
6	0 2 3 4

Key

2 | 3 = 23 years old

c 3

2 a 3.2 s   **b** 21   **c** 4.3 s
  d It shows the shape of the distribution.

3 a 25   **b** 1.91 m   **c** 17
  d

Height, h (cm)	Frequency
$1.5 \leqslant h < 1.6$	3
$1.6 \leqslant h < 1.7$	5
$1.7 \leqslant h < 1.8$	9
$1.8 \leqslant h < 1.9$	7
$1.9 \leqslant h < 2.0$	1
Total	**25**

e

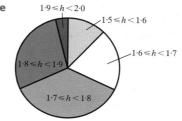

4 a 56–60
  b $45.5 \leqslant g < 50.5$; $50.5 \leqslant g < 55.5$; $55.5 \leqslant g < 60.5$; $60.5 \leqslant g < 65.5$; $65.5 \leqslant g < 70.5$
  c All classes are the same width.
  d Nearly half the eggs are in the middle class.
    Almost quarter of the eggs are in each class on either side of the middle class.
    Very few eggs are in the smallest and largest classes.

5 a

Frequency	Angle
8	$72°$ $(= \frac{8}{40} \times 360°)$
13	$117°$ $(= \frac{13}{40} \times 360°)$
12	$108°$ $(= \frac{12}{40} \times 360°)$
7	$63°$ $(= \frac{7}{40} \times 360°)$

b **Percentage of iron in ore sample**

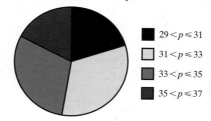

■ $29 < p \leqslant 31$
□ $31 < p \leqslant 33$
■ $33 < p \leqslant 35$
■ $35 < p \leqslant 37$

Exercise 3C

1 a

Height, h (cm)	Cumulative frequency
110 < h ≤ 120	5
120 < h ≤ 130	17
130 < h ≤ 140	52
140 < h ≤ 150	92
150 < h ≤ 160	130
160 < h ≤ 170	150

b

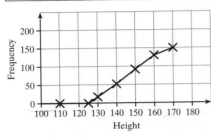

c Accept 14 to 20 boys.

2 a

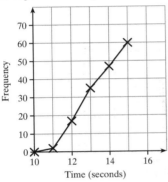

b 54
c Accept 52 to 56.

3 a

Time late, t (min)	Frequency	Cumulative frequency
0 < t ≤ 3	10	10
3 < t ≤ 5	31	41
5 < t ≤ 7	35	76
7 < t ≤ 10	21	97
10 < t ≤ 15	3	100

b

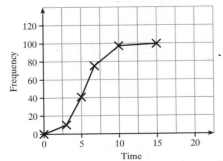

c Accept 14 to 19 times.
d Accept about 40%.

4 a

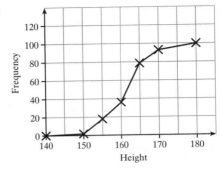

b Accept 51 to 54.

Exercise 3D

1

A frequency polygon to show distances travelled

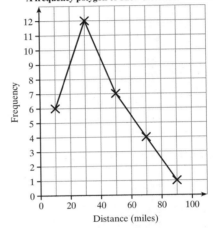

2

A frequency polygon to show plant heights

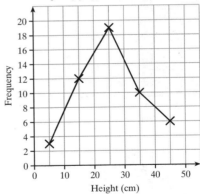

3 a 3   **b** 7   **c** A boy.

d

Time, t (s)	Girls' frequency	Boys' frequency
0 < t ≤ 20	0	3
20 < t ≤ 40	3	4
40 < t ≤ 60	10	8
60 < t ≤ 80	14	9
80 < t ≤ 100	6	8
100 < t ≤ 120	0	1
Total	33	33

e Same number of each.

f Girls. More balanced for over a minute.
OR: Both the same. The modal times are the same.

g

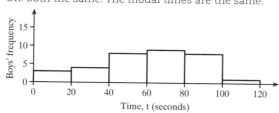

h The distribution has a weak negative skew.

4 a

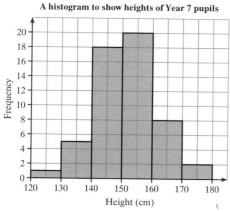

A histogram to show heights of Year 7 pupils

b

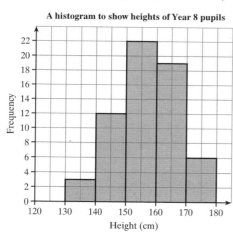

A histogram to show heights of Year 8 pupils

c

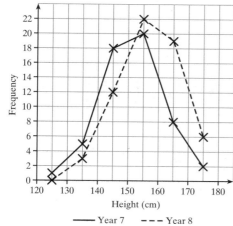

—— Year 7   - - - Year 8

5 a

Time, t (min)	Tally	Frequency
$0.5 \leqslant t < 10.5$	卌 I	6
$10.5 \leqslant t < 20.5$	卌	5
$20.5 \leqslant t < 30.5$	卌 I	6
$30.5 \leqslant t < 40.5$	IIII	4
$40.5 \leqslant t < 50.5$	卌 I	6
$50.5 \leqslant t < 60.5$	III	3
Total		30

b

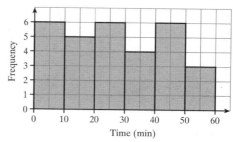

Exercise 3E

1

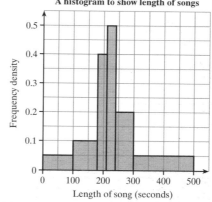

A histogram to show length of songs

2 a and **b**

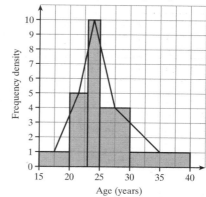

A histogram to show ages of guests

3

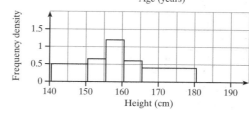

4

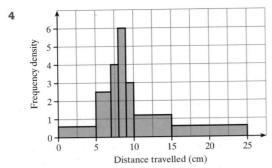

Frequency density vs Distance travelled (cm)

5

Distance, d (m)	Frequency
$0 \leqslant d < 20$	40
$20 \leqslant d < 35$	75
$35 \leqslant d < 45$	110
$45 \leqslant d < 60$	105
$60 \leqslant d < 65$	10
Total	340

6 a

Time, t (s)	Frequency
$0 < t \leqslant 5$	56
$5 < t \leqslant 15$	128
$15 < t \leqslant 20$	104
$20 < t \leqslant 30$	160
$30 < t \leqslant 45$	144
$45 < t \leqslant 70$	120
$70 < t \leqslant 80$	24
Total	736

b A bar between 70 and 80 that is 1.5 squares high.

Exercise 3F

1

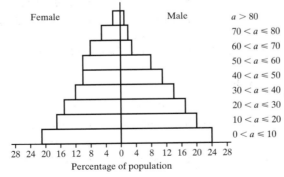

A population pyramid to show percentage of ages

Female	Male	
		$a > 80$
		$70 < a \leqslant 80$
		$60 < a \leqslant 70$
		$50 < a \leqslant 60$
		$40 < a \leqslant 50$
		$30 < a \leqslant 40$
		$20 < a \leqslant 30$
		$10 < a \leqslant 20$
		$0 < a \leqslant 10$

28 24 20 16 12 8 4 0 4 8 12 16 20 24 28
Percentage of population

2 a Country B **b** Females
  c $30 < a \leqslant 40$ **d** Country A
  e It has a high mortality rate and a high birth rate.

3 a Regions A and C **b** Region I
  c Region B **b** Regions D, H and J

4

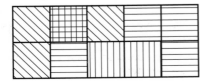

Exercise 3G

1 The scale does not start at 0; the 3-D effect distorts, so heights are difficult to read.

2 No scale; thick line.

3 Two of: Dark colour dominates; pieces cut out make it difficult to see proportions correctly, no sector for 'other pets'.

Chapter 3 review

1 a The frequency polygon shows that 13 boys balanced a dictionary for between 30 and 40 seconds.

b

Time, T (s)	Frequency
$0 < T \leqslant 10$	1
$10 < T \leqslant 20$	6
$20 < T \leqslant 30$	8
$30 < T \leqslant 40$	13
$40 < T \leqslant 50$	2
$50 < T \leqslant 60$	1
Total	31

c

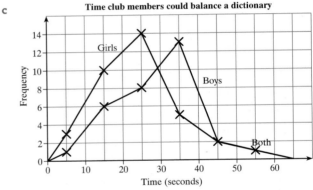

Time club members could balance a dictionary

d Boys generally balanced for longer (e.g. modal class was higher).

2 They would appear firstly to like a high altitude and then a high rainfall rainfall. They do not like heath.

3 a Population pyramid
  b 25–29 **c** 17%–22%
  d Women tend to live longer than men in both; Both have same proportion between 40 and 44; Camden has more in the 20–30 age group.

4 a

```
0 | 0 1 2 5 6
1 | 1 1 3 3 3 7 8 9
2 | 0 1 1 2 2 6 7 8
3 | 0 0 2 3 4 5 6 7 8
4 | 2 2 3 5 7
5 | 2
```

Key
1 | 2 means 1.2

b

Distance, d (cm)	$0 < d \leqslant 1$	$1 < d \leqslant 2$	$2 < d \leqslant 3$	$3 < d \leqslant 4$	$4 < d \leqslant 5$	$5 < d \leqslant 6$
Frequency	5	9	9	7	5	1

5 a

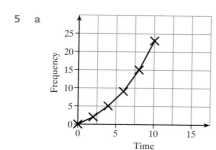

b 13 **c** Accept 6, 7 or 6.5.

d

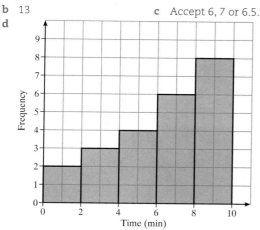

6

Length, l (cm)	Frequency
$3 \leqslant l < 4$	3
$4 \leqslant l < 5$	4
$5 \leqslant l < 8$	14
$8 \leqslant l < 10$	20
$10 \leqslant l < 11$	9

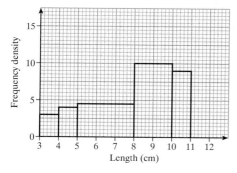

Chapter 3 Test yourself

1 a 35–39 **b** 50–54

c A larger percentage of people in Northern Ireland is aged up to 19 than in the UK.

2

3 Any **six** from:

- Scales that do not start at zero or have parts of them missed out give a misleading impression of the heights of bars etc.
- Scales that do not increase uniformly distort the shape of anything plotted on them.

- Lines on a graph that are drawn too thick make it difficult to read information from them.
- Three-dimensional diagrams make comparisons difficult. Often things at the front of the diagram can appear larger than those at the back (e.g. angled pie charts). Parts at the back may be hidden behind those at the front and appear smaller than they should.
- Sections of the diagram separated from other parts make comparisons difficult (e. g. pie charts with slices pulled out).
- Use of colours can make some parts stand out more than others. Generally dark colours stand out more than light colours and cause things to look bigger.
- Use of different width bars/pictures. To make charts more interesting, the bars can be made up of pictures of the thing they represent. For example, bags of money to illustrate wages: if the bags are different sizes it is unclear whether it is the height of the money bags that should be compared or the area.
- Axes that are not labelled properly.
- Some data may be excluded.

4 25 or 26 **5** 4 cm

Chapter 4

Exercise 4A

1 a 12 min **b** 15 min **c** 15 min

2 a 5 5 7 7 8 9 9 9 9 9 10 11 12

b 9 **c** 9 **d** 8.46

3 a 5.5 s **b** 2 s **c** 6.5 s

4 Mode 33; median 33; mean 33.42

5 a 520 kg **c** 50 kg

Exercise 4B

1 a 13 **b** 13 **c** 13.26 (2 d.p.)

2 a 3 **b** 4 **c** 4.29 (3 s.f.)

3 a 20°C **b** 20°C **c** 20.48°C (2 d.p.)

4 a i) B **ii)** C

b Ratings would need to be numbers, not in letters.

Exercise 4C

1 a $6 < x \leqslant 8$ **b** 7 cm $\left(6\frac{11}{12}\right)$ **c** 6.91 cm

2 a $3 < x \leqslant 6$ **b** 4.75 hours **c** 4.8 hours

3 a $60 < speed \leqslant 70$ **b** 57 mph **c** 54.9 mph

4 a $20 \leqslant age < 30$ **b** 31 years **c** 31.75 years

Exercise 4D

1 104.5 **2** 3004.9 (1 d.p.) **3** 2.153 75

Exercise 4E

1 a 14 **b** 25

c The median because the number 100 distorts the mean making it unrealistic.

2 The mode is £16. This is the minimum amount so it does not well describe the average pay.

3 **a** Ordinary
  b The mode is the only average that can be found from non-numerical data.

4 **a** The mode is the only average that can be found from non-numerical data.
  b Colour is not numerical.

5 The mean would not be satisfactory if there were one or two very high or low values – these would distort the mean.
The mode could be acceptable but it could apply to just the minimum or maximum price.
The median is the middle price and it is not affected by outliers.
Each has points for and against but probably the median would be best.

Exercise 4F

1 Jimmy 54.65 (accept 55);
Sumreen 58.25 (accept 58).

2 £319

3 54.5

Exercise 4G

1 **a** 9   **b** 6   **c** 10   **d** 4

2 **a** 70 km/h **b** 50 km/h **c** 95 km/h **d** 45 km/h

3 **a** Range = 8; LQ = 3; UQ = 9; IQR = 6
  b Range = 57; LQ = 21; UQ = 65; IQR = 44
  c Range = 11; LQ = 6; UQ = 12; IQR = 6

4 **a**
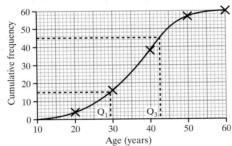

  b LQ = 29; UQ = 44; IQR = 15

5 **a**
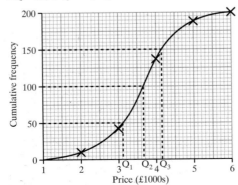

  b **i)** £3600 **ii)** £4200 **iii)** £3100 **iv)** £1100
  c **i)** £4500 **ii)** £3400 **iii)** £1100 **iv)** £3500
    v) £4300

6 **a** £39
  b
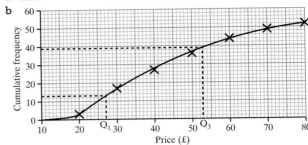

  c **i)** LQ = £27; UQ = £53
    ii) £42.50   **iii)** £23

7 **a**
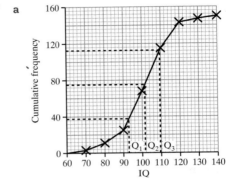

  b **i)** 102   **ii)** 93   **iii)** 109
  c **i)** 104   **ii)** 21

Exercise 4H

1 **a** **i)** 23 m   **ii)** 25 m   **iii)** 26 m
  b
```
    ┌─────┬───┐
├───┤     │   ├───┤
    └─────┴───┘
21    23  25 26  27
        Length (m)
```

2 **a** **i)** 8   **ii)** 12   **iii)** 14
  b
```
  ┌────────┬─────┐
├─┤        │     ├──┤
  └────────┴─────┘
7 8       12    14 15
      Number of eggs
```

3 **a** **i)** 26   **ii)** 42
  b **i)** 40   **ii)** 80
  c
```
        ┌────┬──────┐
    ├───┤    │      ├────┤
        └────┴──────┘
        41  46  57  72  81
```
```
      ┌──────┬──────────┐
  ├───┤      │          ├──────────┤
      └──────┴──────────┘
  19  23    46         65          99
```
  d The girls' marks are have a negative skew and the boys' marks have a positive skew.

4 **a**
```
    ┌──────┬──────┐
├───┤      │      ├────┤
    └──────┴──────┘
750 810   940   1100  1250
       Weight (g)
```
  b Positive skew.

5 **a** **i)** 32   **ii)** 27   **iii)** 41
  b 75
  c
```
      ┌───┬───┐
  ├───┤   │   ├──────┤              ×
      └───┴───┘
  14  27 32  41      52            75
        Age (years)
```

6 a i) 38 **ii)** 32 **iii)** 47 **iv)** 15
b 76
c

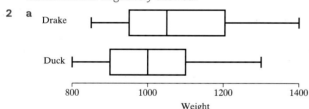

7 a LQ = 160; UQ = 180; median = 172
b 120 and 214 **c** Negative skew.

Exercise 4I

1 a Mean = 8; sd = 2.61 (3 s.f.)
b Mean = $6\frac{2}{3}$; sd = 3.06 (3 s.f.)
c Mean = 4.07 (3 s.f.); sd = 1.77 (3 s.f.)
2 a 2.61 (3 s.f.) **b** 3.06 (3 s.f.) **b** 2.02 (3 s.f.)
3 a Mean = 1.65; sd = 4.66 (3 s.f.)
b Mean = 0.1575; sd = 0.596 (3 s.f.)
4 a 41.6 **b** 110.44 **c** 10.509
5 a 99.7 mph **b** 9.4 mph
6 a 3.48 (3 s.f.) **b** 1.805 (4 s.f.) **c** 1.344 (4 s.f.)

Exercise 4J

1 The median is greater for the males.
The IQRs are the same.
The range is greater for the males.
The males' distribution is symmetrical but the females' distribution is negatively skewed.

2 a

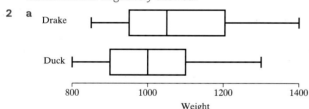

b The median is greater for the drakes.
The IQR is greater for the drakes.
The range is greater for the drakes.
The drakes' distribution is positively skewed and the ducks' distribution is symmetrical.

3 a 1.83 (3 s.f.) **b** 3.25
4 a

	Sean	Theresa	Victoria
History	1.83	−0.5	2.83
Geography	−2	−0.2	1.4

b Victoria is above average in both subjects.
Sean is above average in History, but below average in Geography.
Theresa is below average in both subjects.

Exercise 4K

1 120 **2** 12%
3 Generally it has risen by 67% over the years although it did go down by 1% in 2006.
4 126
5 a 101 102 109 95
b 71 86 81 66

c Flat prices rose in each of the first three years, but they dropped by 5% in the fourth year.
Car prices dropped every year.
6 a 104p and 158p **b** 152 **c** 52%

Chapter 4 review

1 a 72 **b** 69 **c** 64.$\dot{3}$
2 a 28.71°C **b** No values occur more than once.
3 a 32 **b** 32 **c** 30.4$\dot{3}$
4 a i) 69 **ii)** 71 **iii)** 76
b 72 $\left(= \frac{71+73}{2}\right)$
5 a 9 subjects **b** 9 subjects
c 8.48 subjects (3 s.f.)
6 a 20 ≤ age < 30 **b** (almost) 30 years
c 30.38 years (4 s.f.)
7 a 110 $\left(= \frac{7040}{6400} \times 100\right)$
b

Thorpe		
Year	1970	1990
Index number	100	98

8 a 16 minutes (= 35 − 19) **b** 24 minutes
c 17 minutes
d E.g. larger median for A or highest time for A is greater than the highest time for B.
e B: Travel to home; positive skew.
9 a i) 50 mph (= 95 − 45) **ii)** 15 mph (= 70 − 55)
b Affected by extreme values.
c

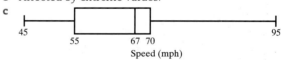

d Some (25%) of the cars were speeding.
OR: A car travelled at 25 mph above the speed limit.
OR: 100 cars were speeding.
10 64
11 a 109 **b** Dropped 4%.
12 83.75 83.58 80.95 75
13 Mean = 12; sd = 2.45 (3 s.f.)
14 a and b

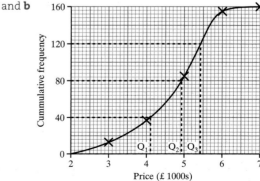

c £1300
d i) £4400 **ii)** £5500
e

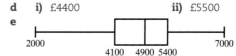

15 a Jennifer: English 1.25, Science 0.5
Samuel: English 0.083, Science 1.83 (3 s.f.)

b She is above average in both subjects.

c He is above average in both subjects, but well above average in Science.

16 a 100 m: $-0.42(0...) \left(= \dfrac{10.85 - 10.95}{0.238}\right)$

400 m: $-0.97(6...) \left(= \dfrac{48.36 - 49.62}{1.290}\right)$

b He performed better in the 400 m race. His standardised score is further from the mean (or greater negative score).

17 a **i)** 6.7 thousand (allow 6700)

ii) 3.6 thousand $\left(= \sqrt{\dfrac{457}{10} - 5.7^2}\right)$ (AWRT)

b Lost fewer days due to strikes. Manufacturing has a lower mean so Public Administration had a greater SD so was more variable (a greater spread) than manufacturing.

Chapter 4 Test yourself

1 Mean = 5.4, mode = 7, median = 6

2 Mean = 5, mode = 5, median = 4

3 Mean = 19.167 (3 d.p.), median = 18.$\dot{3}$

4 Range = 12, IQR = 5, SD = 3.325

5 Q_1 = 3.3, Q_2 (median) = 5, Q_3 = 7.3

6 −2

Chapter 5

Exercise 5A

1 a
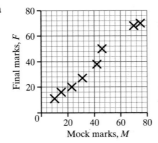

b Yes

2 a
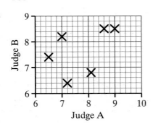

b The is a very slight agreement between the two judges.

3 a
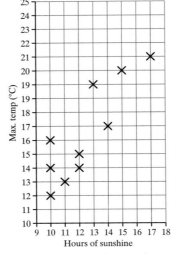

b The temperature rises as the number of hours of sunshine increases.

4 a
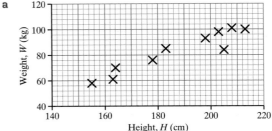

b Height and weight are associated. The taller the person, the more they weigh.

5 a
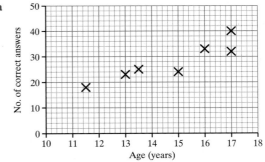

b The older the boy, the more answers he gets correct.

Exercise 5B

1 a No correlation.

b Weak positive correlation.

c Strong negative correlation.

2 a Strong positive correlation.

b Students who do well in the mock examination are likely to do well in the final examination

3 a
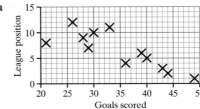

b Weak negative correlation; the more goals a team scores, the lower its position number is likely to be.

4 a

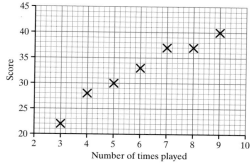

b Strong positive correlation.

c The score is likely to increase at the next attempt.

5 a

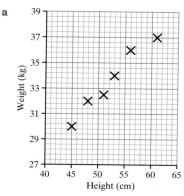

b Strong positive correlation.

c The greater the dog's height, the more it weighs.

Exercise 5C

1 A and D (In C, low temperature and snowfall may occur together, but one does not cause the other.)

2 a £3500   **b** 5 years old

c Fairly strong negative correlation.   **d** Yes

3 a and b

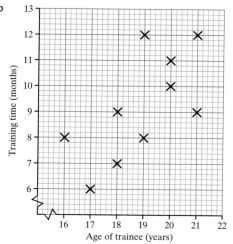

c Evidence of weak positive correlation.   **d** No

4 a

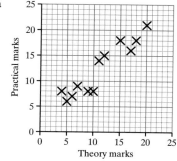

b Fairly strong positive correlation.   **c** No

5 a

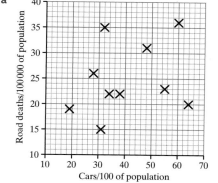

b No correlation.   **c** Yes

Exercise 5D

1 a

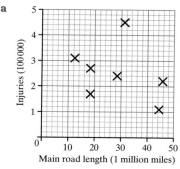

b No. There is no correlation.

2 a (41, 46)

b and c

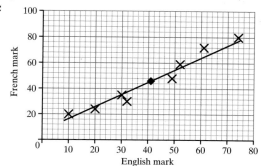

3 **a**, **b** and **d**

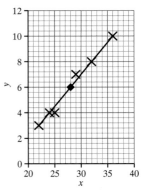

c (28, 6)

4 **a**, **b** and **c**

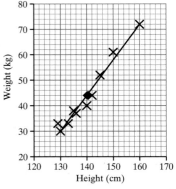

b (140, 44) **d** Strong positive correlation.

Exercise 5E

1 **a**

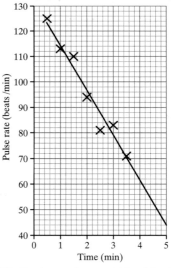

b 44 beats/min

c No. 5 minutes is a considerable time after the last measurement. His normal pulse rate is likely to be about 71 beats per minute, the last measurement.

2 **a**

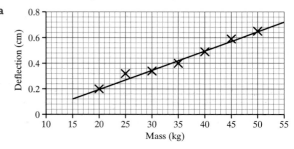

b About 0.32 cm **c** About 0.12 cm, 0.72 cm

d The first because this is the only one that uses interpolation.

3 **a** and **b**

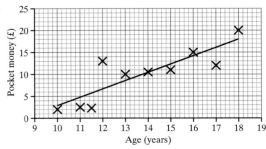

b £9.50

c People of 25 don't usually get pocket money because they have a job.

4 **a** and **b**

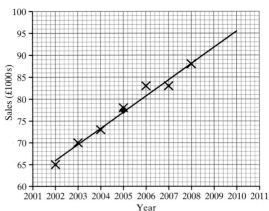

b £95 000–£96 000. This should be treated with caution because it is an extrapolated value.

5 **a** and **b**

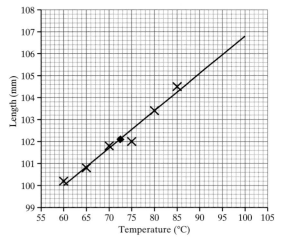

c 101.4 mm – interpolated so reasonable reliable.
106.8 mm – extrapolated so should be treated with caution.

Exercise 5F

Note that your answers will depend on how accurately you can read from the graph. Also for answers to questions 3– the lines of best fit are drawn by eye so the values of a and b might differ from those given here.

1 a −1.6   **b** 9 years   **c** $y = -1.6x + 9$

2 $y = 27x - 3$

3 a Student's own answer.   **b** $y = 1.2x + 54$

4 a and **b**

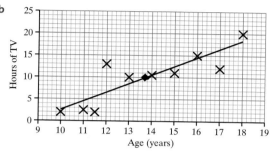

c $y = 1.9x - 16$; When $x = 16.5$, $y = 15$.
d People of 40 are well outside the ranges of ages given.

5 a 2002 = year 1, then $y = 3.8x + 62.1$.
b When $x = 9$, $y = 95.89$ (£95 890), but extrapolation has been used, so treat with caution.

6 a $y = 0.17x + 90$
  i) 101.56 mm
  ii) 107 mm
    The first uses interpolation and so is more reliable than the second which uses extrapolation.
b a is the amount the bar expands for each 1°C rise in temperature. b is the length of the bar at 0°C.

7 a

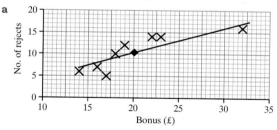

b Weak positive correlation
c $y = 0.63x - 2.2$
d £17.80

8 a and **b**

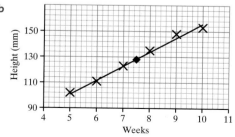

c $y = 10.8x + 48$
d a is the increase in height of the seedling in one week. b is the seedling height (in mm) when the experiment began.
e 264 mm. 20 weeks is a long way outside the range of given values so it is unlikely to be very accurate (the plants might never grown this high).

Exercise 5G

1 a $y = a\sqrt{x} + b$ **b** $y = ax + b$ $(a > 0)$
c $y = \frac{a}{x} + b$
d $y = -ax + b$ $(a > 0)$ or $y = ax + b$ $(a < 0)$

2 a and **b**

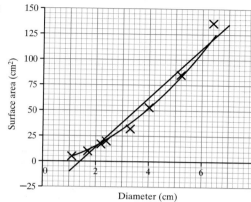

c The curve.

3 a and **b**

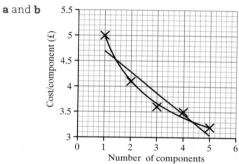

c The curve.

4 a

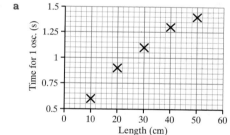

b Equation i (though if $a < 0$, equation ii is possible)

5 a

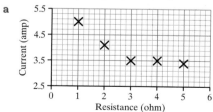

b A curve.   **c** $y = \frac{a}{x} + b$ and $y = ka^x$

Exercise 5H

Rankings from lowest to highest are equally valid.

1

Number	49	44	43	36	40	39	29	28	30	33	26
Rank	1	2	3	6	4	5	9	10	8	7	11

2

Marks after, x	12	22	40	33	18	25	14	4
Rank	7	4	1	2	5	3	6	8

Marks before, y	10	30	45	12	28	18	19	4
Rank	7	2	1	6	3	5	4	8

3

Lung damage, x	5	3	7	8	1	4	2	6
Years smoked, y	6	4	7	3	2	8	5	1

4   **a**   0.9371 (4 s.f.)   **b**   0.6   **c**   −0.1905 (4 s.f.)

5   −0.3714 (4 s.f.)

6   0.7714 (4 s.f.)

7   **a**

		Skater				
	A	**B**	**C**	**D**	**E**	**F**
Judge 1	5	2	1	6	3	4
Judge 2	4	3	2	6	1	5

0.7714 (4 s.f.)

Exercise 5I

1   −0.1152 – a very weak negative correlation.

2   The two judges are in almost complete disagreement.

3   **a**

		Song				
	1	**2**	**3**	**4**	**5**	**6**
Judge A	5	2	1	6	3	4
Judge B	6	4	1	5	2	3

$r_s = 0.7714$

b   There is fairly strong positive agreement between the judges.

4   **a**   0.4303

     b   There is weak positive correlation between choices of girls and boys.

     c   There is moderate agreement about the boys' and girls' choices of subjects.

5   **a**   0.8810

     b   There is fairly strong positive correlation between the tasters' rankings. The two tasters agree fairly well.

6   **a**   0.4061

     b   Gemma's belief is false. There is positive correlation between the ranks, implying that people with long surnames are likely to have long first names.

7   **a**   0.9524

     b   There is strong correlation. The weight after 25 days depends on the original weight.

Chapter 5 review

1   **a**

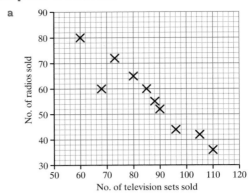

b   Negative

c   No. You can have both. Ownership of one does not affect ownership of the other.

2   **a and c**

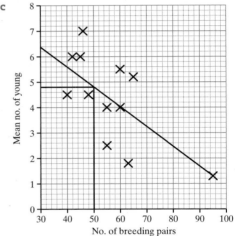

b   Weak negative   **d**   4.8

3   **a and b**

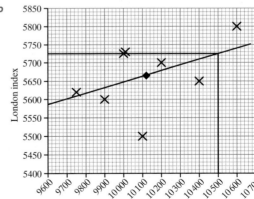

b   (10 120, 5665)   **c**   Weak positive

d   5725

e   No. This is far outside the range and the correlation is weak.

f $y = 0.155x + 4375$

4 a

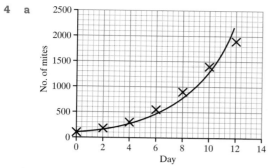

b $y = ka^x$ (The Law of Growth)

5 a

x	y	Rank x	Rank y	d	d²
280	90	1	1	0	0
290	95	2	2	0	0
297	110	4	3	1	1
300	125	5	4	1	1
310	140	6	5	1	1
295	145	3	6	3	9
311	155	7	7	0	0
330	170	9	8	1	1
320	175	8	9	1	1
				Total	14

b $r_s = 0.8833$

c Strong positive correlation. The number of people manufacturing vehicle accessories is closely related to the number manufacturing vehicles.

6 a

	Country						
	Niger	Rwanda	India	Oman	China	Cuba	UK
HDI rank	7	6	5	4	3	2	1
GNP rank	7	4	5	2	6	3	1
Difference in ranks (d)	0	2	0	2	−3	−1	0
d²	0	4	0	4	9	1	0

b 0.678 (3 d.p.)

c There is some positive correlation (or association/agreement).
PLUS EITHER: The higher the HDI the higher the GNP.
OR: The lower the HDI the lower the GNP.
(Accept wealth and quality of life.)

Chapter 5 Test yourself

1 (3.5, 3250)

2 b £2700 or £2800   **d** £1500 or £1600

e The interpolated answer **c** is more reliable because it is within the range of the data. The extrapolated answer **d** is not so reliable because age-related factors, such as the availability of spare parts and reliability, might make the price drop sharply, or the item's rarity might make the price increase.

f There is a causal relationship. Items generally deteriorate with age thereby lessening the value.

3 40 mm was the unloaded length of the spring. 0.2 mm was the length the spring increased for every 1 g mass added.

4 a −0.94 (2 d.p.)

b Strong positive correlation. The older they are, the less time they take to reach the level of proficiency.

Chapter 6

Exercise 6A

1 a 3 hours   **b** Wednesday   **c** It is the weekend.

2 a

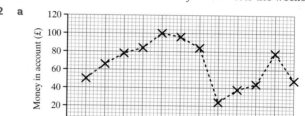

b August and December   **c** May

3 a

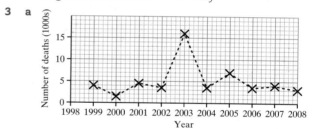

b There were many deaths in 2003. There must have been an epidemic.

4 a

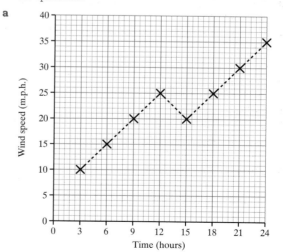

b The wind speed rose steadily except for the period between 12 00 and 15 00 when it dropped slightly.

5 a

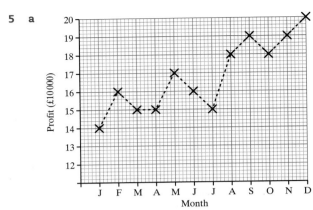

b Late summer (August/September). Local vegetables are growing well so production costs, transport and storage costs are low giving maximum profit.

Exercise 6b

1 a and **c** can be shown on a time series graph because they both alter with time. **b** does not.

2 a

b Rainfall seems to be lowest in Quarter 3 and highest in Quarter 1 of each year. Rainfall seems to be decreasing from year to year.

3 a

b Sales are highest in the fourth quarter and lowest in the first quarter of each year. Fourth quarter sales are decreasing from year to year, but first quarter sales are increasing.

4 a

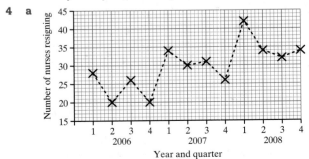

b First quarter.
c Yes. Resignations seem to be increasing each year.

5 a

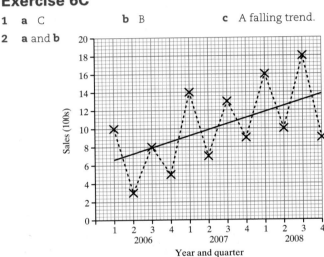

b The manager is wrong. Profits are rising.

Exercise 6C

1 a C   **b** B   **c** A falling trend.

2 a and **b**

c The general trend is for sales to increase.

3 a and **b**

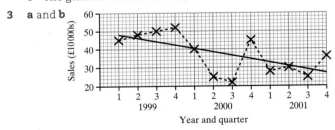

c The trend is for sales to decrease.

4 a and **b**

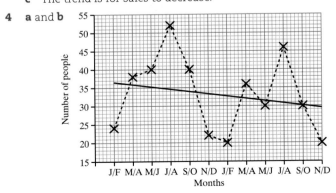

c The trend is for the number of people to decrease. The tours are getting less popular.

Exercise 6D

1 **b** and **c**. The number of swimsuits and the number of hours of sunshine rise in the summer and fall in the winter. Bank accounts and breakfast cereals are unaffected by the season.

2 **a** Seasonal variation. **b** First quarter.

3 **a** 500 **b** 1200
   c The trend is rising. More people seem to be taking the tour each year.
   d The tour is most popular in the second quarter of a year and least popular in the fourth quarter of a year.

4 **a** and **b**

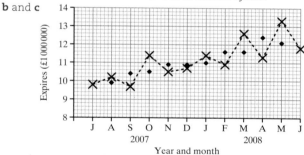

   c The long-term trend is downwards. Sales are lowest in the third quarter and generally highest in the first quarter.

Exercise 6E

1 97, 97, 106.3, 111.3, 110.7, 108, 109, 113.3, 120.3, 119, 119.3, 119.7

2 **a** A four-point moving average is the average of the four observations from successive four-season cycles.
   b and **c**

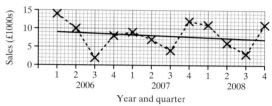

3 **a** and **b**

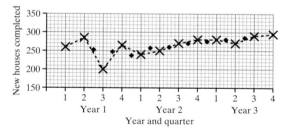

   b 280, 283.75
   c The trend is rising. Numbers of new builds have increased over the three years.

4 **a**

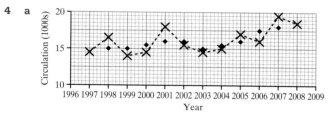

   b 16, 17.5, 18
   c The trend is rising slightly. Between 1997 and 2008 circulation has increased.

5 **b** 20, 20, 22, 24, 24, 25, 26, 27, 30
   a, **b** and **c**

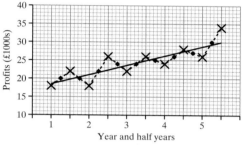

   c The trend is rising. Over the 5 years profits have increased.

6 **a**, **c** and **d**

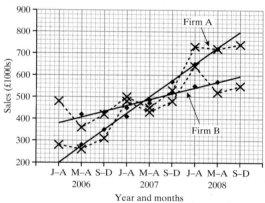

   b Firm A: 283.3, 350, 413.3, 486.7, 570, 656.7, 726.7
       Firm B: 420, 426.7, 450, 470, 516.7, 546.7, 570
   e Between April and August 2007.

7 **a** and **b**

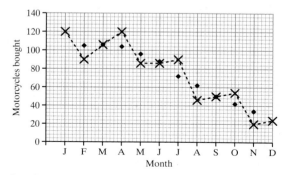

   b 105, 105, 103.3, 96.7, 86.7, 73.3, 61.7, 50, 41.7, 33.3
   c The trend is falling. Over the year sales have fallen.

Exercise 6F

Trend lines are drawn by eye, so your readings might vary either side of those given.

1 £16.59

2

Year	Seasonal variation			
	Quarter 1	Quarter 2	Quarter 3	Quarter 4
1	0.0	−6.2	8.1	4.0
2	3.1	−4.0	5.0	3.5
Total	3.1	−10.2	13.1	7.5
Mean seasonal variation	1.55	−5.1	6.55	3.75

3

Year	Quarter	Actual value	Trend	Seasonal variation
1	1	26	22	4
	2	38	24	14
	3	14	18	−4
	4	10	14	−4
2	1	20	19	1
	2	32	23	9
	3	13	17	−4
	4	10	10	0

4 Estimated mean seasonal variations 2.5, 11.5, −4, −2

5 **a** and **b** The last four moving averages are 79, 87, 97 and 106.

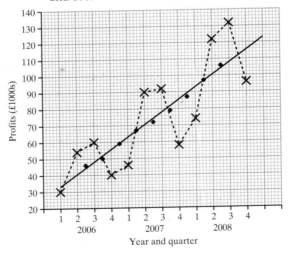

c £108 000

6 **a**, **b** and **c**

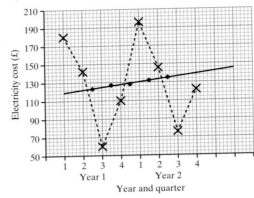

b 123, 127, 128, 132, 135   **d** 59, 20
e £201, £164

7 **a** The way in which the y-variable rises and falls with the seasons (e.g. the sales of ice cream will be at a peak in the summer and low in the winter.

b **i)** and **ii)**   111, 114, 112, 113, 109, 108, 106, 111, 111, 114, 113, 115, 117

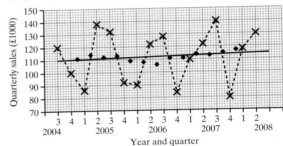

c −£11 750, £17 000, £17 500, −£23 500
d £131 500, £91 500

8 **a** 799, 822

b and **c**

d

686	630	56			
590	650		−60		
660	670			−10	
720	690				30
754	710	44			
642	730		−88		
732	750			−18	
808	770				38
842	790	52			
738	810		−72		
808	830			−22	
900	850				50
	Total	152	−220	−50	118
	Mean	$50\frac{2}{3}$	$-73\frac{1}{3}$	$-16\frac{2}{3}$	$39\frac{1}{3}$

e 921, 817

Exercise 6G

1 a £1125/quarter
  b The rate at which sales are going up each quarter.
2 a and b

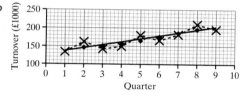

  b i) 146, 151, 157, 165, 176, 186, 196
  c $y = 8.54x + 124$ **d** (in £1000s): −4.3, 16, −8.3
  e £218 000, £202 000
3 a, c and d

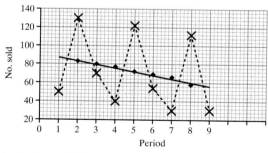

  b Each period is one third of a year.
  c 83, 80, 77, 72, 69, 66, 58
  e −4. The rate at which sales of size 6 trainers decrease.
  f −36, 49, −16   **g** 16, 97, 28

Chapter 6 review

1

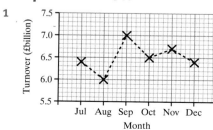

2 a and b

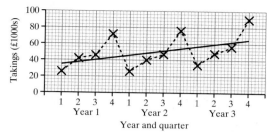

  c The general trend is for takings to rise.
  d Yes. Fourth quarter. More post sent before Christmas.
3 a and b

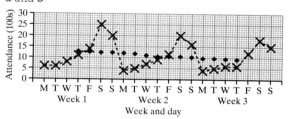

  b 12.86, 12.57, 12.43, 12.29, 12.00, 11.71, 11.00, 10.43, 10.43, 10.43, 10.29, 9.86, 9.86, 9.57, 9.43
  c The trend is down. The highest attendance is on a Saturday. The lowest attendance is on a Monday.
4 a i)

Year	Period	Number of houses	Three-point moving average
2006	1	9	
	2	26	18
	3	19	19
2007	1	12	21
	2	32	22
	3	22	23
2008	1	15	25
	2	38	26
	3	25	

  ii)

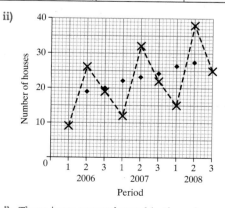

  b i) There is an upward trend in the sales.
  ii) The increase in sales may be due to other factors (e.g. a new manager, better publicity etc.).

5 a i) £152, £155

ii)

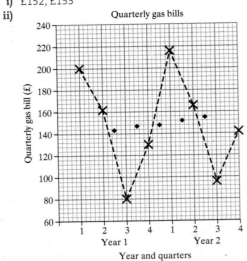

Quarterly gas bills

Year and quarters

b The gas bills are increasing.

c i) Seasonal variations.

   ii) E.g. variations are linked to the time of year. More gas is used in winter to heat the house.

6 a, .b and c

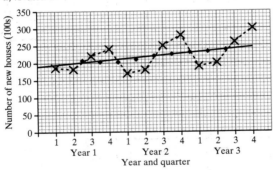

Year and quarter

b 207, 203, 202.5, 210, 220, 225, 230, 232.5, 237.5

d $y = 4.675x + 188$

e −29.38, −28.72, 22.61, 47.93   **f** 225 houses

Chapter 6 Test yourself

1 12.8°C   **2** 8.8°C

3 Level   **4** +6°C

5 −4.3°C (1 d.p.)   **6** 8.4°C

7 4.1°C

Chapter 7

Exercise 7A

1 a rains, does not rain   **b** 1, 2, 3, 4, 5, 6, 7, 8

   c point up, on side (point down)

2 Outcomes that have the same chance of happening.

3 a Certain   **b** Evens

   c Unlikely or very unlikely

4

5 a 0   **b** 1

   c The probability of event W happening is a half (or 0.5 or evens).

6 0.8

7

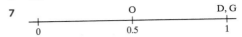

Exercise 7B

1 a $\frac{1}{6}$ **b** $\frac{2}{3}\left(=\frac{4}{6}\right)$ **c** $\frac{1}{3}=\frac{2}{6}$

2 a $\frac{1}{4}\left(=\frac{2}{8}\right)$ **b** $\frac{3}{8}$ **c** $\frac{1}{4}\left(=\frac{2}{8}\right)$

   d $\frac{1}{8}$ **e** $\frac{3}{8}$

3 a i) $\frac{1}{4}\left(=\frac{13}{52}\right)$ **ii)** $\frac{1}{52}$ **iii)** $\frac{1}{2}\left(=\frac{26}{52}\right)$ **iv)** $\frac{1}{13}\left(=\frac{4}{52}\right)$

   b If it is not picked at random, then all outcomes are not equally likely.

4 a $\frac{1}{3}\left(=\frac{2}{6}\right)$ **b** $\frac{1}{6}$ **c** $\frac{1}{3}\left(=\frac{2}{6}\right)$

5 a $\frac{1}{2}\left(=\frac{25}{50}\right)$ **b** $\frac{8}{25}\left(=\frac{16}{50}\right)$ **c** $\frac{8}{25}\left(=\frac{16}{50}\right)$

6 a

	Male	Female	Total
Jam A	10	13	23
Jam B	2	20	22
Total	12	33	45

   b i) $\frac{4}{9}\left(=\frac{20}{45}\right)$ **ii)** $\frac{2}{9}\left(=\frac{10}{45}\right)$ **iii)** $\frac{23}{45}$

7 a

	Boys	Girls	Total
Raven	11	40	51
Eagle	35	19	54
Total	46	59	105

   b i) $\frac{59}{105}$ **ii)** $\frac{8}{21}\left(=\frac{40}{105}\right)$ **iii)** $\frac{18}{35}\left(=\frac{54}{105}\right)$

8 a $\frac{3}{8}$ **b** $\frac{5}{8}$

Exercise 7C

1 $\frac{3}{5}\left(=\frac{30}{50}\right)$ **2** $\frac{17}{20}$

3 When an experiment or survey is difficult to carry out.

4 Repeated simulations will give the chef a good idea of the maximum daily demand for each dish. He will then know how many of each to prepare so as to satisfy all possible requests from workers.

5 a Student's own random numbers.

   b Allocate numbers 1–5 to A, 6 and 7 to B, 8 to C, 9 and 10 to D. The fifty number would then represent the number of visitors wanting each activity one rainy day.

   c Other simulations will produce different results but several simulations will give an idea of the maximum daily demand for each activity.

6 a We need 30 numbers (30 = 2 × 15 because 100 = 2 × 50). The first 64 numbers have been used (00 to 63 is 64 numbers).

   b Frequency column: 7, 24, 17, 2.

   c They are not a good match.

   d More trains are cancelled and late than during the first week.

7 Student's own results.

Exercise 7D

1 $\frac{2}{75}\left(=\frac{4}{150}\right)$ **2** $\frac{1}{40}\left(=\frac{20}{800}\right)$ **3** 4

4 £48   **5** £100

Exercise 7E

1 a

		Coin 1	
		H	**T**
Coin 2	**H**	HH	HT
	T	TH	TT

   b $\frac{1}{2}\left(=\frac{2}{4}\right)$ **c** $\frac{1}{4}$

2 a

		Dice 1					
		1	**2**	**3**	**4**	**5**	**6**
Dice 2	**1**	2	3	4	5	6	7
	2	3	4	5	6	7	8
	3	4	5	6	7	8	9
	4	5	6	7	8	9	10
	5	6	7	8	9	10	11
	6	7	8	9	10	11	12

   b $\frac{1}{36}$ **c** $\frac{7}{12}\left(=\frac{21}{36}\right)$ **d** $\frac{1}{12}\left(=\frac{3}{36}\right)$ **e** $\frac{1}{2}\left(=\frac{18}{36}\right)$

3 a

		Men			
		A	**B**	**C**	**D**
Women	**X**	XA	XB	XC	XD
	Y	YA	YB	YC	YD

   b $\frac{1}{4}\left(=\frac{2}{8}\right)$ **c** $\frac{1}{8}$ **d** $\frac{3}{8}$

4 a

		Dice					
		1	**2**	**3**	**4**	**5**	**6**
Spinner	**1**	2	3	4	5	6	7
	2	3	4	5	6	7	8
	3	4	5	6	7	8	9
	4	5	6	7	8	9	10

   b $\frac{1}{8}\left(=\frac{3}{24}\right)$ **c** 0   **d** $\frac{7}{12}\left(=\frac{14}{24}\right)$

5 a

	Blue	Red	Green
Blue	BB	BR	BG
Red	RB	RR	RG
Green	GB	GR	GC

   b $\frac{1}{9}$ **c** $\frac{1}{3}\left(=\frac{3}{9}\right)$ **d** $\frac{1}{9}$

6 a

```
1 1 1
1 1 2   1 2 1   2 1 1
1 2 2   2 1 2   2 2 1
2 2 2
```
   b $\frac{1}{8}$ **c** 0   **d** $\frac{3}{8}$ **e** $\frac{7}{8}$

7 a

```
ABC   ACB
BAC   BCA
CAB   CBA
```
   b $\frac{1}{2}$ **c** $\frac{1}{3}$

Exercise 7F

1 a x = 28. The number of students who study Spanish but not French.

   b x = 18. The number of students who study both Geography and History.

   c x = 3. The number of students who study neither Maths nor Science.

2 a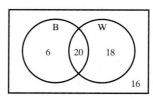

   b x = 18   **c** 26   **d** $\frac{4}{15}\left(=\frac{16}{60}\right)$

3 a 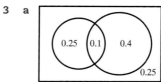   **b**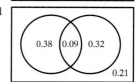

   c 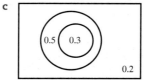   **d**

4 a 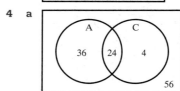   **b** $\frac{7}{15}\left(=\frac{56}{120}\right)$

5 a   **b** 52%

   c 26%

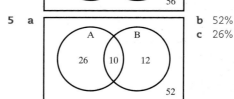

6 a   **b** 17

   c $\frac{7}{30}$

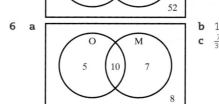

7 a   **b** 0.25

   c 9

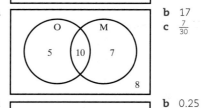

8 a   **b** $\frac{3}{25}\left(=\frac{12}{100}\right)$

   c $\frac{1}{5}\left(=\frac{20}{100}\right)$ or 0.2 or 20%

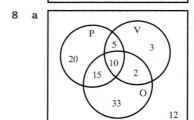

9 $x = 0.2$, $y = 0.2$, $z = 0.05$

10 a

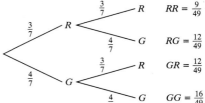

b **i)** $\frac{3}{5}$

ii) $\frac{12}{25}$

iii) $\frac{2}{5}$

Exercise 7G

1 A and C

2 a 0.6   **b** 0.5   **c** 0.7

3 A and C

4 B

5 a $\frac{1}{8}$ **b** $\frac{1}{8}$ **c** 1

6 0.4

7 a 15   **b** $\frac{1}{3}\left(=\frac{10}{30}\right)$ **c** $\frac{1}{6}\left(=\frac{5}{30}\right)$

  d $\frac{1}{2}$ **e** $\frac{5}{6}$ **f** $\frac{2}{3}$

8 a 0.75   **b** 0.95   **c** 0.8

Exercise 7H

1 A and B

2 a 0.06   **b** 0.08   **c** 0.12

3 a Yes   **b** 0.42

4 a $\frac{4}{35}$ **b** $\frac{6}{7}$ **c** $\frac{6}{35}$

5 a $\frac{1}{480}$ **b** $\frac{1}{18}$

6 0.006

7 a 0.225   **b** 0.8   **c** 0.24

Exercise 7I

1 0.9

2 77 people

3 a **i)** $\frac{5}{12}$ **ii)** $\frac{7}{12}$ **iii)** $\frac{1}{6}\left(=\frac{2}{12}\right)$

  b $\frac{5}{6}\left(=\frac{10}{12}\right)$

4 $P(A \cup B) = 1$

5 a 0.98   **b** 0.02

Exercise 7J

1 a $\frac{1}{2}$ **b** $\frac{1}{2}$

  c

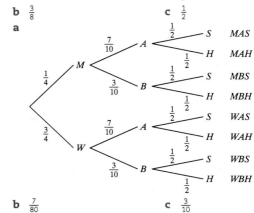

  d 0.25

2 a

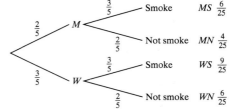

b $\frac{9}{49}$ **c** $\frac{24}{49}$

3 a $\frac{3}{4}\left(=\frac{6}{8}\right)$ **b** $\frac{1}{2}\left(=\frac{6}{12}\right)$

  c

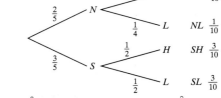

d $\frac{3}{10}$ **e** $\frac{2}{5}$

4

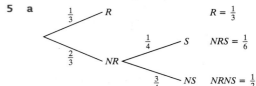

5 a

R $= \frac{1}{3}$

NRS $= \frac{1}{6}$

NRNS $= \frac{1}{2}$

b $\frac{1}{6}$ **c** $\frac{1}{2}$

6 a

111 $\frac{1}{8}$
116 $\frac{1}{8}$
161 $\frac{1}{8}$
166 $\frac{1}{8}$
611 $\frac{1}{8}$
616 $\frac{1}{8}$
661 $\frac{1}{8}$
666 $\frac{1}{8}$

b $\frac{3}{8}$ **c** $\frac{1}{2}$

7 a

MAS, MAH, MBS, MBH, WAS, WAH, WBS, WBH

b $\frac{7}{80}$ **c** $\frac{3}{10}$

Exercise 7K

1 a $\frac{5}{12}$ **b** $\frac{4}{11}$

2 a $\frac{5}{19}$ **b** $\frac{1}{114}$ **c** $\frac{5}{114}$

3 a

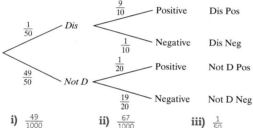

b $\frac{3}{13}$ $\left(= \frac{81}{351} = BBB + RRR\right)$ **c** $\frac{28}{65}$ $(= RRB + RBR + BRR)$

4 a $\frac{7}{145}$ **b** $\frac{69}{580}$ $\left(= \frac{483}{4060}\right)$

5 a

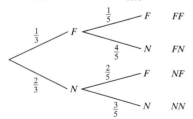

b i) $\frac{7}{15}$ $\left(= \frac{14}{30}\right)$ **ii)** $\frac{7}{15}$ $\left(= \frac{42}{90}\right)$

6 a

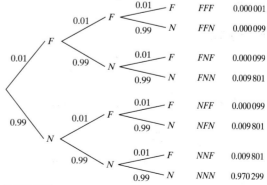

b i) $\frac{49}{1000}$ **ii)** $\frac{67}{1000}$ **iii)** $\frac{1}{50}$

7 a

b i) $P(FF) = \frac{1}{15}$ $\left(= \frac{2}{30}\right)$

ii) $P(FF) + P(NF) + P(FN) = 1 - P(NN) = \frac{3}{5}$ $\left(= \frac{18}{30}\right)$

Chapter 7 review

1 a

```
    A      C  B
  +--+-----+--+-----------+
  0        0.5            1
```

b For example: She takes an even number or a number greater than 5.

2 a 0.25 or $\frac{1}{4}$ **b** 0.75 or $\frac{3}{4}$

3 a i) $\frac{7}{10}$ $\left(= \frac{70}{100}\right)$ **ii)** $\frac{1}{10}$ $\left(= \frac{10}{100}\right)$

b Yes. E.g. a larger proportion of those vaccinated did not get foot rot.

c E.g. 'control group' or 'to compare the vaccinated sheep with the non-vaccinated sheep'.

4 a

	Smokers	Non-smokers	Total
One or more teeth taken out	7	13	20
No teeth taken out	15	65	80
Total	22	78	100

b i) $\frac{39}{50}$ $\left(= \frac{78}{100}\right.$ or $\left. 0.78\right)$ **ii)** $\frac{3}{20}$ $\left(= \frac{15}{100}\right.$ or $\left. 0.15\right)$

c More likely to have teeth out.

5 a

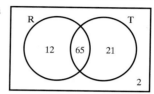

b 33% (0.33)   **c** 14% (0.14)   **d** 84% $\left(\frac{65}{77}\right.$ or $\left. 0.84\right)$

6 a

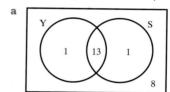

b i) $\frac{1}{23} = 0.043$ (3 d.p.)   **ii)** $\frac{1}{23} = 0.043$ (3 d.p.)

7 a The outcome of one event does not affect the outcome of the other event.

b i) 0.15   **ii)** 0.85

c 0.84

8 a

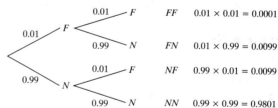

Engine 1	Engine 2	Outcome	Probability
	0.01 F	FF	0.01 × 0.01 = 0.0001
0.01 F	0.99 N	FN	0.01 × 0.99 = 0.0099
0.99 N	0.01 F	NF	0.99 × 0.01 = 0.0099
	0.99 N	NN	0.99 × 0.99 = 0.9801

b 0.9999

c

Engine 1	Engine 2	Engine 3	Outcome	Probability
	0.01 F	0.01 F	FFF	0.000 001
		0.99 N	FFN	0.000 099
0.01 F	0.99 N	0.01 F	FNF	0.000 099
		0.99 N	FNN	0.009 801
0.99 N	0.01 F	0.01 F	NFF	0.000 099
		0.99 N	NFN	0.009 801
	0.99 N	0.01 F	NNF	0.009 801
		0.99 N	NNN	0.970 299

d 0.999 702

e Two engines. The probability of a successful flight is greater for a two-engine plane.

9 a $\frac{4}{49}$ = 0.082 (3 d.p.)   **b** $\frac{5}{49}$ = 0.102 (3 d.p.)

c

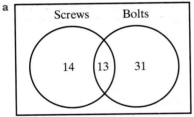

	Resistor 1	Resistor 2	Outcome	Probability

0.082 F   FF   0.1 × 0.082 = 0.0082

0.1 F

0.918 N   FN   0.1 × 0.918 = 0.0918

0.102 F   NF   0.9 × 0.102 = 0.0918

0.9 N

0.898 N   NN   0.9 × 0.898 = 0.8082

d   0.9918

Chapter 7 Test yourself

1 X is $\frac{2}{7}$ of the way along the line from the left (or on the probability scale of 0 to 1, X is at approximately 0.29).

2 $\frac{1}{2}$ or 0.5 (50%)

3 a

	Boys	Girls	Total
Art	10	12	22
Music	6	6	12
Total	16	18	34

b   **i)** $\frac{8}{11}\left(=\frac{16}{34}\right)$ or 0.47 (47%)   **ii)** $\frac{3}{8}\left(=\frac{6}{16}\right)$ or 0.375 ($37\frac{1}{2}$%)

4 $\frac{9}{14}$ or 0.64 (64%)

5 a H1   H2   H3   H4   T1   T2   T3   T4

   b $\frac{1}{4}\left(=\frac{2}{8}\right)$ or 0.25 (25%)

6 a

Screws   Bolts

14 13 31

b $\frac{13}{58}$ or 0.22 (22%)

7 A and B

8 0.1 × 0.5 = 0.05

9 0.9 + 0.5 − 0.7 = 0.7

10 a   **Throw 1**   **Throw 2**   **Outcome**

H = Hit
M = Miss

0.5 H   HH

0.5 H

0.5 M   HM

0.5 H   MH

0.5 M

0.5 M   MM

b   P(one hit) = (0.5 × 0.5) + (0.5 × 0.5)
              = 0.25 + 0.25
              = 0.5 (= $\frac{1}{2}$)

11   P(B ∩ A) = P(B | A) × P(A)
                = 0.1 × 0.3
                = 0.03

Chapter 8

Exercise 8A

1 $\frac{1}{4}$ or 0.25   **2**   0.3 (30%)

3 a

x	Fault 1	Fault 2	Fault 3	Fault 4	Fault 5
p(x)	0.2	2k	k	2k	0.3

   b   0.1

4 a

y	Sector 1	Sector 2	Sector 3	Sector 4	Sector 5	Sector 6	Sector 7
p(y)	s	2s	3s	4s	s	2s	3s

   b   0.0625   **c**   0.1875

Exercise 8B

1 B and C. People are equally likely to be born on any day of the week and any of the numbers 1 to 9 are equally likely to come at the end of a telephone number. However, the heights of people cluster around a mean value.

2 Discrete uniform distribution.

3 a

x	1	2	3	4	5	6	7	8
p(x)	$\frac{1}{8}$	$\frac{1}{8}$	$\frac{1}{8}$	$\frac{1}{8}$	$\frac{1}{8}$	$\frac{1}{8}$	$\frac{1}{8}$	$\frac{1}{8}$

   b $\frac{1}{8}$ **c** $\frac{1}{4}$

4 There is not an equal probability of hitting all the numbers if you aim for the 20. The distribution is unsuitable.

5 a Discrete uniform distribution   **b** $\frac{1}{50}$

6 $\frac{1}{5}$

7 a Discrete uniform distribution   **c** $\frac{1}{3}\left(=\frac{2}{6}\right)$

Exercise 8C

1 a Binomial
   b 12 is the number of trials. 0.325 is the probability of success.

2 a 0.6   **b** 0.3456

3 a 0.614 125   **b** 0.057 375

4 a 0.156 25   **b** 0.968 75   **c** 0.031 25

5 a 0.2   **b** 0.896

6 a 0.001   **b** 0.729

7 a 0.0256   **b** 0.2688

8 a 0.000 03   **b** 0.999 (3 d.p.)

9 a 0.000 01   **b** 0.918 54

10 a 0.614 125   **b** 0.057 375

Exercise 8D

1 B, C and D. They are continuous variables and natural occurrences, so they will be norally distributed. The number of accidents is not continuous.

2 a They are all equal.
   b The distribution is symmetrical about the mean. 95% of observations lie within two standard deviations, and 99.8% lie within three standard deviations of the mean.

3 0.9 cm and 5.7 cm.

4

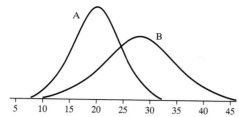

5 a 2
  b i) 14   **ii)** 24

6 a 70.5 km/h   **b** 34.5 km/h
  c 25.5 km/h   **d** 79.5 km/h

7 a 97.5%   **b** 2.5%   **c** 47.5%

8 a 95%   **b** 47.5%   **c** 97.4%   **d** 2.4%

9 2.4%

10 a i) 95%   **ii)** 2.5%
  b 570

11 a i) 95%   **ii)** 99.8%
  b i) 950   **ii)** 998
  c 97.4%

12 a 29 days   **b** 1 day   **c** 29 days

13 a i) 0.025 or 2.5%   **ii)** 0.95 or 95%
  b 5000 hours

14 $7\frac{1}{2}$ cm

15 25

Exercise 8E

1 A control chart is used to check that a process mean, range, variance etc. stay within set limits (i.e. under control).

2 Components being manufactured may be outside the set limits if either the mean is too large or too small, or if the range is too large.

3 a

Sample	1	2	3	4	5	6	7	8
Mean	150.4	150	149.8	150.1	150	149.75	149.75	150
Range	0.8	1.7	1.0	1.7	1.4	0.8	1.1	1.4

b

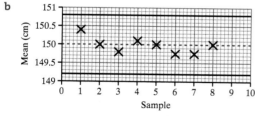

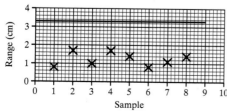

4 a

Sample	1	2	3	4	5	6	7	8
Mean	49.58	49.98	49.61	49.93	50	49.42	49.9	49.77
Range	1.0	0.29	0.85	0.25	0.08	1.62	0.4	0.62

b

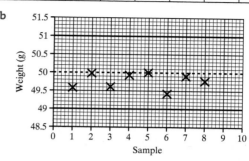

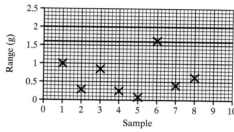

5 Warning: 64.4 mm and 65.6 mm
  Action: 64.1 mm and 65.9 mm

6 a i) 37.1 cm and 38.9 cm   **ii)** 37.4 cm and 38.6 cm
  b Machine should have been stopped and reset after Sample 8.

Chapter 8 review

1 a 10   **b** 6
  c Discrete uniform distribution

2 a 0.970 299   **b** 0.000 297

3 a 0.815 (3 d.p.)   **b** 0.000 006 25   **c** 0.986 (3 d.p.)

4 a i) Binomial   **ii)** $n = 5, p = \frac{5}{7}$
  b 0.1447...
  c $P(5) = \left(\frac{5}{7}\right)^5 = 0.186$

    $P(4) = \frac{5 \times 5^4 \times 2}{7^5} = 0.372$

    $P(3) = \frac{10 \times 5^3 \times 4}{7^5} = 0.297$

    $P(2) = \frac{10 \times 25 \times 8}{7^5} = 0.119$

    $P(1) = \frac{5 \times 5 \times 16}{7^5} = 0.024$

    $P(0) = \frac{32}{7^5} = 0.002$

    The probability of 4 sunny days is highest.

5 a i) 95%   **ii)** 99.8%
  b 97

6 a 47.5% = 0.475   **b** 97.5% = 0.975

7 **a** 25.61 cm and 26.39 cm

b

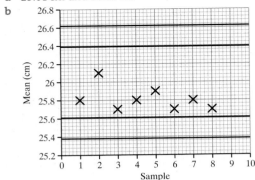

c

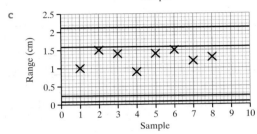

d Both the mean and the range are under control. No action is taken.

Chapter 8 Test yourself

1 **a**

X	A	B	C	D	E
p(x)	$\frac{1}{5}$	$\frac{1}{5}$	$\frac{1}{5}$	$\frac{1}{5}$	$\frac{1}{5}$

  b Discrete uniform distribution

  c $\frac{1}{6}$ **d** $\frac{1}{3}$

2 **a** Binomial **b** $n = 5, p = \frac{4}{9}$

  c 0.339 **d** 0.397

3 **a** 0.95 **b** 0.974 **c** 0.025

4 **a** Another sample should be taken.

  b The machine should be stopped and reset.

  c **i)** 394 g–406 g **ii)** 391 g–409 g

  d Warning: 394 g and 406 g

  Action: 391 g and 409 g

Examination practice papers

Foundation

1 **a**

Time	Number of fires
00 00 – 05 59	♨ ♨ ♨

  b 33 **c** Between 12:00 and 17:59.

  d It is night-time.

2 **a**

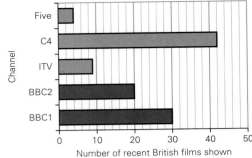

  b 42 **c** Five

3 **a** Any **one** from: it will be difficult to get replies from everyone; it will take along time; it could be expensive.

  b It will get everyone's opinion.

  c It is a bad question because it is open. It could be improved by making it a closed question, for example 'Which of the following colours should the waiting room be painted: cream, green, blue or pink?'

4 **a**

	Prefer a theatre	Prefer a dance	Total
Men	65	70	135
Women	20	30	50
Total	85	100	185

  b **i)** $\frac{13}{37} \left(= \frac{65}{185} \right)$ **ii)** $\frac{10}{37} \left(= \frac{50}{185} \right)$

5 **a** 2.8% **b** Submerged

  c Many more people were involved in fatal accidents where no object was collided with than there were people involved in a collision with a crash barrier. (2.8% of 2409 is 67 people. 1.5% of 41 110 is 616. 616 > 67.)

6 **a** 69.01 $\left(= \frac{760}{11} \right)$

  b 70 **c** 70 **d** 5 ($= 71 - 66$)

  e Two of them are the same as the expected number 70. The mean (69) is lower than the expected value.

7 **a** The more trials, the more accurate the answer. The last one is after 50 throws. The others are after less than 50.

  b The dice is probably not biased. 0.167 is very close to the graph answer of 0.17.

8 **a**

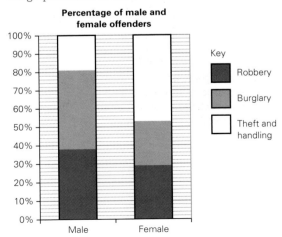

  b Burglary

  c There were more males than females because 405.6 (= 47% of 863) < 2986 (= 19% of 15 716).

9 a

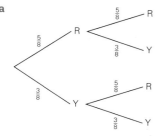

b $\frac{15}{32} \left(= \frac{30}{64}\right)$

10 a

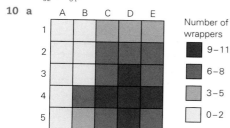

b D4. It has the most wrappers and is surrounded by sectors that have a lot of wrappers.

11 a Either: H_1: Men answer calls more quickly than women.
Or: H_1: Women answer calls more slowly than men.

Plus

Either: H_2: Women give better answers than men.
Or: H_2: Men give worse answers than women.

b All the telephone operators working at the call centre.

c Taking a sample is quicker than taking a census.
It is cheaper. It is easier.
But it does not get information on everyone and it could be biased.

12 a 48, 49

b and c

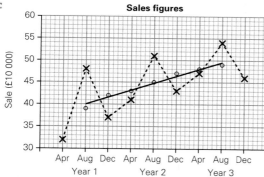

d Rising or going up.

e The product is seasonal in sales or it sells more in May to August each year. Plus: It is a product that sells better in warm weather.

13 a

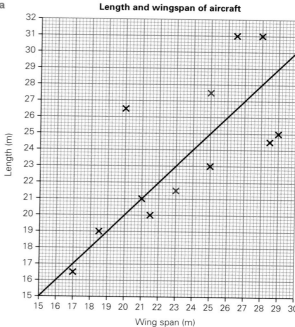

b A trend line that goes through (15, 15) and (25, 25). See answer **a**.

c Not very well. Several aircraft are a long way from the line of best fit.

d Any figure in the range 23 m to 27 m.

e 66.7 is well outside the data given, so the line of best fit would need to be extrapolated. Also the data do not fit the line of best fit well.

f This is fairly close to the student's belief. To be in complete agreement the aircraft would need a wingspan of 66.6 m. Its wingspan is only 2.2 m too short.

Higher

1 a i) Crosses for quantitative and continuous.
ii) Crosses for discrete and quantitative.

b 143.5 cm to 144.5 cm

2 a Any **one** of: it is three dimensional; it is at an angle; it is a bird's-eye view.

b

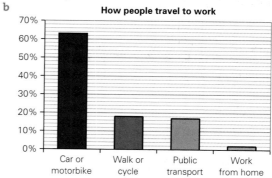

c Car or motor bike.

3 a 8.82 million tonnes

b Decreasing or going down.

c Rising/increasing or going up.

d People are changing the fuel they use for transport from motor spirit to diesel.

4 a 205, 202
b
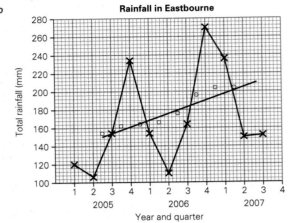

c The second quarter.
d See answer **b**.
e Rising/going up.

5 a
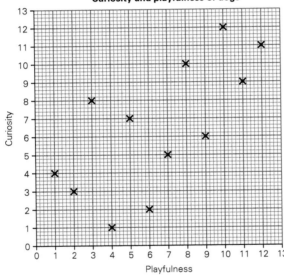

b 0.69 (2 d.p.)
c Positive
d The greater the playfulness, the greater the curiosity.
OR
The less the playfulness, the less the curiosity.

6 a
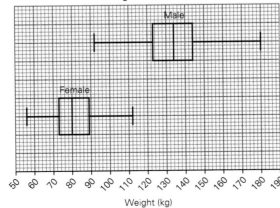

b 16 kg

c The male red deer has either a greater range, a greater IQR or a greater spread. Plus, the male red deer has a greater median. Over 75% of the male deer were heavier than the heaviest female deer.

7 a 1.93 **b** 1.20
c The mean is now lower suggesting a decrease in accidents. The standard deviation is a very small amount higher suggesting a slightly larger spread of accidents.

8 a

Method of reception	Satellite	Cable	Digital terrestrial	None of these
Frequency	7.2	3.6	7.2	2

b 36% of the households get satellite. One number represents each per cent. There are 36 numbers between 00 and 35.

c

Method of reception	Satellite	Cable	Digital terrestrial	None of these
Frequency	8	6	6	0

d Satellite and cable are higher than the expected value. Digital, Terrestrial and None are less than expected.
e No **f** No

9 a
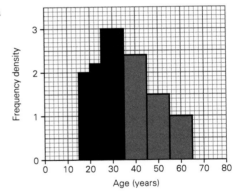

b It is positive. **c** 31.8
10 a 1.02 kg and 1.1 kg **b** 1.0 kg and 1.12 kg
c
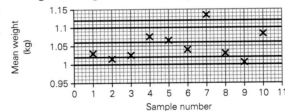

d See answer to **c**.
e Another sample would be taken after the 2nd sample. Production would be stopped and the machines reset after the 7th sample.

11 a All the students at the university.
b A closed question because it gives a clear answer.
c E.g. Tick the box that best describes the amount of rent you pay each week.
This should be followed by up to five non-overlapping boxes that cover all possibilities (e.g. under £100, £100 up to £124.99, £125 up to £150, over £150).

d Any **five** of the following: it will identify problems; it shows the possible responses; it finds errors; it checks the questions are clear; it gives feedback so things can be changed; it costs money, but might save it in the long run; it takes time, but might save it in the long run.

12 a They are randomly spread in the lake.
  b 240

13 a Binomial   **b** It is constant.
  c **i)** 0.256
     ii) Three or four times are equally likely.
  d Unlikely. She should have got three hits for her claim to be true.

14 a

	Given pocket money	Had to earn pocket money	Did not get pocket money	Total
Girls	35	15	5	55
Boys	35	8	2	45
Total	70	23	7	100

  b **i)** $\frac{7}{10}\left(=\frac{70}{100}\right)$ **ii)** $\frac{3}{20}\left(=\frac{15}{100}\right)$

  c The following points need to be made: equal numbers of boys and girls were given pocket money but more girls than boys had to earn their pocket money and more girls than boys did not get any pocket money.

  d 232 (2006); 222.5 (2007)   **e** 132%

  f Increased yearly until the last year when it went down.

Index